# ALGORITHMS, COMPLEXITY ANALYSIS AND VLSI ARCHITECTURES FOR MPEG-4 MOTION ESTIMATION

# ALGORITHMS, COMPLEXITY ANALYSIS AND VLSI ARCHITECTURES FOR MPEG-4 MOTION ESTIMATION

*by*

**Peter Kuhn**

*Technical University of Munich, Germany*

KLUWER ACADEMIC PUBLISHERS

BOSTON / DORDRECHT / LONDON

A C.I.P. Catalogue record for this book is available from the Library of Congress.

ISBN 978-1-4419-5088-8

Published by Kluwer Academic Publishers,
P.O. Box 17, 3300 AA Dordrecht, The Netherlands.

Sold and distributed in North, Central and South America
by Kluwer Academic Publishers,
101 Philip Drive, Norwell, MA 02061, U.S.A.

In all other countries, sold and distributed
by Kluwer Academic Publishers,
P.O. Box 322, 3300 AH Dordrecht, The Netherlands.

*Printed on acid-free paper*

# CONTENTS

# PREFACE

The emerging multimedia standard MPEG-4 combines interactivity, natural and synthetic digital video, audio and computer-graphics. Typical applications are: video conferencing, mobile videophone, multimedia cooperative work, teleteaching and, last but not least, games. With MPEG-4 the next step from block-based video coding (ISO/IEC MPEG-1, MPEG-2, ITU-T H.261, H.263) to object-based video coding is taken. The step from block-based to object-based video coding requires a new methodology for VLSI design, as well as new VLSI architectures with considerable higher flexibility, as the motion estimation (ME) of the encoder is the computationally most demanding tool within the MPEG-4 standard.

The MPEG-4 encoder with its support of several modes and video objects suggests at first glance a programmable processor based implementation. The question arose, whether processor based or dedicated VLSI-based solutions are better suited for low power and high throughput motion estimation. Processors offer high flexibility and programmability, which has, however, to be paid for by higher power consumption, higher area demands and lower throughput, compared to dedicated ASICs (application specific integrated circuits). Dedicated ASICs, however, usually have the disadvantage of limited flexibility.

Therefore, within this work VLSI architectures were developed, which offer the flexibility to support several motion estimation algorithms. Block-matching motion estimation algorithms usually depict relatively simple arithmetic operations, but have high memory bandwidth. Therefore, for low-power and high throughput optimization of motion estimation at algorithm and system level, the minimization of the memory bandwidth from the data path to local buffers is performed here.

The dedicated, but flexible VLSI architectures which support several motion estimation algorithms developed within this book, are compared to processor based implementations. It is shown by means of an Area-Time-Bandwidth (ATB) metric, that the flexible VLSI architectures presented within this book are favorable in terms of magnitudes of order, compared to processor based solutions, both in terms of high throughput and low-power constraints.

In the following, the methodology and the results are summarized. An overview and classification of fast motion estimation algorithms for MPEG-4 is given first. Then software complexity metrics and software based complexity analysis tools are discussed within this work. As existing tools for software based computational complexity analysis are limited in terms of portability and analysis of memory bandwidth, a new portable tool, iprof, was developed within this work. A detailed time-dependent complexity analysis of a software implementation of the MPEG-4 video part is described and a complexity and visual quality analysis of several fast motion estimation (ME) algorithms within the MPEG-4 framework is presented.

An overview of and discussion on VLSI complexity metrics and the basic components of ME architectures for MPEG-4 is provided. With reference to this design space of basic components, algorithm-specific, flexible and programmable VLSI architectures are discussed for a number of ME algorithms. Based on this discussion, the basic architectures for a high throughput MPEG-4 ME architecture, supporting optionally luminance-corrected full-search variable block size ME (Search Engine I) and a low-power ME architecture, are derived for several fast ME algorithms (Search Engine II).

A VLSI implementation of the flexible high throughput ME architecture (Search Engine I) is described supporting: 1.) fixed and 2.) variable block size full-search ME, 3.) segment-matching, 4.) support of arbitrarily-shaped video objects (MPEG-4) and 5.) luminance-corrected variable block size ME. It is shown that, using this architecture the additional area costs for the support of block-matching with arbitrarily-shaped objects (MPEG-4) are very small and luminance-corrected variable block size ME can be implemented with about 15 $mm^2$ additional chip size using 0.35 μm CMOS-technology, 100 Mhz (min.). The total chip size area is 54.1 $mm^2$ without wiring. The presented VLSI architecture was implemented using synthesizable VHDL and allows variable block size ME at 23,668 32x32 blocks per second with +/- 32 pel search area. Parallelization of an iterative partial quadtree ME algorithm was shown with less than 1 dB PSNR loss. The results were significantly superior, when compared with an Area-Time-Bandwidth (ATB) metric to a software implementation of a luminance-corrected quadtree ME algorithm.

A VLSI implementation of the flexible low-power MPEG-4 ME architecture (Search Engine II), developed within this work is performed. The presented VLSI architecture supports, besides full-search ME with [-16, 15] and [-8, +7] pel search area, MPEG-4 ME for arbitrarily-shaped objects, advanced prediction mode, 2:1 pel subsampling, 4:1 pel subsampling, 4:1 alternate pel subsampling, Three Step Search, preference of the zero-MV, R/D-optimized ME and half-pel ME. The VLSI architecture resulted into a size of 22.8 Kgates (without RAM), 100 MHz (min.) using 0.25 μm CMOS-technology. As the advanced prediction mode and half-pel motion estimation is also part of the ITU-T H.263 recommendation, H.263 can be also supported beneficially by this architecture. Finally, the presented VLSI architecture showed superior performance in comparison to other dedicated ASIC and processor based solutions, using the above mentioned ATB metric.

**Acknowledgments.** It is a pleasure to acknowledge the invaluable help received from a number of people, institutions and organizations, especially from Prof. Ingolf Ruge. Special thanks are due to the former students for software (E. Haratsch, A. Keil) and for VHDL synthesis (U. Niedermeier, R. Poppenwimmer, and A. Weisgerber). I acknowledge the co-working and discussions with the following persons: W. Bachhuber, L.-F. Chao, G. Diebel, M. Eiermann, Prof. G. Färber, A. Hutter, S. Herrmann, P. Kindsmüller, A. Kobell, Prof. M. Lang, H. Mooshofer, W. Stechele, M. Stimpfl, T. Wild, A. Windschiegl, M. Zeller, the SARE project team, the M4M project partners, and the MPEG comittee. The author acknowledges funding of the European MEDEA M4M project. The author thanks James Finlay and his staff from Kluwer Academic Publishers for their support. Finally I dedicate this book to my parents Ruth and Franz.

# 1

# INTRODUCTION

## 1.1 VISION: A MOBILE VIDEO COMMUNICATOR

Modern technology like telephone and e-mail offers the possibility of an efficient information exchange between people. These technologies enable *verbal communications*, but interpersonal communications also consists of a *non-verbal* part, like visual gestures, expressing, e.g., the mood of the person you are dealing with or making the subject of discussion clear to your partner. For example, a business man needs to see the face of his business partner, to be able to decide whether his offer seems to be reasonable for his partner or not. Therefore, visual communications is regarded to be essential to avoid misunderstandings between people, occurring, e.g., because of the anonymous style of an e-mail.

To satisfy the ever increasing demand towards visual communications in an increasingly mobile society, a mobile video communicator with Internet access is envisaged as depicted in Fig. 1.1. The video communicator consists of a display, a video camera, and a radio link for connection with a partner within the Internet or a corporate network. To enable intuitive usage for a broad range of users, the video communicator employs a pen for natural user interaction [Kaplan 94]. Typical software installed on the video communicator is, for example, an Internet browser supporting MPEG-4.

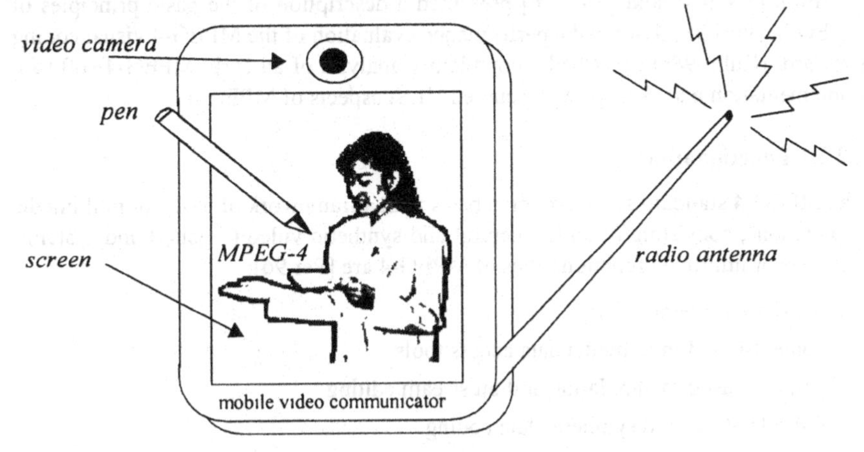


*Figure 1.1: Vision: A mobile MPEG-4 video communicator*

MPEG is an acronym for Moving Pictures Experts Group and describes the International Standardization working group ISO/IEC JTC1/SC29/WG11. MPEG-4 enables video object manipulation, which means that - as an example - with the pen a particular visual object could be marked, manipulated or tracked during a visual sequence. Other user-interaction includes asking for additional information for the marked object (like price, availability of a product, etc.). MPEG-4 only delivers and standardizes the coding and decoding tools for the audio-visual objects. The user-interface capabilities and implementation is completely determined by the application and differentiates the MPEG-4 based products. However, research and development on these user-interfaces for MPEG-4 has just begun.

## 1.2 ENABLING FACTORS

### 1.2.1 MPEG-4 Standardization

One of the enabling factors for mobile multimedia communications is the emerging MPEG-4 audiovisual standard, which is expected to become an International Standard (IS) for MPEG-4 version 1 in January 1999. MPEG-4 version 2 is expected to become an International Standard (IS) in February 2000. A common standard guarantees interoperability between products from different manufacturers. The standardization process also guarantees that the intellectual property of a new technology is available to several companies and organizations, thus leading to interoperability of products and to a healthy competitive environment. Note that MPEG basically standardizes only the bit stream format and the tools of the decoder. Therefore, more freedom in terms of an optimization for the implementation of an MPEG-4 encoder is guaranteed.

Only a short description and personal view on MPEG-4 can be given here. Detailed and up-to-date information on MPEG-4 version 1 and version 2 can be found at [MPEG]. [Sik 97a] and [Sik 97b] presented a description of the basic principles of MPEG-4, [Puri 98] described a performance evaluation of the MPEG-4 visual coding standard, [Kuhn 98a] described a complexity analysis of an early MPEG-4 software implementation and [Kneip 98] discussed VLSI aspects of MPEG-4.

#### 1.2.1.1 Functionalities

The MPEG-4 standard sets a common basis with a framework of tools for multimedia applications, consisting of audio, natural and synthetic video (=visual) and systems. The new or improved functionalities of MPEG-4 are [Per 96]:

Content-based interactivity:

- Content-based multimedia data access tools
- Content-based manipulation and bit stream editing
- Hybrid natural and synthetic data coding
- Improved temporal random access

Compression:

- Improved coding efficiency
- Coding of multiple concurrent data streams

Universal access:

- Robustness in error-prone environments
- Content-based scalability

### 1.2.1.2 Video coding

The MPEG-4 video compression scheme is based on a block-based hybrid coding concept [Mus 85], [Net 88], Fig. 1.2a, as used within the ISO/IEC MPEG-1, MPEG-2, and the CCITT H.261 / ITU-T H.263 video compression standards and recommendations, which have been extended within the MPEG-4 standardization effort to support arbitrarily-shaped video objects. The arbitrarily-shaped video objects of MPEG-4, Fig. 1.2b, are split up into macro-blocks (MBs, 16x16 pel) within a bounding box, cf. Fig. 2.6, and are coded on MB and block (8x8 pel) basis similar to block based video compression schemes, but with the coding tools of MPEG-4.

The shape of a video object, is represented by a so-called alpha-plane with pel resolution and is gained by an application-dependent method which is beyond standardization, cf. chapter 2. Visual objects can be translucent, change their size, they can be of natural video object (VO) type or they may be synthetic (computer generated) objects which can be manipulated by the user. Every object is encoded and decoded by a different encoder and decoder instance and may use different coding options. The video object representation at a specific time instance is called video object plane (VOP) which is the equivalent to "frame" for block-based video.

### 1.2.1.3 Applications

Typical applications of MPEG-4 include: search in multimedia databases, teleworking, interpersonal real-time video communications with low bit rate, broadcast video distribution with high bit rate, manipulation of video sequences for home video production, mobile multimedia, infotainment, multimedia games, surveillance, DVD (digital versatile disk), content-based storage and retrieval, streaming video on the Internet/intranet, digital set-top box and many more [N 2195].

**a) Block-based video coding: MPEG-1, MPEG-2, H.261, H.263**

**b) Object-based video coding: MPEG-4**

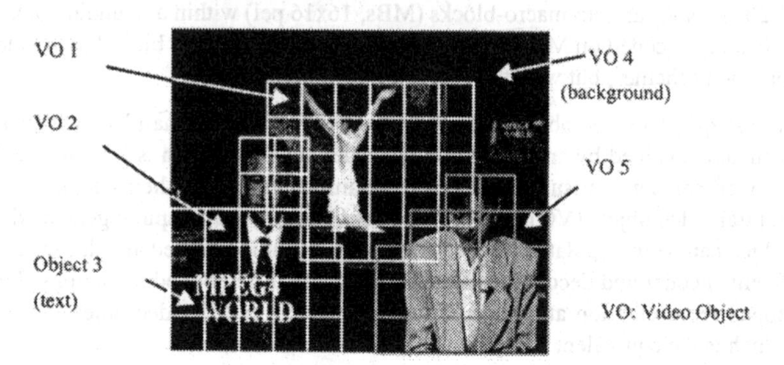


*Figure 1.2: MPEG-4: Comparison of block-based video coding and object-based video coding*

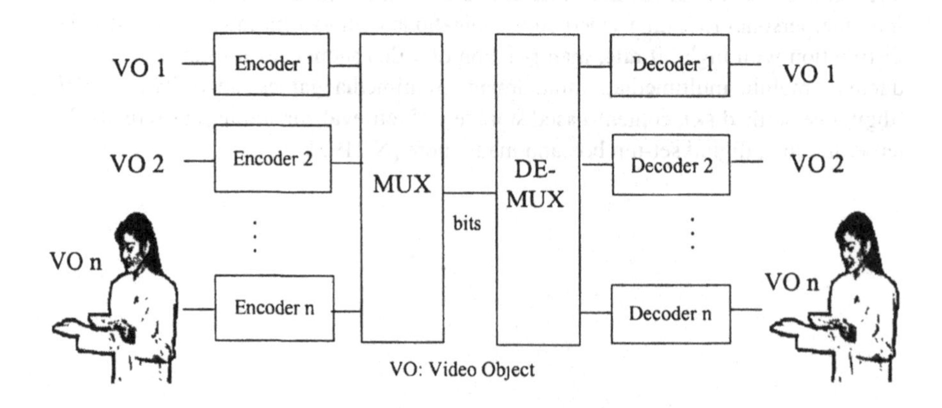


*Figure 1.3: MPEG-4: Encoding, multiplex, bit stream transmission, demultiplex, and decoding of several audio-visual objects*

### 1.2.2 Advances in mobile communications

For a real-time video communications application the minimum network bandwidth is estimated to be at least 24 kbit/s for low motion video with small session size and to be at least 48 kbit/s for generic video telephone applications. However, by adding bandwidth the quality increases.

At the moment, in Europe the GSM (Global System for Mobile Communications) standard is widely used and supports a maximum of 9.6 kbit/s for data transmission. GSM is expected to be enhanced to GSM 2+ by the end of 1998 to support transparent data services with 14.4 kbit/s. The next step may be HSCSD (High Speed Circuit Switched Data) which is currently being standardized by ETSI (European Telecommunication Standards Institute) and is expected to be finalized in 1999 providing a data rate support of 76.8 kbit/s (8 * 9.6 kbit/s), which is sufficient for real-time video communications. DECT (Digital European Cordless Telecommunications) supports a data rate of above 24 kbit/s and was basically developed for home usage, covering a small area. DECT also offers a lower delay compared to GSM and may therefore be of interest for real-time video applications within intranets. A disadvantage of DECT may be that it cannot be used for mobile applications that move with a speed higher than 7 km/h.

For mobile video communications a key factor in MPEG-4 is the enhanced error robustness and the use of audio-visual objects. The object-based coding methodology allows priorisation of objects, e.g. for an object which requires high quality (foreground), a better transmission channel and/or a higher bit rate could be selected, than for other objects (background). However, this feature has to be supported by the mobile network.

Until mobile networks offer a widespread support of higher data rates, MPEG-4 usage may grow in the non-mobile accessed Internet, as currently available data modems for POTS (Plain Old Telephone Service) support enough bandwidth (56.8 kbit/s) for real-time video applications. The next generation of high speed modems based on ADSL (Asymmetric Digital Subscriber Line) or subsequent technologies, which have to be supported by the physical network access provider may provide the required bandwidth of up to 1 Mbit/s enabling the broadcast of video applications.

### 1.2.3 Market push

Not only does every new step in VLSI technology enable exciting new and computationally more demanding applications, but VLSI technology also needs for every new development step new mass applications, which sell a high number of silicon chips to return the development investments. The highest return of investment occurs, when products which are ahead of the market (e.g. because of high performance), can be sold to a large number of customers with high profit. For example, for medium-performance CPUs (Central Processing Unit) the profit margin is relatively small, and may eventually be too small to earn the money already invested in the development of the next generation of processor and silicon technology.

Therefore, processor manufacturers, e.g., promote applications which require the latest processor of their brand. The need to find and promote these new "selling" applications grows more and more important, as every new step in silicon technology requires higher monetary investitions, which have to be returned.

Fig. 1.4 depicts the processor performance evolution and some "VLSI-selling" applications. For example, at around 1990 Microsoft massively promoted their graphical user interface and operating system "Windows". This operating system was targeted for the Intel processor architecture and resulted in a high number of sold high-performance processors from Intel (80286 and 80386) at that time, leading to a significant growth of Intel.

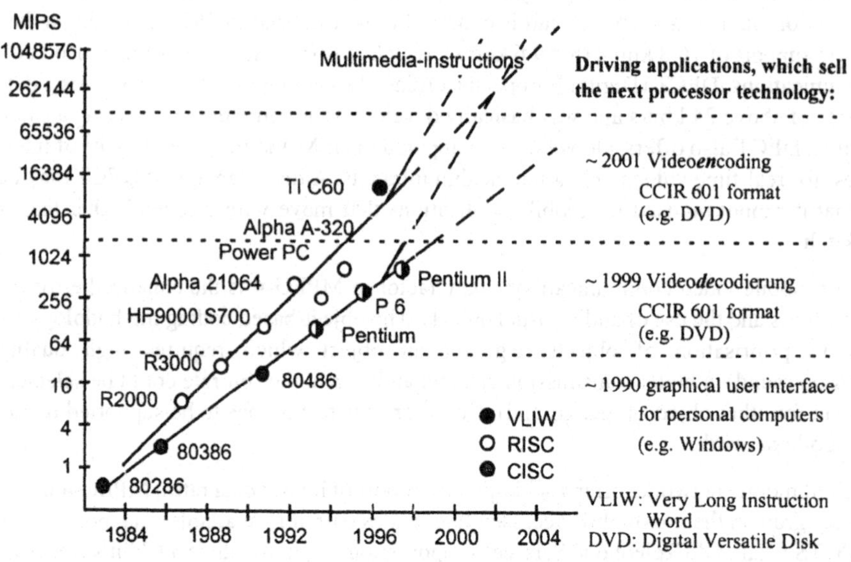


*Figure 1.4: Processor performance development and enabling applications*

Today, in 1999, the personal computer market finds itself in a situation in which the computational power of state-of-the-art CPUs, which have become affordable for the mass-market, are sufficient for typical applications like word processing, Internet, etc. This results into dropping demands for high performance CPUs as well as in dropping profit margins, as the processor technology for the required computational power can be delivered by several competing companies.

Now video decoding for formats comparable to CCIR 601 (720x480 pel PAL TV format or 720 x 576 NTSC TV format) comes into the range of the processing capabilities of the current high-performance CPU generation, especially where special multimedia processor instructions are available. Therefore, processor manufacturers currently promote video applications (e.g. by free software distribution), to create a demand for their high-performance CPUs.

However, as far as video encoding is concerned, the processing power of currently available CPUs is not sufficient for real-time applications beyond CIF (common intermediate format, 352 x 288 pel) session size, due to the high computational amount (60-80%) for motion estimation [Pir 95:1], [Kuhn 98a]. Therefore, until programmable processors for the mass market reach the computational power to perform motion estimation in real-time, a high demand for cost effective ASIC (application specific integrated circuit) solutions for motion estimation is expected to emerge in the meantime, especially for applications, for which, e.g., low power consumption or high throughput is required.

# 1.3 TECHNICAL ISSUES TO SOLVE

In the previous section the opportunities which lie in mobile MPEG-4 video communications have been described. However, there are some technical issues to be solved which are described within this section. Hardware aspects for video coding are described in [Pir 95:1], [ZKuhn 97a], [ZKuhn 97b] and [Kneip 98].

## 1.3.1 Computational power requirements

### 1.3.1.1 Video encoder

The computational power requirements for real-time operation of an early implementation of the MPEG-4 video encoder for the QCIF (Quarter Common Intermediate Format, 176 x 144 pel) format are depicted in Fig. 1.5. However, a fully optimized software implementation can reduce the total computational requirements significantly, but usually VLSI designers start with an unoptimized code as used here. For details and test conditions cf. chapter 4. Note that for the video encoder the motion estimation shares about 66% of the total computational complexity. Therefore, it seems to be sensible to reduce the computational power requirements of motion estimation by advanced fast motion estimation algorithms, which is the central scope of this book.

Fig. 1.6 depicts the computational complexity of an early MPEG-4 decoder at QCIF session size. Note that the optional deringing and deblocking postfilters demand a very high computational power and are suited for VLSI implementation. In the section of "others", the RVLD (reversal variable length decoding) also shares a significant part of the computational power and is also a good candidate for VLSI acceleration. However, as the MPEG-4 decoder depicts far less computational complexity requirements than the MPEG-4 encoder, it may also be suitable for processor based implementation.

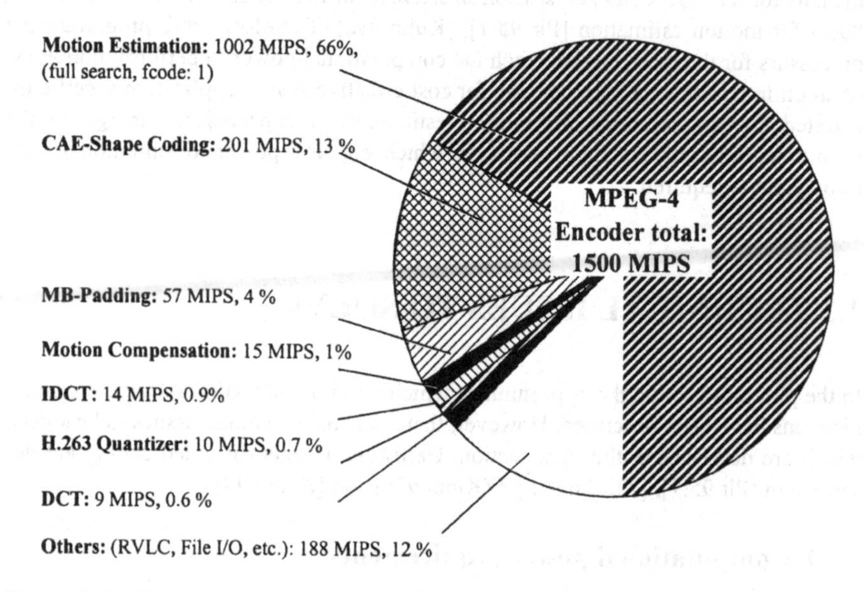


*Figure 1.5: Computational power distribution among single tools for MPEG-4 video encoding*

### 1.3.1.2 Video decoder

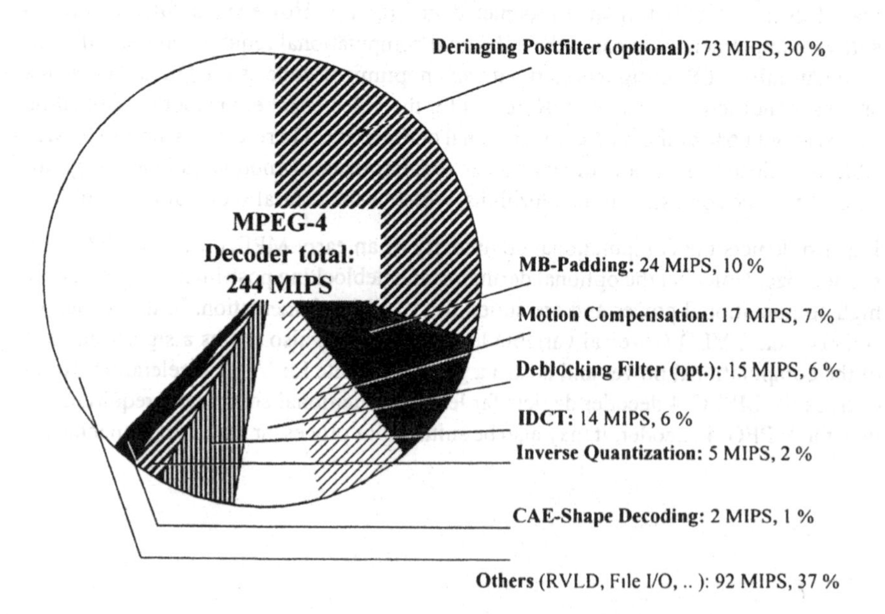


*Figure 1.6: Computational power distribution among single tools for MPEG-4 video decoding*

### 1.3.1.3 Variability of the complexity over time

As MPEG-4 video objects change their size and number over time, the encoder and decoder depicts a variable complexity over time, which leads to a new design methodology developed within this book (chapter 3). Fig. 1.7 depicts as an example the computational complexity of the sequence "weather" over time (video object 1), cf. chapter 4 for details.

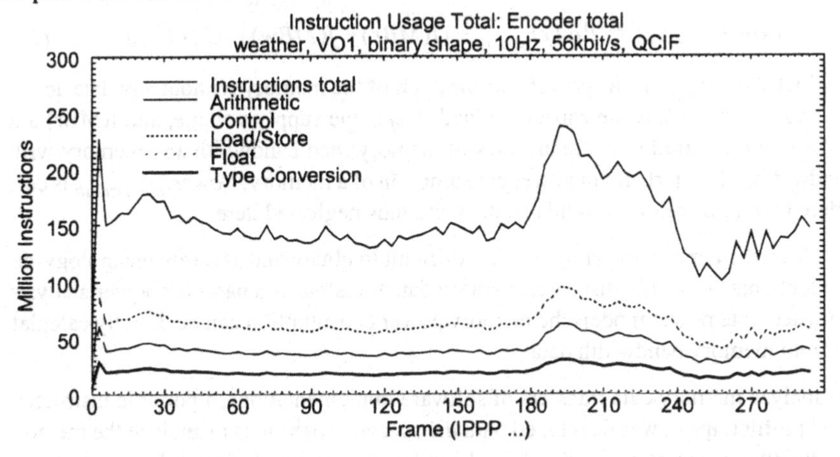


*Figure 1.7: MPEG-4: variable computational complexity over time (encoder)*

## 1.3.2 Power Consumption

### 1.3.2.1 VLSI design

For mobile video communications the requirement emerges to provide the necessary computational power under low power constraints. Therefore, an aim of this book is to analyze and to optimize the power consumption of motion estimation algorithms.

***Power analysis***

There exists a number of power analysis tools on transistor and gate level, providing very detailed results. However, to determine the power requirements of a given algorithm implemented as a software program for a specific processor is very time consuming with these tools and requires the low level circuit details of that processor, which are often unavailable. The exact power consumption of a dedicated VLSI architecture for a specific algorithm can be (at the time being) only determined after laborous VLSI design.

It was observed by [Nacht 98] that the IDCT, as the computationally most dominant tool in the H.263 decoder recommendation, depicts a power consumption in the range of about two magnitudes lower than the power consumption of the (unoptimized) memory access of this IDCT module. Therefore, the analysis and minimization of memory power consumption for low-power video architectures is given high priority here. The memory access bandwidth determines mainly the power consumption in most modern low-power RAMs, and not the clock rate of the RAM [Itoh 95].

The power consumption of a CMOS circuit in general is given as:

$$Power = Power_{static} + Power_{dynamic} + Power_{shortcircuit} \qquad (1.1)$$

in which $Power_{static}$ is the static power consumption, $Power_{dynamic}$ is the dynamic power consumption, and $Power_{shortcircuit}$ is the short circuit power consumption [Rab 96]. The memory power consumption can therefore be defined as:

$$Power_{memory} = Power_{static} + (f(MBr) + f(MBw)) \cdot C_0 \cdot V^2_{dd} \qquad (1.2)$$

in which $Power_{static}$ is the power consumption of the memory without any data access (standby-mode), $C_0$ is the capacitive load, $V_{dd}$ is the supply voltage, and f(MBr) and f(MBw) are referred to as frequencies of memory read bandwidth and memory write bandwidth. The short circuit power consumption of a memory, $Power_{shortcircuit}$, is considered not to be relevant in this context and thus neglected here.

As accurate memory power models are difficult to obtain and severely technology-dependent, this book only discusses memory data transfers as a basis for power analysis. With accurate power models the memory power consumption can be directly calculated from memory bandwidth data.

To analyze memory data transfers of software implementations, a portable instruction level profiler, iprof, was developed with the special capabilities to analyze the memory bandwidth consumption for 8/16/32/64-bit data words, cf. chapter 3. However, to analyze the memory data transfer for VLSI implementations of fast motion estimation algorithms, a VLSI architecture has to be developed and the data flow of the fast motion estimation algorithm has to be mapped to this architecture. This was performed for a two-dimensional processor element array (2D PE array) in chapter 7 and an one-dimensional processor element array (1D PE array) architecture within chapter 8.

***Optimization for low power***

Once having determined a basic VLSI architecture supporting several fast motion estimation algorithms, this architecture can be further optimized in terms of low power consumption. As the motion estimation algorithm consists of relatively simple arithmetic operations but a huge amount of power intensive memory accesses, the optimization of memory access has high priority. Therefore, for motion estimation the required data is usually stored locally within a motion estimation accelerator.

However, this book describes how the local memory access by the data path can be further optimized using special data flow designs described within chapter 8. Another point of optimization is the possible reduction of the number of memory modules, as every single memory module requires an address generation unit, address decoding unit and row and column drivers. A low number of memory modules may also eliminate the need for an interconnection network to link the memory modules with the arithmetic units. A further point of optimization is the combination of several address generation units, which otherwise consume significant area in case several fast motion estimation architectures are supported.

### 1.3.3 Other issues

Besides the VLSI circuits of a mobile video communicator, the display unit consumes a considerable amount of power for backplane illumination. However, display technology is beyond the scope of this book. Other issues of work for the envisaged mobile video communicator are image stabilization, battery technology, radio technology, and a proper user interface which are also beyond the scope of this book.

## 1.4 VLSI DESIGN METHODOLOGY FOR MPEG-4

### 1.4.1 New VLSI design requirements for MPEG-4

***Block-based video***
The computational complexity of previously standardized block-based video coding schemes is predictable for I-, P- and B- frames over time. Therefore, an analytical calculation of the computational requirements as a function of the number of MBs was feasible. Typical VLSI designs are based on the calculation of the worst case computational power requirements, using a pipeline of function specific modules, e.g. for motion estimation (ME), DCT/IDCT, etc. and a control processor.

***Object-based video: MPEG-4***
However, the support of arbitrarily-shaped visual objects, as well as various coding options within MPEG-4 introduce now content- (and therefore time-) dependent computational requirements with significant variance, cf. chapter 4. For the encoder an average load design on a statistical basis may often be feasible instead of a worst case design, reducing hardware costs significantly. Therefore, a new methodology for time-dependent computational complexity analysis is required, which was developed within this book and is presented in chapter 3. MPEG-4 supports multiple video objects, which now define new requirements: task scheduling, memory allocation, rate control, etc. for multiple VOs. A dedicated accelerator module for an MPEG-4 tool has now to be made usable for multiple video objects.

### 1.4.2 Objectives, design methodology, organization, and contributions of this book

#### 1.4.2.1 Objectives

To realize the vision of a portable MPEG-4 video communicator, low power consumption of the VLSI circuits of the communicator is mandatory. Within an MPEG-4 system the video encoder requires a significantly higher computational power than the decoder, and it is therefore assumed, that the encoder is mainly responsible for high power consumption. As far as the MPEG-4 encoder is concerned, it has already been depicted in Fig. 1.5 that the motion estimation (ME) tool shares the highest computational power consumption among the MPEG-4 tools.

Therefore, this book investigates ME algorithms and architectures.

While for previously standardized video coding schemes usually a single rectangular video object and a single motion estimation algorithm was employed, MPEG-4 introduces now the support of multiple arbitrarily-shaped video objects (VOs), where for every single VO different quality constraints and therefore different motion estimation algorithms can be selected. This is especially useful for handheld applications, where for example the coding options and the motion estimation algorithm for the background VO (e.g. a landscape) can be optimized for low complexity. For the foreground VO (e.g. a person talking) a high visual quality can be obtained by spending more computational power on motion estimation.

Therefore, for a MPEG-4 ME architecture the requirements arise to support:

1. multiple VOs (video objects),
2. several motion estimation algorithms with different complexity/distortion trade-off, and
3. the alpha-plane defined within MPEG-4.

It has already been discussed (cf. 1.3.2) that power consumption is difficult to optimize on algorithm and system architectural level, and that for video applications the memory access power consumption is dominant. The latter is especially true for motion estimation, which is based on relatively simple arithmetic operations but high memory access demands. Therefore, to become independent from VLSI technology, the power consumption is optimized within this book by minimizing the memory access bandwidth of the motion estimation.

The MPEG-4 encoder, with its support of several modes and video objects, suggests at first glance a processor based implementation. However, at this point it is not clear, whether processor based or dedicated VLSI-based solutions are better suited for low power motion estimation. Processors offer a high flexibility and programmability, which is, however, paid for by higher power consumption, higher area demands and lower throughput compared to dedicated ASICs (application specific integrated circuits). Dedicated ASICs usually have the disadvantage of limited flexibility. Therefore, within this book, the effort has been made to develop VLSI architectures with low area and low computation time demands for motion estimation which are flexible enough to support several algorithms, hence being favorable to programmable processors.

Therefore, the basic thesis of this book is formulated as follows: Dedicated, but flexible VLSI architectures, supporting several motion estimation algorithms are favorable within the MPEG-4 context to programmable processors in terms of: area (A), computation time (T), memory bandwidth (B), and are therefore advantageous in terms of power consumption and throughput. Within this book this thesis is proved for a flexible high throughput MPEG-4 motion estimator (Search Engine I, e.g. for the base station of the portable MPEG-4 video communicator) and for a flexible low-power MPEG-4 motion estimator (Search Engine II, for the communicator itself).

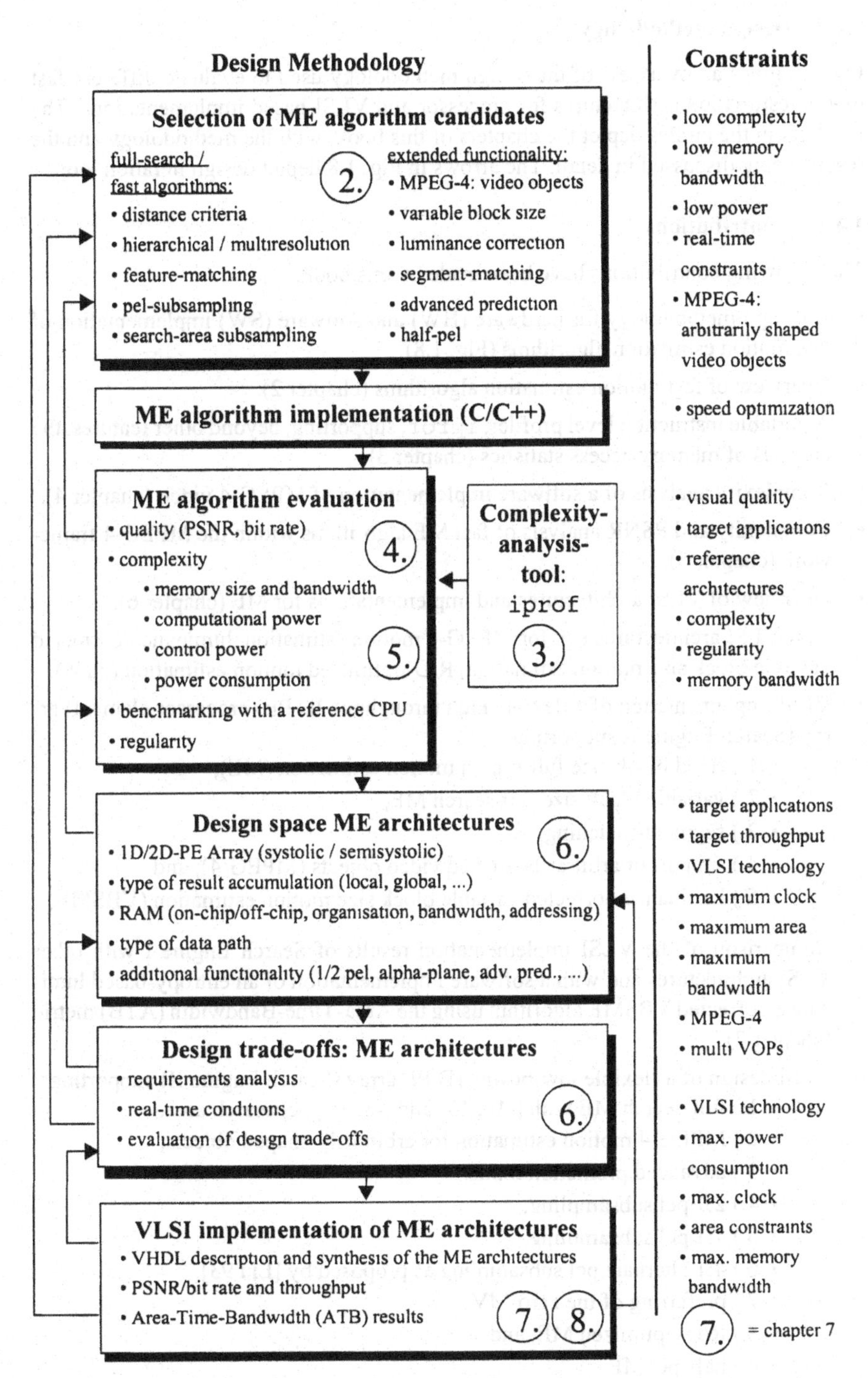


*Figure 1.8: Design space exploration of motion estimation VLSI architectures*

#### 1.4.2.2 Design methodology

Fig. 1.8 gives an overview of the design methodology used to evaluate different fast motion estimation architectures for processor and VLSI-based implementations. The numbers in the circles depict the chapters of this book, with the methodology and the results being discussed in detail. The arrows in Fig. 1.8 depict design iteration loops.

#### 1.4.2.3 Contributions

The following contributions have been made by this book:

- A design methodology for hardware (HW) and software (SW) implementation of fast motion estimation algorithms (Fig. 1.8).
- Overview of fast motion estimation algorithms (chapter 2).
- A portable instruction level profiler, `iprof`, supporting, beyond other features, the analysis of memory access statistics (chapter 3).
- Complexity analysis of a software implementation of MPEG-4 video (chapter 4).
- Complexity and PSNR analysis of fast ME algorithms within the MPEG-4 framework (chapter 5).
- Overview of VLSI architectures and implementations for ME (chapter 6).
- New VLSI architectures e.g. for MPEG-4 motion estimation, luminance-corrected variable block size motion estimation, R/D-optimized motion estimation (ch. 6).
- VLSI implementation of a flexible high throughput 2D PE (processor element) array (Search Engine I) supporting:
    - 1.) fixed block size full-search motion estimation (ME),
    - 2.) variable block size full-search ME,
    - 3.) segment-matching,
    - 4.) support of arbitrarily-shaped video objects (MPEG-4), and
    - 5.) luminance-corrected variable block size motion estimation (VBSME).
- Comparison of the VLSI implementation results of Search Engine I with other VLSI architectures and with a software implementation of an entropy-based luminance-corrected VBSME algorithm using the Area-Time-Bandwidth (ATB) metric (chapter 7).
- VLSI design of a flexible low-power 1D PE array (Search Engine II) supporting:
    - 1.) full-search ME with [-16, 15] and [-8, +7] pel search area,
    - 2.) MPEG-4 motion estimation for arbitrarily-shaped objects,
    - 3.) advanced prediction mode,
    - 4.) 2:1 pel subsampling,
    - 5.) 4:1 pel subsampling,
    - 6.) 4:1 alternate pel subsampling as proposed by [Liu 93],
    - 7.) preferring of the zero-MV,
    - 8.) R/D-optimized ME, and
    - 9.) half-pel ME.

- Comparison of the VLSI implementation results of Search Engine II with a software implementation of the algorithms and other VLSI architectures using an Area-Time-Bandwidth (ATB) metric (chapter 8).

### 1.4.2.4 Organization of the book

The organization of this book is as follows:

- **Chapter 1** gives an introduction to this book.
- **Chapter 2** gives an overview of MPEG-4 motion estimation together with an overview and classification of fast motion estimation algorithms.
- **Chapter 3** describes software complexity metrics and software based complexity analysis, as well as the development of a new tool for this task, `iprof`, a portable instruction level profiler, capable of memory bandwidth analysis.
- **Chapter 4** describes a detailed time-dependent complexity analysis of a software implementation of the MPEG-4 video part.
- **Chapter 5** describes a complexity and visual quality analysis of several fast motion estimation algorithms within the MPEG-4 framework.
- **Chapter 6** provides an overview of and discussion on VLSI complexity metrics and the basic components of motion estimation (ME) architectures for MPEG-4. Based on these basic components dedicated, flexible and programmable VLSI architectures are discussed for a number of motion estimation algorithms. Based on this discussion, the basic architectures for a high throughput MPEG-4 motion estimation architecture supporting optionally luminance-corrected full-search variable block size motion estimation (Search Engine I) and a low-power MPEG-4 ME architecture for several fast motion estimation algorithms (Search Engine II) are determined.
- **Chapter 7** describes the VLSI implementation of a flexible high throughput ME architecture: Search Engine I and the comparison with an Area-Time-Bandwidth (ATB) metric with a software implementation.
- **Chapter 8** describes the VLSI implementation of a flexible low power ME architecture (Search Engine II) and the comparison by an Area-Time-Bandwidth (ATB) metric with a software implementation.
- **Chapter 9** concludes this book.
- **Annex** gives additional information on the evaluation methodology, and the used video test sequences.

# 2

# MOTION ESTIMATION ALGORITHMS

## 2.1 INTRODUCTION

Motion estimation (ME) has proven to be effective to exploit the temporal redundancy of video sequences and is therefore a central part of the ISO/IEC MPEG-1, MPEG-2, MPEG-4 and the CCITT H.261 / ITU-T H.263 video compression standards. These video compression schemes are based on a block-based hybrid coding concept, [Mus 85], which was extended within the MPEG-4 standardization effort to support arbitrarily-shaped video objects.

Motion estimation algorithms have attracted much attention in research and industry, because of these reasons:

1. It is the computational most demanding algorithm of a video encoder (about 60-80% of the total computation time) which limits the performance of the encoder in terms of encoding speed.
2. The motion estimation algorithm has a high impact on the visual performance of an encoder for a given bit rate.
3. Finally, the method to extract motion vectors from the video material is not standardized, thus being open to competition.

Motion estimation is also used for other applications than video encoding like image stabilization, computer vision, motion segmentation and video analysis. However, the requirements for these applications differ significantly for video encoding algorithms: motion vectors have to reflect the real motion within the image sequence, otherwise the algorithms will not show the desired results. In video encoding the situation is different. Motion vectors are used to compensate the motion within the video sequence and only the remaining signal (prediction error) has to be encoded and transmitted. Therefore, the motion vectors have to be selected in order to minimize the prediction error and to minimize the number of bits required to code the prediction error. Within this book the investigations on motion estimation algorithms are focused on video coding. However, similar concepts are applicable to other areas as well. Motion estimation algorithms can be classified into time-domain and frequency-domain algorithms (Fig. 2.1), comprehensive overviews are given in [Mus 85], [Duf 95], [Ohm 95], [Tek 95], and [Mit 97].

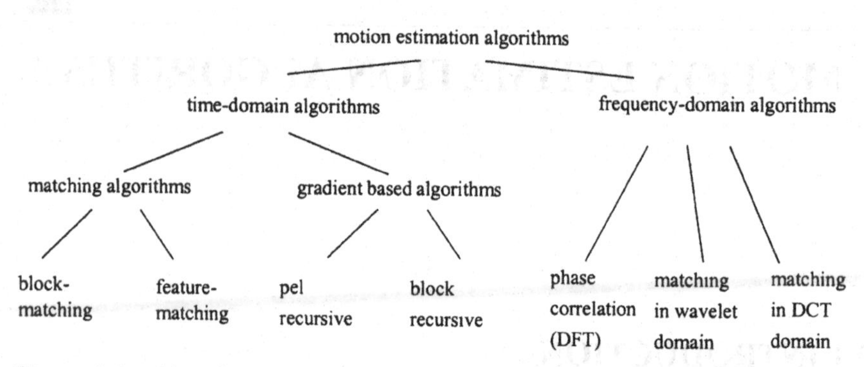


*Figure 2.1: Classification of motion estimation algorithms*

The time-domain algorithms comprise matching algorithms and recursive (gradient-based) algorithms, whereas the frequency-domain algorithms comprise phase correlation algorithms [Girod 93] and algorithms using e.g. the wavelet transformation or DCT. As block-matching algorithms (BMA) and feature-matching algorithms (which attracted more attention recently, cf. 2.4.6) are most commonly used to implement standardized block-based coding schemes (due to their block-based nature and to their inherent parallelism), these algorithms and their variants are described in detail here.

## 2.2 MPEG-4 MOTION ESTIMATION

### 2.2.1 Full-search block-matching algorithm (FSBMA)

Motion estimation is in most cases based on a search scheme which tries to find the best matching position of a 16x16 macro-block (MB) of the current frame with a 16x16 block within a predetermined or adaptive search range in the previous frame, Fig. 2.2. However, as the ME algorithm by itself is not standardized, there exist several variations on the depicted scheme. The matching position relative to the original position is described by an motion vector, which is (after subtraction of the predictor and variable length coding) transmitted in the bit stream to the video decoder. Fig. 2.2 describes the basic principle of the block-matching algorithm (BMA), which is most popular for motion estimation in standardized video compression schemes. $I_k(x, y)$ is defined as the pixel intensity (luminance or Y component) at location (x, y) in the *k*-th frame (*k*-th VOP in MPEG-4 parlance), and $I_{k-1}(x, y)$ is the pixel intensity at location (x, y) at the *k-1*-th frame. For block-matching motion estimation, $I_{k-1}(x, y)$ usually represents a pel located in the search area of the size $R^2 = R_x \times R_y$ pel of the reference frame and $I_k(x, y)$ belongs to the current frame. The maximum motion vector displacement is defined here as *[-p, p-1]*. The block size is defined as $N^2 = N \times N$ pel, with $N=16$ being used for generic standardized motion estimation, and $N=8$ being used for the advanced prediction mode (cf. 2.2.6) resulting in eq. (2.1). The restrictions on these two block sizes will be released later in 2.5.2.

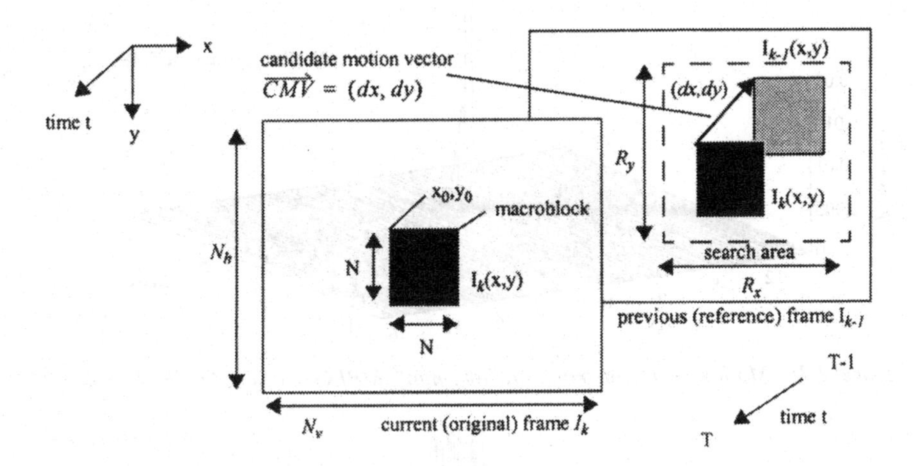


*Figure 2.2: Block-matching algorithm*

$$N \in [8, 16] \; \textit{block size} \tag{2.1}$$

Each individual search position of a search scheme is defined by $\overrightarrow{CMV} = (dx, dy)$ representing the candidate motion vector in horizontal and vertical direction. The distance criteria to determine the best motion vector position is in most cases the SAD (Sum of Absolute Differences), cf. eq. (2.2), (2.3):

$$SAD(dx, dy) = \sum_{m=x}^{x+N-1} \sum_{n=y}^{y+N-1} \left| I_k(m, n) - I_{k-1}(m + dx, n + dy) \right| \tag{2.2}$$

$$\overrightarrow{MV} = (MV_x, MV_y) = \min_{(dx, dy) \in R^2} SAD(dx, dy) \tag{2.3}$$

Other distance criteria are discussed in 2.4.2. The motion vector $\overrightarrow{MV} = (MV_x, MV_y)$ represents the displacement of the best block with the best result for the distance criterion, after the search procedure is finished. Fig. 2.3 and Fig. 2.4 depict example distributions of motion vectors on MPEG-4 test sequences (cf. Annex B on simulation details).

Motion vector distributions are often center-based (Fig. 2.3) and the horizontal motion is often the preferred motion direction (Fig. 2.4). Note that the block-matching technique is depicted in Fig. 2.2 in terms of pixels of a MB of the current frame (for which the predicted frame is to be approximated), which is compared to the pixels of a block of the search area of the previous frame. However, the predicted frame (denoted as VOP in case of MPEG-4) can also be the previous frame, which is the case for bidirectional, B-frame (or denoted as B-VOP in case of MPEG-4) motion estimation. Note that in literature sometimes the *current* frame is referred to as reference frame and in other publications sometimes the *previous* frame is referenced to as reference frame).

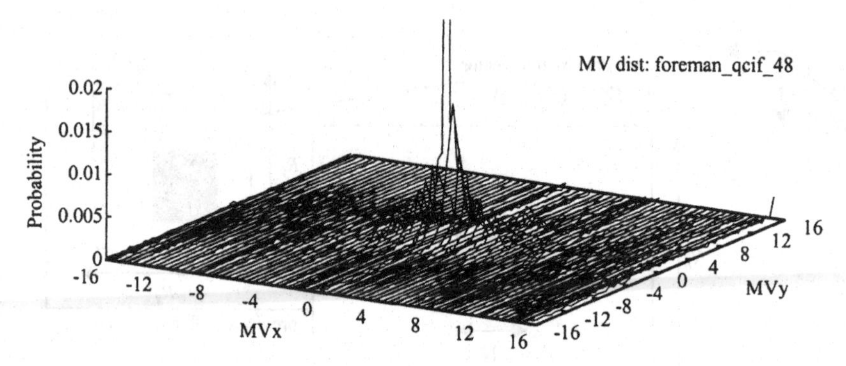


*Figure 2.3: Motion vector distribution "foreman", MPEG-4 VM 8, QCIF, 48 kbit/s*

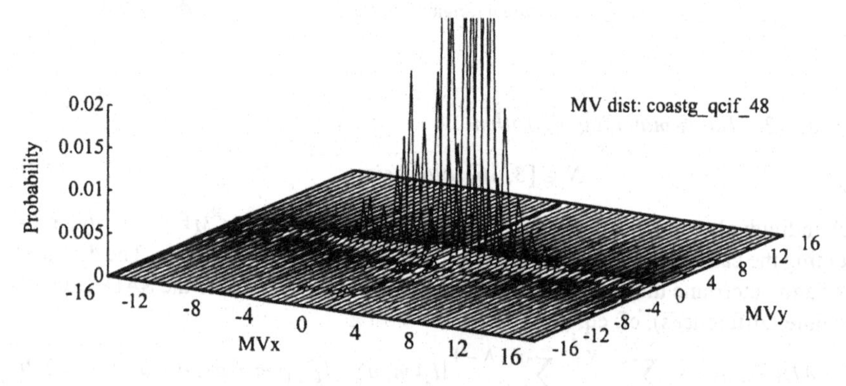


*Figure 2.4: Motion vector distribution "coastguard", MPEG-4 VM 8, QCIF, 48 kbit/s*

## 2.2.2 Prerequisites

Block-matching algorithms assume that all pels within a block have the same motion. Block-matching algorithms result in proper behavior provided that following prerequisites are met:

- Object displacement is constant within a 8x8 or 16x16 block of pels (this condition will be released in section 2.5.2).
- Pel illumination between successive frames is spatially and temporally uniform (released in 2.5.4)
- Motion is restricted to the type translational.
- Matching distortion increases (monotonically) as the displaced candidate block moves away from the direction of the exact minimum distortion. This is especially important for most of the fast motion estimation algorithms, cf. section 2.4.

For real-life video sequences most of these conditions are usually not met, but a large number of motion estimation algorithms still give proper results. The resulting higher prediction error is coded by DCT/IDCT which increases the bit rate.

### 2.2.3 Computational Complexity

Full-search (or exhaustive search) motion estimation is often employed for selection of the best motion vector according to the minimum SAD by iterating over all candidate motion vectors of the search area. However, considering all possible candidate motion vectors and calculating a distortion measure at every search position exhibits a high computational burden of typically 60 - 80% [Pir 95:1], [Kuhn 98a] of a video encoder's computational load. The SAD distance criterion (Sum of Absolute Differences) consists of basically 3 operations: SUB, ADD, ABS, cf. eq. (2.2). Eq. (2.2) gives the number of operations per second (to calculate the distance criterion only) for the full-search motion estimation algorithm:

$$Op = 3 \cdot 2p \cdot 2p \cdot N_h \cdot N_v \cdot f \tag{2.4}$$

with the horizontal/vertical image size being $N_h$ and $N_v$ in pel, and $f$ representing the frame rate (frames per second, fps). The memory bandwidth (only to access every pel for the distance criterion calculation) is given by eq. (2.5):

$$Mem = 2 \cdot 2p \cdot 2p \cdot N_h \cdot N_v \cdot f \tag{2.5}$$

For a typical real-time video application with p=16, f=30, $N_h$=352, and $N_v$=288 (CIF) eq. (2.6) and (2.5) result in a computational load of 9.34 billion integer arithmetic operations (with 8 and 16 bit data) per second, and a memory bandwidth of 6.22 billion 8 bit accesses per second. These numbers have not taken into account the implementation-dependent number of operations and memory accesses for address calculation, result comparison, coding decisions, and control. Computational complexity of various fast motion estimation algorithms is investigated theoretically, as well as experimentally by a new method [Kuhn 98a] for processor and VLSI implementation in the next chapters.

### 2.2.4 Motion Vector Prediction

Within the MPEG-1/MPEG-2/MPEG-4 or H.26X framework, non-linear MV prediction, based on the median calculation, is used to efficiently code the motion vector, [MPEG-4], with the predicted motion vector $\vec{P} = (P_x, P_y)$ being denoted as:

$$P_x = median(MV1_x, MV2_x, MV3_x) \tag{2.6}$$

$$P_y = median(MV1_y, MV2_y, MV3_y) \tag{2.7}$$

in which the neighboring MBs are numbered as follows (the borders are treated specially):

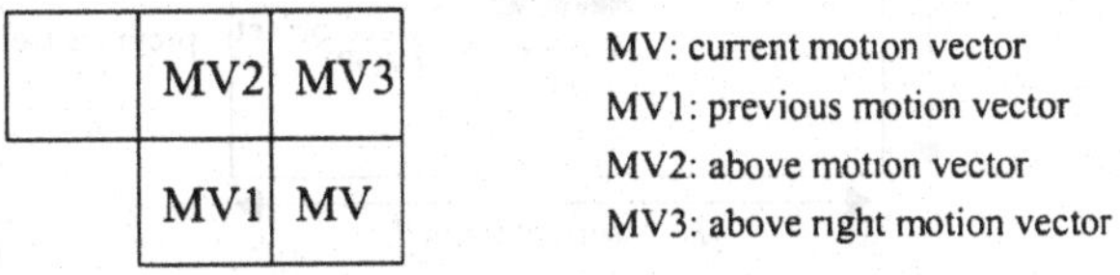

|  | MV2 | MV3 |
|---|---|---|
|  | MV1 | MV |

MV: current motion vector
MV1: previous motion vector
MV2: above motion vector
MV3: above right motion vector

*Figure 2.5: MPEG / ITU-T H.263 method for MV prediction*

In MPEG/H.26x the predicted motion vector $\vec{P}$ is subtracted from the actual motion vector $\overrightarrow{MV}$ and the resulting difference, $\overrightarrow{MVD}$ (motion vector difference), is coded using a variable length code.

$$\overrightarrow{MVD} = \overrightarrow{MV} - \vec{P} \tag{2.8}$$

### 2.2.5 MPEG-4: Block-matching for arbitrarily-shaped objects

Natural video scenes consist (for example) of stationary background and moving foreground objects, which are of arbitrary shape. In previously standardized video coding schemes the error measure criterion (e.g. SAD) for motion estimation is calculated on all pels, regardless whether they belong to the background or to a foreground object, causing the resulting motion vector often not to represent the real motion of the object's pels. To avoid this (but mainly for other reasons like enabling of object manipulation, static background memory, sprites, etc...) MPEG-4 introduces an object shape description of a video object.

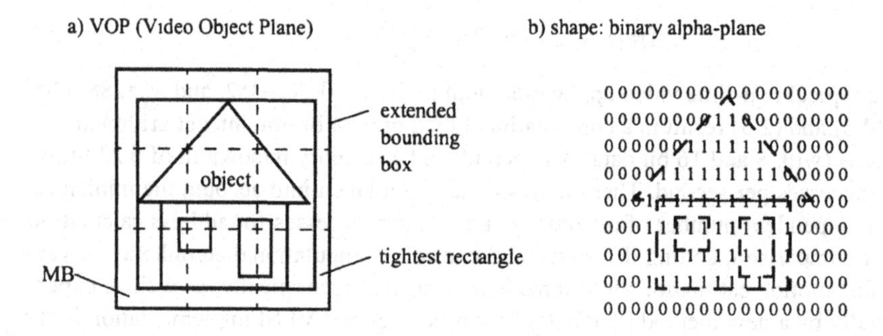


*Figure 2.6: Coding of arbitrarily-shaped objects within MPEG-4: a) VOP (Video Object Plane) and b) binary alpha-plane*

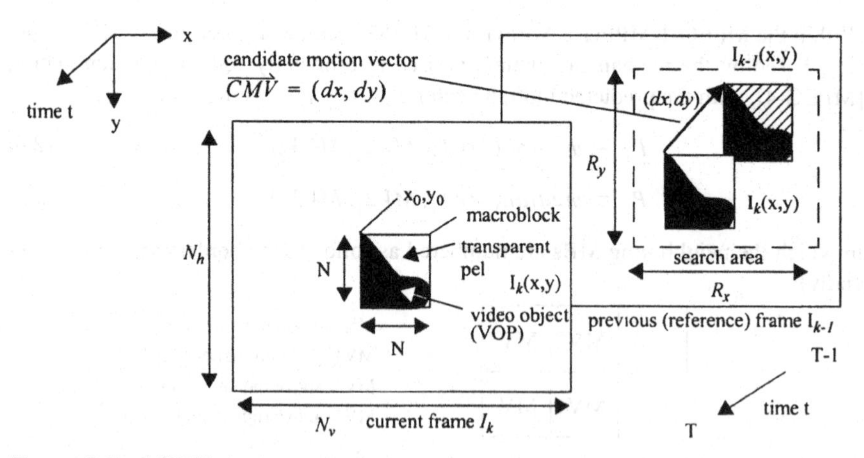


*Figure 2.7: MPEG-4 motion estimation*

In MPEG-4 the object shape description is called alpha-plane. The alpha-plane of a video object can be provided by (semi-) automatic segmentation of the video sequence. This technique is not covered by the MPEG-4 standardization process and depends on the application. The alpha-plane refers to the pel of the current VOP at time instance $k$, and contains the information which of the pixels form an object ($alpha_k > 0$), and which pixels are outside of an object ($alpha_k = 0$). The greyscale alpha-plane allows translucent objects, whereas the binary alpha-plane, as depicted in Fig. 2.6, is restricted to values of 0 and 1. At the encoder the shape information allows to restrict the error measurement of the motion estimation process to pixels inside objects. For transmission of the object, the object's shape is coded by special shape coding techniques, which are beyond the scope of this section. The SAD is calculated only for pixels with non zero alpha-plane [MPEG-4] as depicted by eq. (2.9):

$$SAD(dx, dy) = \sum_{m=x}^{x+N-1} \sum_{n=y}^{y+N-1} |I_k(m, n) - I_{k-1}(m+dx, n+dy)| \cdot (alpha_k \neq 0) \tag{2.9}$$

$$(MV_x, MV_y) = \min_{(dx, dy) \in R^2} SAD(dx, dy) \tag{2.10}$$

However, the pels of the previous frame, which are located outside of the VOP but still inside the bounding box of the VOP, have to be replicated by a *repetitive padding* method [MPEG-4]. For MPEG-4, the motion vector range is defined as:

$$\left[-2^{fcode+3}, -2^{fcode+3} - \frac{1}{2}\right] \tag{2.11}$$

with

$$0 \leq fcode \leq 7 \tag{2.12}$$

Within MPEG-4 the zero motion vector for $N=16$ is favored by reducing the $SAD(0, 0)$, when there is no significant difference:

$$SAD_{N=16}(0, 0) \equiv SAD_{N=16}(0, 0) - \left(\frac{N_B}{2} + 1\right) \tag{2.13}$$

with $N_B$ being the number of pels inside the VOP multiplied by $2^{(bits\_per\_pel-8)}$. The $(dx, dy)$ pair resulting from eq. (2.10) and (2.11) is chosen as the integer pel motion vector MV0 for the 16x16 macro-block.

## 2.2.6 Advanced prediction motion estimation (4MV)

For H.263 and MPEG-4 an advanced prediction motion estimation mode exists, Fig. 2.8, where for the four 8x8 blocks the four motion vectors MV1, MV2, MV3, MV4 are found, using the same motion estimation technique as described before. In a first step the 16x16 motion vector, MV0, is determined and in the successive step the four 8x8 MVs are searched in a [-2, +2] pel search range around MV0. The SAD results of the four 8x8 blocks being summarized and compared with the SAD for the 16x16 block. The smaller SAD value decides, whether this displaced MB is coded either by MV0 or by MV0, MV1, MV2, MV3.

The main advantages of this method in terms of visual quality are: 1.) improved representation of inhomogeneous motion fields and 2.) OBMC (Overlapped Block Motion Compensation) resulting in smoothed block boundaries. However, the coding of four MV in the bit stream is more costly in terms of bit rate than coding only one single MV. It will be shown later, that this method causes irregular data flow requiring specific motion estimation VLSI architectures.

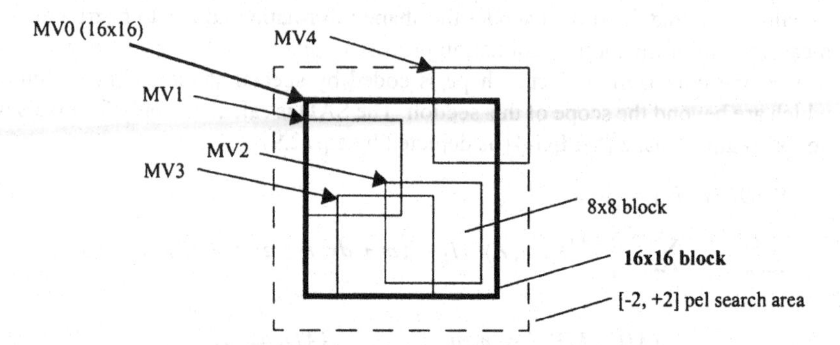


*Figure 2.8: Advanced prediction motion estimation (four 8x8 blocks)*

To decide between using one 16x16 MV (MV0) or four 8x8 MVs, (MV0, MV1, MV2, MV3), the $SAD_{N=8}$ of these four 8x8 blocks are summarized and compared with the $SAD_{N=16}$. Finally the mode with the smallest SAD is chosen.

## 2.2.7 Fractional-pel motion estimation

Fractional-pel motion estimation is advantageous to gain an improved prediction image. A general analysis on the benefits of fractional-pel motion estimation algorithms can be found e.g. in [Girod 93] and [Bha 95] p122.

***Half-pel motion estimation***

The half-sample search (i.e. half-pel ME) is performed for the best 16x16 MB or the best 8x8 blocks (in case of advanced prediction) of the previous reconstructed frame (or VOP), [H.263], [MPEG-4]. The search area is [-0.5, +0.5] pel around the target matrix pointed to by MV0, ... MV4. Pel values for half-pel interpolation at the block boundaries are gained by mirroring the pel values inside the block boundaries.

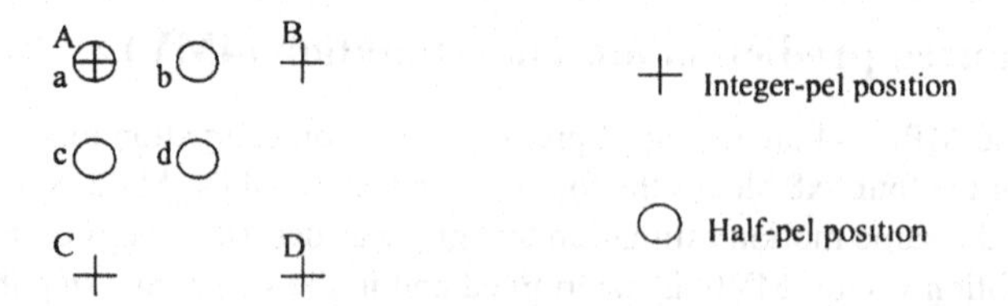


*Figure 2.9: Half-pel interpolation*

The pel values at the half-pel locations are calculated by bilinear interpolation, where "/" denotes the division by truncation:

$$a = A \tag{2.14}$$

$$b = (A+B+1\text{-}rounding) / 2 \tag{2.15}$$

$$c = (A+C+1\text{-}rounding) / 2 \tag{2.16}$$

$$d = (A+B+C+D+2\text{-}rounding) / 2 \tag{2.17}$$

***Quarter-pel motion estimation***

Within the algorithm evaluation for the emerging MPEG-4 standard ("Core experiment process") a quarter-pel motion estimation method was proposed [Benz 97]. This technique shows advantages for CCIR 601 size video sequences at bit rates in the range of 500 kbit/s - 2 MBit/s

## 2.3 RATE/DISTORTION-OPTIMIZED MOTION ESTIMATION

Rate/distortion (R/D)-optimized ([Berg 71], [Bha 95] p93) motion estimation algorithms [Gir 94], [LKLiu 97], [Kos 97], [Sri 98] are employed to find the optimum motion vectors, not only by using a distance criteria like the SAD, but also by taking into account the resulting number of bits in the bit stream, using this particular motion vector candidate.

R/D-optimization is usually performed employing the Lagrange optimization technique, which is subject to be applied using only the bits of the variable length coded (VLC) motion vector difference (MVD) bits as "rate", or (better) also using the number of bits contributed by the resulting quantized DCT coefficients in the bit stream. However, gaining the exact number of bits after DCT coding and quantization, would require a DCT transformation with every search step of the motion estimation algorithm. Currently this is computationally not feasible. Therefore, either bit rate models for the DCT were developed (e.g. [Chu 96]), or the DCT coefficients were generally not taken into account for rate calculation. Neglecting the bit rate for prediction error coding in R/D-optimized ME is only suitable for very low bit rate video coding, as the number of bits for the DCT coefficients increases significantly with higher bit rates, compared to the VLC coded MVD bits. R/D-optimized motion estimation algorithms offer the advantage of improved image quality at the same bit rate and are still compliant with the MPEG / ITU-T video coding standards.

***Lagrange optimization***

Eq. (2.18) depicts the cost function for R/D-optimization, *J(dx, dy)*, for a motion vector candidate $\overrightarrow{CMV} = (dx, dy)$. *D(dx, dy)* is the distortion for this $\overrightarrow{CMV}$, $\lambda$ is the predetermined Lagrange operator (which is usually in some relation with the inter-quantization parameter in case of R/D-optimized motion estimation) and *R(dx, dy)* is the calculated number of bits in the bit stream for this motion vector candidate. The goal is now to minimize the cost function *J(dx, dy)* by employing the full-search algorithm or one of the search area subsampling algorithms to obtain the best $MV(dx,dy)$.

$$J(dx, dy) = D(dx, dy) + \lambda \cdot R(dx, dy) \qquad (2.18)$$

Rate-distortion optimization has some tradition in the selection of the optimal block size in variable block size motion estimation (VBSME) algorithms, cf. section 2.5.2. Application of R/D for *fast* motion estimation algorithms (cf. 2.4.9) is an emerging new area of research.

***Note on VLSI implementation***

VLSI implementation of entropy-constrained motion estimation for low bit rate video can be performed with little effort, as basically only the precalculated number of bits in the bit stream have to be provided by a lookup-table and added to the distortion measure $D$. As usually a predetermined $\lambda$ is used, the multiplication of eq. (2.18) has to be performed only once per frame / VOP, when the lookup-table is updated. With every search step of the motion estimation algorithm only one additional addition operation is required therefore.

# 2.4 FAST MOTION ESTIMATION ALGORITHMS: AN OVERVIEW

For battery powered real-time visual communications applications, there are stringent requirements on low power consumption. The number of 9.34 billion arithmetic operations per second and the memory bandwidth of 6.22 GByte/s (as calculated before, eq. (2.4) and (2.5)) for the full-search motion estimation algorithm to encode a CIF sized video at 30 fps, is not feasible at low power constraints using today's processor technology.

Therefore, fast motion estimation algorithms with reduced computational complexity, as well as their advantages for VLSI design in terms of area and power consumption, are considered here. However, reduced computational complexity gained by these fast motion estimation algorithms has often to be paid for with losses in terms of visual quality and/or by irregularities in data flow. This makes it difficult to achieve efficient VLSI implementations, a problem that is discussed in detail in the next chapters.

In recent years, an abundant number of fast motion estimation algorithms was proposed in literature, some being targeted on software implementation and others on VLSI implementation. In a recently published book [Furht 97] five well-known (and a proposed new one) fast motion estimation algorithms have been compared in terms of visual quality. However, an in-depth overview of the state of research, classification, MPEG-4 applicability, complexity analysis, and VLSI architecture mapping of currently used fast motion estimation algorithms cannot be found there.

Therefore, the aim of this book is:

1. To give an overview of the state-of-the-art of fast motion estimation algorithms, cf. the rest of this chapter.

2. To analyze the trade-off of complexity-reduced motion estimation versus image quality degradation with respect to the PSNR/bit rate for rectangular video objects, as well as for binary-shaped video objects supported by MPEG-4.

3. To evaluate the implementation complexity on typical processor architectures and for (parallelized) VLSI implementations, cf. chapter 5 and 8.

## 2.4.1 Classification of fast motion estimation algorithms

Fig. 2.10 gives an overview of fast motion estimation algorithm classes and their specific implementations which are described in detail in this chapter.

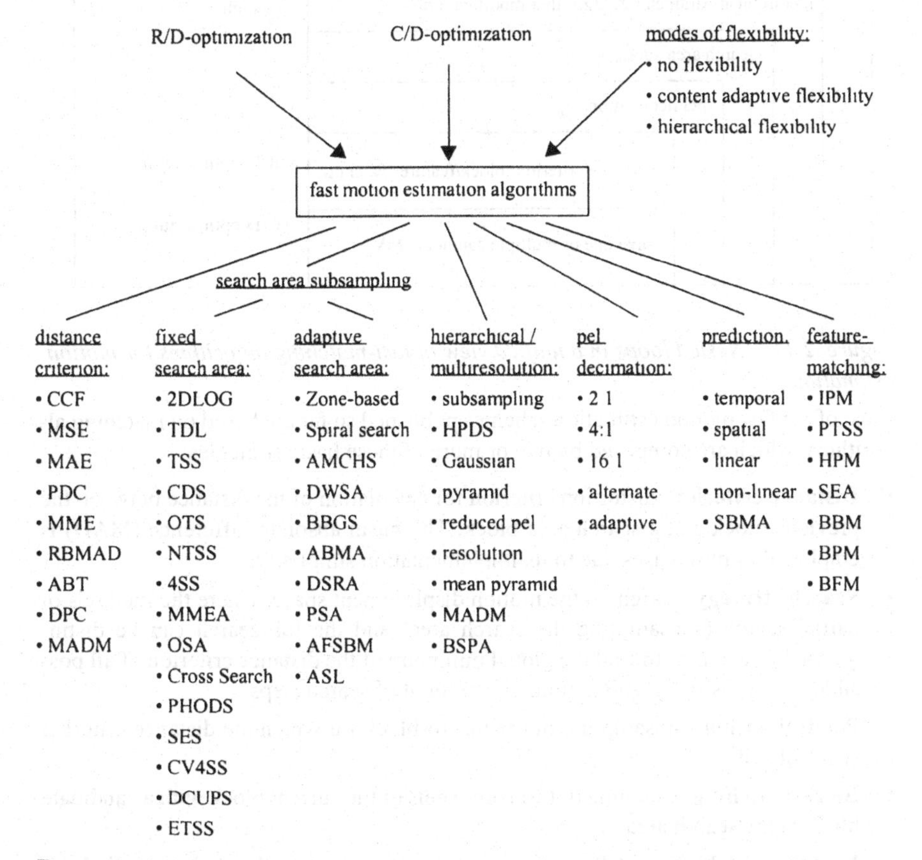


*Figure 2.10. Overview of fast motion estimation algorithms (acronyms are described within the next pages)*

The techniques depicted here are "standard conformant" and had been mainly developed for the H.261, H.263, MPEG-1, MPEG-2 video compression schemes, i.e. for 16x16 block-based video. These algorithms are applied solely on the luminance (Y) component of the signal. In one of the following chapters the use of some of these basic algorithmic elements is extended for arbitrarily-shaped video objects as supported within MPEG-4.

Fig. 2.11 depicts an unified view of the nested loops of fast-matching algorithms for motion estimation (other classification schemes can be found, e.g. in [Mit 97] p307).

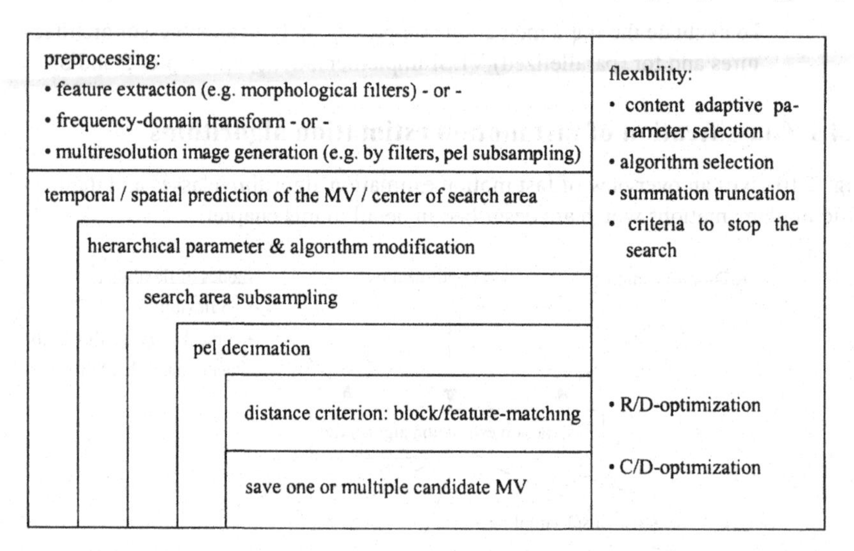


*Figure 2.11: Nested loops in a unified view of fast-matching algorithms for motion estimation*

Most of the fast motion estimation schemes published so far are based on matching algorithms, which are composed by one or more of these basic strategies:

- **Distance criterion**: distortion criterion for calculation of the distance between the previous block and a search area block. The mean absolute difference (MAD) is employed in most cases due to its implementation simplicity.
- **Search strategy**: search on the motion displacement space, where the methods of partial search (subsampling the search area) and the full-search can be distinguished. The aim is to find the global minimum of the distance criterion of all possible block positions with a minimum amount of search steps.
- **Pel decimation**: subsampling of the macro-block for which the distance criterion is calculated.
- **Block-matching**: matching of (all/some) pels of the current block with a candidate block in the search area.
- **Feature-matching**: matching of meta information extracted (e.g. by morphological filters) from the current block and the search area pels.

- **Temporal prediction**: prediction of the candidate MV and/or the center of the search area using temporal correlation techniques.
- **Spatial prediction**: prediction of the candidate MV and/or the center of the search area employing spatial correlation techniques.
- **Multiple candidate MV**: during the search procedure several candidate MVs are saved, from which the final MV is selected at the end of the search procedure.
- **Frequency-domain search**: motion estimation after transformation of the current block and search area pixels into the frequency-domain (e.g. by DCT, Wavelet).
- **Multi-resolution algorithm:** The motion estimation algorithm works on image representations with different resolutions, gained by preprocessing (e.g. subsampling, filtering).

The above basic strategies can be combined with the following different options of flexibility:

- **Fixed algorithm**: fixed parameters and algorithms.
- **Adaptive algorithm**: content-adaptive parameters and/or algorithms.
- **Hierarchical algorithm**: hierarchically-adaptive parameters and/or algorithms.

Apart from these parameters for fast motion estimation algorithms recent publications show growing interest in:

- **R/D-optimized ME**: rate/distortion-optimized selection of the motion vectors by taking into account distortion, coding costs of the candidate motion vector, and the coding of the prediction error.
- **C/D-optimized ME**: computational complexity/distortion optimized selection of the motion vectors within fast motion estimation algorithms.

In the forthcoming sections these parameters for fast motion estimation algorithms are discussed in detail.

### 2.4.2 Distance Criteria

Block-matching motion estimation algorithms gain the motion vector by minimizing a cost function, which is generally based on a distance criterion. Various distance criteria have been proposed and analyzed in literature. All of them they vary in terms of implementation complexity and efficiency, concerning the approach used to find the global optimum. This section gives an overview of several distance criteria. In the next chapters these distance criteria will be evaluated with regard to: 1.) quality (in terms of the PSNR/bit rate), 2.) processor implementation, and 3.) VLSI implementation.

Basically the following distance criteria can be used in any fast motion estimation algorithm.

***CCF (Cross-Correlation Function)***

The CCF (Cross-Correlation Function) is derived from the correlation between the two random variables $I_k(m, n)$ and $I_{k-1}(m + dx, n + dy)$, and is depicted in eq. (2.19) and (2.20) [Rao 90] p242, [Fuhrt 97] p58. As the CCF suffers from high computational complexity, the practical importance for real-time video codecs is negligible.

$$CCF(dx, dy) = \frac{\sum_{m=x}^{x+N-1}\sum_{n=y}^{y+N-1} I_k(m, n) I_{k-1}(m+dx, n+dy)}{\sqrt{\sum_{m=x}^{x+N-1}\sum_{n=y}^{y+N-1} I_k^2(m, n)}\sqrt{\sum_{m=x}^{x+N-1}\sum_{n=y}^{y+N-1} I_{k-1}^2(m+dx, n+dy)}} \quad (2.19)$$

$$(MV_x, MV_y) = max_{(dx, dy) \in R^2} CCF(dx, dy) \quad (2.20)$$

***MSE (Mean Square Error Function)***

The MSE (Mean Square Error Function) is known to produce superior results, as the MSE can be interpreted as Euclidean distance between two MBs, which is close to the human visual perception [Haus 94]. However, with one multiplication per pel difference the MSE suffers from high computational complexity.

$$MSE(dx, dy) = \frac{1}{N \cdot N}\sum_{m=x}^{x+N-1}\sum_{n=y}^{y+N-1} [I_k(m, n) - I_{k-1}(m+dx, n+dy)]^2 \quad (2.21)$$

$$(MV_x, MV_y) = min_{(dx, dy) \in R^2} MSE(dx, dy) \quad (2.22)$$

***MAE (Mean Absolute Error)***

The MAE (Mean Absolute Error), which is also known as MAD (Mean Absolute Difference), is a very popular cost function for block-matching algorithms, due to it's simplicity and straightforward VLSI implementation. The MAE tends to overemphasize small differences over large differences, and may therefore cause in some cases sub-optimal results in terms of convergence of fast ME algorithms and MV accuracy, compared to the more complex MSE [Ghar 90].

$$MAE(dx, dy) = \frac{1}{N \cdot N}\sum_{m=x}^{x+N-1}\sum_{n=y}^{y+N-1} |I_k(m, n) - I_{k-1}(m+dx, n+dy)| \quad (2.23)$$

$$(MV_x, MV_y) = min_{(dx, dy) \in R^2} MAE(dx, dy) \quad (2.24)$$

***SAD (Sum of Absolute Differences)***

For practical purposes the constant division by $N \cdot N$ in eq. (2.23) (and other distance criteria) is often omitted, resulting in the SAD criteria (Sum of Absolute Differences):

$$SAD(dx, dy) = \sum_{m=x}^{x+N-1}\sum_{n=y}^{y+N-1} |I_k(m, n) - I_{k-1}(m+dx, n+dy)| \quad (2.25)$$

$$(MV_x, MV_y) = min_{(dx, dy) \in R^2} SAD(dx, dy) \quad (2.26)$$

***SAD summation truncation***

Typical sequential implementations of the SAD calculation iterate over every single pixel of the MB. The computational complexity of the SAD calculation can be reduced significantly by stopping the SAD calculation, when the current interim SAD value is higher than the minimum SAD calculated so far [Kap 85]. This process is called summation truncation, for which [Eck 95] reported a speed improvement for MPEG-2 encoding of approximately a factor two.

***SAD estimation***
[Leng 98] observed that a partially computed SAD is highly correlated with the actual SAD of an entire block at a specific candidate MV position. The authors proposed a SAD estimation technique for a complexity-distortion-optimized (cf. 2.4.10) fast motion estimation algorithm. The SAD estimation of successive pels (which were gained from subsampling) of a particular block is refined until there is enough confidence in the SAD of that particular block.

***PDC (Pixel Difference Classification)***
PDC (Pixel Difference Classification), [Ghar 90], [Fuhrt 97] p61, is a simple block-matching criterion with reduced complexity.

$$PDC(dx, dy) = \sum_{m=x}^{x+N-1} \sum_{n=y}^{y+N-1} T(dx, dy, m, n) \tag{2.27}$$

with *T (dx, dy, m, n)* being a binary representation of the pel difference defined in eq. (2.28) and the *threshold* being a predefined value:

$$T(dx, dy, m, n) = \begin{cases} 1 & if \quad |I_k(m, n) - I_{k-1}(m + dx, n + dy)| \le threshold \\ 0 & otherwise \end{cases} \tag{2.28}$$

$$(MV_x, MV_y) = max_{(dx, dy) \in R^2} PDC(dx, dy) \tag{2.29}$$

However, the *threshold* selection depends on the scene itself and may be difficult to choose [Chen 95:1].

***MME (Minimized Maximum Error Function)***
In [Chen 95:1] the MME (Minimized Maximum Error Function) is presented with the advantage of reduced area for VLSI implementation, as only an 8-bit comparator unit is necessary for MME compared to a 16-bit accumulator for the MAE.

$$MME(dx, dy) = max_{(m, n) \in R^2} |I_k(m, n) - I_{k-1}(m + dx, n + dy)| \tag{2.30}$$

$$(MV_x, MV_y) = min_{(dx, dy) \in R^2} MME(dx, dy) \tag{2.31}$$

[Chen 95:1] reports a 15% chip-size reduction at 0.6 - 4% PSNR degradation using MME, compared to the MAE criterion using the full-search method. A label-updating scheme for an efficient bit level maximum/minimum circuit that may be applicable for MME is described in [Lee 94a].

***RBMAD (Reduced Bit Mean Absolute Difference)***
In [Baek 96] the RBMAD (Reduced Bit Mean Absolute Difference) aims to reduce VLSI implementation costs. RBMAD reduces the number of bits used in an absolute difference calculation (i.e. bit truncation) and therefore reduces the VLSI area and power consumption, while it enables a higher operating speed, as the difference unit and the adder can be of reduced wordlength.

For any pixel value $I_k(m,n)$ let $I_k(m, n)_{s, t}$ be the $s$-$t$+$1$ bits $A_s \ldots A_t$ of the binary representation of $I_k(m, n) = A = A_{u-1} \cdot 2^{u-1} + \ldots + A_2 \cdot 2^2 + A_1 \cdot 2^1 + A_0$, where $u$ is the number of bits of a pixel representation. [Baek 96] uses a fixed number of 1 ... 7 bits which are subtracted. Since RBMAD uses only the upper *k* bits of pixels, RBMAD is equivalent to MAD in case of $s = u\text{-}1$ and $t = 0$.

$$RBMAD(dx, dy) = \sum_{m=x}^{x+N-1} \sum_{n=y}^{y+N-1} \left| I_k(m, n)_{u-1, u-t-1} - I_{k-1}(m+dx, n+dy)_{u-1, u-t-1} \right| \quad (2.32)$$

$$(MV_x, MV_y) = \min_{(dx, dy) \in R^2} RBMAD(dx, dy) \quad (2.33)$$

***ABT (Adaptive Bit Truncation)***

[He 97] extended the RBMAD approach by using an adaptive pixel truncation scheme, aiming at low power motion estimation. This scheme determines the number of pixels compared on a frame-per-frame basis, controlled by the mean quantizer step-size of a frame. The authors showed that on average more than four bits can be truncated without affecting picture quality.

***DPC (Different Pixel Count)***

In [LeeC 96] the DPC (Different Pixel Count) scheme was proposed. Each pixel value of the current block and the search window is normalized using the *mean* of the current block pixel values followed by a quantization to a 2-bit resolution. $\hat{I}_k(x, y)$ is the quantized pixel luminance at the position *(x,y)* in the *k*-th frame. The DPC calculation per pixel is a simple 2-bit comparison, which can be implemented by two XOR gates and a single OR gate.

$$DPC(dx, dy) = \sum_{m=x}^{x+N-1} \sum_{n=y}^{y+N-1} \delta[\hat{I}_k(m, n), \hat{I}_{k-1}(m+dx, n+dy)] \quad (2.34)$$

$$\delta = \begin{cases} 1 & if \quad x \neq y \\ 0 & otherwise \end{cases} \quad (2.35)$$

$$(MV_x, MV_y) = \min_{(dx, dy) \in R^2} DPC(dx, dy) \quad (2.36)$$

***MADM (Mean Absolute Difference of the Means)***

The MADM (Mean Absolute Difference of the Means) was applied by [Chun 94] p117 in a hierarchical framework consisting of *l* layers. In this context, every layer is subdivided into *n* x *n* blocks until the smallest block size of 1x1 pel is reached. The MADM is defined at each layer as the mean of the absolute frame differences of the means, obtained from subblocks in the blocks under comparison. The approximated MADM criterion of the *l*-th layer of the quadtree partitioning can be obtained as:

$$MADM^l(dx, dy) = \frac{1}{2^{2(l-1)}} \sum_{m=x}^{x+N-1} \sum_{n=y}^{y+N-1} \left| \bar{I}_k^l(m, n) - \bar{I}_{k-1}^l(m, n, dx, dy) \right| \quad (2.37)$$

with the $\bar{I}_k^l$ and the $\bar{I}_{k-1}^l$ being the mean values of the (m, n)-th subblocks of the corresponding blocks in the current frame and the preceding frame. With a successively refined matching criterion in [Chun 94], a successive refinement of motion vector candidates had been achieved.

***Note on VLSI-implementation***

Among the above mentioned distance criterion functions, the CCF and MSE require many multiplications and are thus costly to implement in real-time processing. The MAE (or MAD, SAD) is the most widely used due to its lower complexity and superior results. The matching criteria with reduced complexity are tested for their suitability in terms of PSNR in chapter 4.

### 2.4.3 Search area subsampling strategies

For full-search block-matching the current MB is shifted to every integer-pel motion vector position of the search area for comparison. Fast motion estimation techniques, however, often employ search strategies which perform a subsampling of the motion displacement space. Generally, these search strategies have to cope with the problem of eventually falling into local minimum of the distance criteria function. Search strategies can be regarded as a hierarchical level above the distance calculation. Numerous search strategies have been developed so far; the early Three Step Search algorithm (TSS), the two-dimensional logarithmic search algorithm (2DLOG), and the conjugate direction search (CDS) have gained some influence on further developments. A selection of some well-known search strategies is presented here.

#### 2.4.3.1 Fixed search area

***2DLOG: 2D Logarithmic Search***

2D logarithmic search (2DLOG) or TDL (two-dimensional logarithmic search) was proposed by [Jain 81]. 2DLOG is a fast search algorithm based on minimum distortion, in which the distortion metric MSE is only calculated for sparse sampling of the full-search area. The step size of the search area is reduced by n/2 with every search step. Logarithmic search had been reported to be efficient and suitable for texture dominated sequences of low complexity with a small search window, such as head- and shoulder-type of video conferencing applications [Netra 88] p335. [Xlee 96] observed a significant performance drop for sequences with complex textures and noisy pictures, e.g. 2DLOG is reported to work fine on the "football" sequence, but showed poor performance for the "flower garden" sequence.

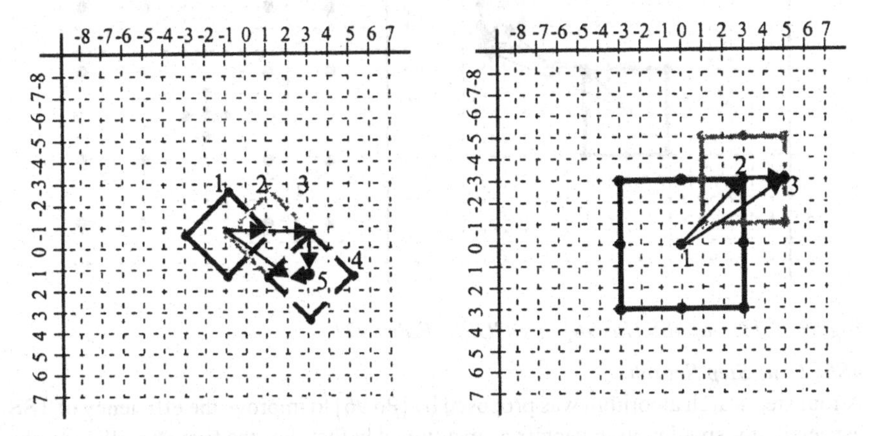


*Figure 2.12: a) 2DLOG (2D logarithmic search), b) TSS (Three Step Search)*

***TSS: Three Step Search***

TSS (Three Step Search, also: 3SS) was proposed by [Koga 81] using a similar structure as 2DLOG, but with the use of MAD instead of the MSE. TSS is one of the most popular fast motion estimation algorithms requiring a fixed number of 25 search steps and is often used as reference. TSS is recommended in RM8 of H.261 and SM3 of MPEG-2. Experimental results showed [Jlu 97], that TSS still provides near-optimal performance, even if the assumption of an unimodal error surface could not be exactly confirmed.

***NTSS: New Three Step Search (= N3SS)***

The NTSS algorithm was proposed by [RLi 94] in order to improve the well-known TSS algorithm especially for video content with slow motion as found in typical head-and shoulder-sequences. As TSS uses a uniform search area subsampling pattern with 2 pel between the search locations in the first step, TSS was regarded to be inefficient in terms of computational complexity for the estimation of small motion [Po 96]. Experimental results in [RLi 94] showed that the block motion field of real world image sequences was usually gentle, smooth, and slowly varying, resulting in a center-biased global minimum motion vector distribution, instead of an uniform distribution. Based on the characteristics of a center-biased motion vector distribution, the NTSS algorithm enhanced TSS by adding additional checking points, which are around the zero MV of the first step of TSS. According to [Po 96] and [Jlu 97], NTSS showed a better performance for small motion than TSS, it is much more robust, and in comparison with the TSS it produces smaller motion compensation errors. However, for larger MV distances, NTSS (worst case: 33 matches) is more complex than TSS (always 25 matches). [LKLiu 96] reported, that NTSS yields significant SNR gains over TSS for head-and shoulder-sequences.

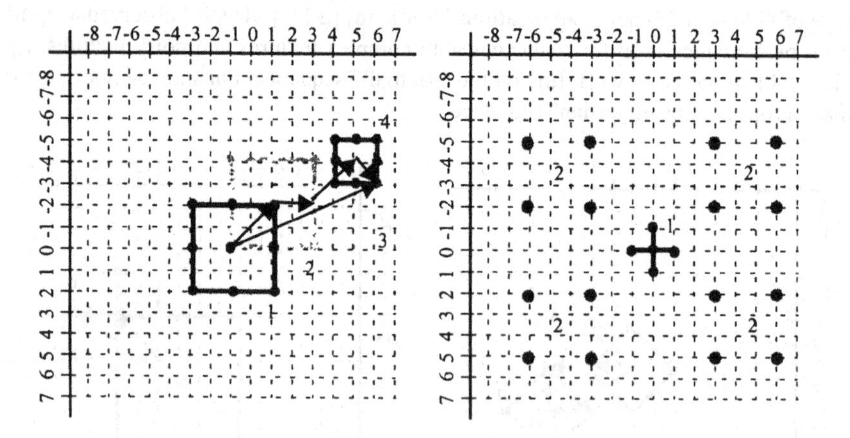


*Figure 2.13: a) 4SS [Po 96], b)DCUPS [Fuhrt 97]*

***4SS: Four Step Search***

A four step search algorithm was proposed by [Po 96] to improve the efficiency of TSS especially for small motion vectors as mentioned before. For the first step, 4SS checks

9 search positions with 1 pel distance in a 5x5 search window. Then this 5x5 search window is moved in 2 pel steps in the direction towards the position with the best match. In the case of the center of the 5x5 search window being the best match, a local search at every integer-pel position in a 3x3 pel window is performed. In the worst case 4SS requires to check 27 matching positions, but the number of average search points lies between 17 and 22.

***Other search area subsampling algorithms:***

- **CDS** (Conjugate Direction Search) was proposed by [Sri 85] and performs a joint horizontal and vertical direction search until a minimum is found.
- **OTS** (One-at-a-Time Search) was also proposed by [Sri 85] and is basically a simplified version of CDS.
- **MMEA** (Modified Motion Estimation Algorithm) was developed by [Kap 85] and is similar to TSS, but computational more efficient.
- **OSA** (Orthogonal direction Search Algorithm) was proposed by [Puri 87], in which with a logarithmic step size, four new locations are searched at each iteration. With this method, at every step 2 positions are searched alternately in vertical and horizontal directions.
- **Cross Search** was proposed by [Gha 90] and uses a cross-formed search pattern. The performance is reported to be comparable to the other search area subsampling algorithms and the computational complexity is reduced.
- **PHODS** (Parallel Hierarchical One-Dimensional Search) was developed by [Chen 91] and is based on an independent search of the x- and y-component of the motion vector. This independent search can be exploited in an advantageous way for VLSI implementations.
- **SES** (Simple and Efficient Search) was proposed by [JLu 97] and is similar to TSS in terms of performance and regularity, while having only about half the complexity of the TSS. This is achieved by selecting at each of the three search steps one of the four quadrants of the search window, depending on the distance criterion value, compared to the neighbor quadrants.
- **CV4SS**: (Candidate Vector-based Four Step Search). CV4SS was proposed by [HeL 97] together with an efficient 1-D systolic array implementation.
- **DCUPS** (Densely-Centered Uniform-P Search) was proposed by [Fuhrt 97] p144 and is based on the observation that in the "cheerleader" sequence 50% of all motion vectors are in the range [-1, 1]. Therefore, these motion vectors are tested in the first step of the search scheme. A VLSI realization of a genetic modification [Chow 93] of DCUPS was given by [Dogi 97].
- **ETSS** (Enhanced Three Step Search) was presented together with a VLSI implementation in [Lak 97]. This modified TSS scheme has for the first step of the TSS four search positions with three pel distance and four search positions +/-1 pel around the center.

#### 2.4.3.2 Scene-adaptive search area

In addition to search strategies with fixed search range, search strategies with scene-adapted search range have been developed recently.

***Zone-Based Motion Estimation***

The **ASA** (Annular Search Algorithm) was presented by [Jung 96] and is depicted in Fig. 2.14 a. In the proposed annular search algorithm, the motion vector search area is partitioned into subregions, called annuli, and each subregion is sequentially examined for the optimal motion vector, which results in the minimum prediction error. The average computational load is reported to be 20% of the computational load of the full-search algorithm, and the PSNR/bit rate ratio is reported to outperform the TSS. This kind of algorithm is also advantageous in a rate/distortion sense, as smaller motion vectors are preferred.

[Zhe 97] proposed a refined scheme for nested zone-based motion estimation, which is depicted in Fig. 2.14 b. The first search area is the innermost zone of a 3x3 or 5x5 square. The next search zone is around the first one and so on. If the minimum point of a search zone is at the center, the matching process stops, otherwise the search continues in the next zone until the distortion criterion reaches a certain threshold.

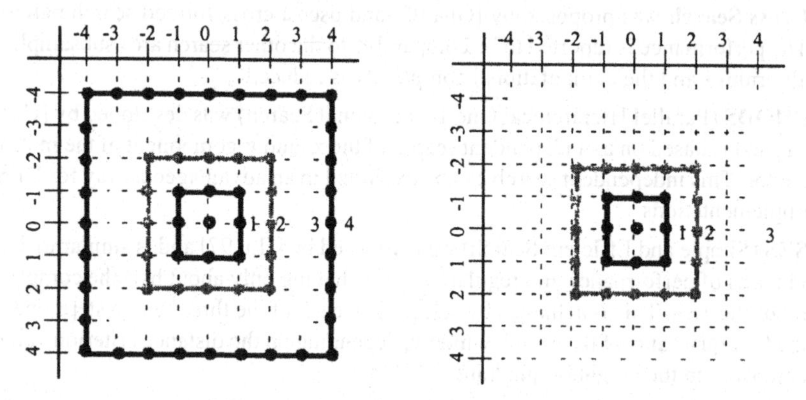


*Figure 2.14: a) ASA (Annular Search Algo.) [Jung 96], b) zone-based ME [Zhe 97]*

***AMCHS (Adaptive Multiple-Candidate Hierarchical Search)***

The AMCHS algorithm was proposed by [Chan 95], basically to eliminate the problem of local minima in coarse-to-fine hierarchical block-matching algorithms. The approach is similar to the TSS, however, after each step more than one candidate motion vector is examined further within a smaller search area. The computational complexity is slightly higher than TSS, which is rewarded by an increase in PSNR performance towards TSS.

***Dynamic Search Window Adjustment (DSWA)***

In [LWLS 93] another dynamic extension of the TSS algorithm, called DSWA, was proposed. The window size is varied depending on a search window convergence ratio, which is defined as the search window size in the present stage, proportionate to that in the next stage. For window size variation three window convergence modes (fast, normal, and slow) are defined. The selection criterion is MAD, and if the smallest and second smallest MAD values are very close, a larger search window is used for the next step. This algorithm is regarded to be beneficial mainly for large search areas, the authors report an average of 24-44% lower computational complexity than TSS.

***Global / local incompensability analysis***
Fast motion estimation based on global and local incompensability analysis was proposed by [Fan 97] and basically determines the search area of the full-search method by the scene complexity of the predicted frame. The scene complexity of the predicted frame is determined by the number of local and global uncompensable MBs.

A local uncompensable MB is an MB where the motion field is inhomogenous over it's four (quadtree partitioned) subblocks. Global uncompensable MBs are MBs, which contain uncovered background or new objects. The goal is to reduce the computational complexity of motion estimation (and bits for false MV) on uncompensable blocks. Uncompensable blocks are determined in the predicted frame by segmentation into four types: a) stationary background, b) uncovered background, c) new objects, and d) moving edges, based on a variance-based scene complexity measure. An overview on variance-based scene complexity measures is given in [Zhou 97].

***Gradient based methods***
**BBGS** (Block-Based Gradient decent Search) was proposed by [LKLiu 96] and relies on the assumption of center-based motion, cf. Fig. 2.15a. A square of 3x3 neighbor integer-pel search positions is moved in the direction of the lowest SAD. The search is stopped, when the center position of the 3x3 square offers the lowest SAD. The authors suggest to stop this search procedure after a certain number of steps (e.g. after 4 steps).

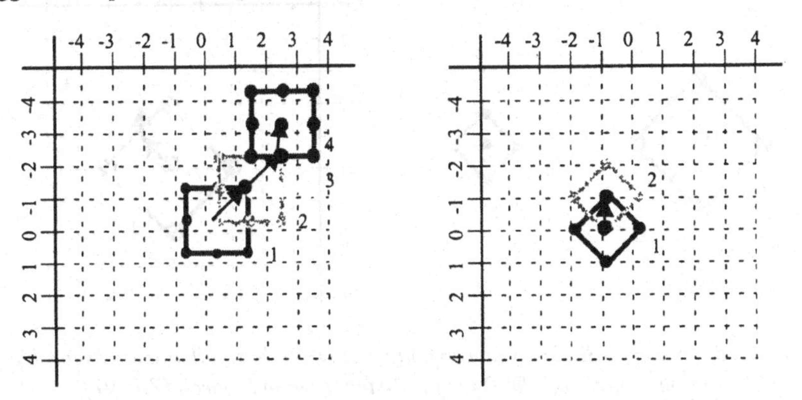


*Figure 2.15: a) BBGS [LKLiu 96], b) Diamond Search by [Cote 97]*

In [Cote 97] a two-dimensional median-predictor-based computation constrained motion estimation algorithm is described, which was developed for the emerging H.263+ recommendation, cf. Fig. 2.15b. The search pattern in the first and in all of the following steps resembles a diamond shape. The diamond search pattern is moved horizontally and vertically in pel steps towards to the lowest SAD. The search is stopped if: 1) all candidate MV of one diamond shape have been considered, and 2) the minimum SAD for the current diamond pattern is larger than the one of the previous shape. [Cote

97] also proposes for H.263+ a new method of motion vector coding, based on semi-fixed length codes.

A rate/distortion-constrained diamond search algorithm based on the mean predictor, is described in [Lee 97], and [Lee 97a] showed similarities to [Cote 97]. The diamond-shaped search area is determined by a predicted motion model, differing from the MPEG MV prediction model. In [Lee 97] two probabilistic models were evaluated, both of which permit the contraction or expansion of the search area as a function of the local statistics of the motion flow. Linear prediction (mean and weighed mean), as well as non-linear prediction (median) are analyzed. A rate/distortion criterion, based on Lagrange optimization is given in [Lee 97a].

A two-step gradient-based diamond search algorithm was proposed by [Zhu 97] and [Zhu 98]. This algorithm is based on the observation, that 53% (in large motion case) - 99% (in small motion case) of the motion vectors are enclosed in a circular area with a radius of 2 pels around the zero motion vector. This algorithm uses two diamond search patterns, depicted in Fig. 2.16. The large diamond search pattern (LDSP) is used for the gradient-based coarse search. When the centered search position of the LDSP shows the minimum SAD, the small diamond search pattern (SDSP) is chosen for fine search. This algorithm is not bound by the use of a stop criterion, but as the minimum block distortion values are found along the search pattern, the algorithm converges. This algorithm has also been tested successfully by the proposers within the framework of the emerging MPEG-4 standard [M 3299].

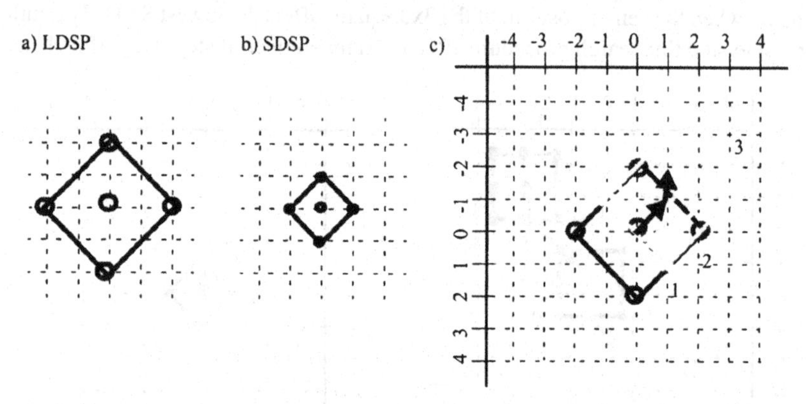


*Figure 2.16: a) Large diamond search pattern (LDSP), b) small diamond search pattern (SDSP), c) movement of LDSP for the 2-step diamond search [Zhu 97]*

***Other scene-adaptive search algorithms:***

- **ABMA** (Adaptive search area Block-Matching Algorithm) was proposed by [Chain 95] and uses the spatial correlation of motion vectors to adjust the size of the search area, based on the previous MV in the MB scan line, and the maximum motion velocity defined by two thresholds in x- and y-direction. ABMA is reported to be about 61 times faster than full-search.

- **DSRA** (Dynamic Search Range Algorithm) is similar to ABMA and was proposed by [In 97], with which a full-search is performed around the feasible motion area. Here the search range can be dynamically adjusted so that the estimated performance will not be significantly reduced, though the computational complexity will be largely reduced. The search range is adjusted based on the correlations of the motion vectors between the neighboring blocks.
- **Spiral search** (Fig. 2.17) is started at the motion vector predictor location and is stopped, when the distance criterion passes a predetermined threshold, e.g. [Ohm 95] p152.
- **PSA** (Prediction Search Algorithm) was proposed by [Lull 97]. The predicted MV is calculated by linear weighting of the MVs of the three adjacent blocks. A 3x3 pel search area is moved similar to BBGS in one pel steps to the minimum, until the local minimum point of the distance criterion is in the center of the 3x3 block.
- **AFSBM** (Adaptive Full-Search Block-Matching) was proposed by [Fang 95a]. Depending on the motion classes of the MB, different search ranges for the full-search algorithm are applied. The motion classes are divided into low-, medium- and high-motion, depending on the SAD value of the MB at the predicted MV location falling within predetermined thresholds. These threshold values are updated from frame to frame, taking the number of high and low motion macro-blocks into account. The bit-plane matching (BPM) technique, which is discussed in more detail in the "feature matching" section of this chapter, was used in [Feng 95] for the AFSBM.
- **ASL** (Adaptive Search Length) was proposed by [Pick 97] and adaptively varies the number of positions searched for each block, while still maintaining control of the average number of searches per block for each frame, thus a trade-off can be adjusted on-line between computation requirements and prediction quality. With this algorithm a small number of searches is performed in case of simple motion and a larger number in case of complex motion. The number of search steps is adaptively determined and updated on the basis of the number of local minima occuring in the search range within each single search step.

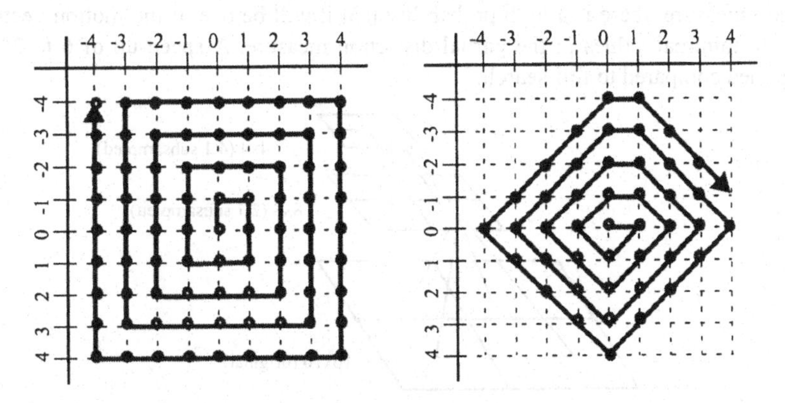


*Figure 2.17: a) Spiral search [Ohm 95] p 152, b) diamond -shaped spiral search*

***Note on VLSI-implementation***
As there exist many search strategies and research in scene-adaptive search techniques is continuing, one or several search strategies can be successfully implemented with high flexibility, using a (sequential) programmable processor. However, (parallel) VLSI support of one or several search schemes with high efficiency is extremely difficult.

## 2.4.4 Hierarchical and multiresolution fast block-matching

The basic idea behind hierarchical and multiresolution fast block-matching algorithms is to predict an approximate large-scale MV in a coarse resolution representation, and to refine the predicted MV in a multiresolution fashion in order to obtain the MV in a finer resolution. One of the first algorithms of this class had been proposed by [Bier 86], and [Bier 88]. An overview of recent work on hierarchical and multiresolution fast block-matching algorithms was given e.g. by [Duf 92], [Duf 95], and [Chali 97]. In general, the process of generating multiresolution images results in a smooth motion vector field, and therefore offers an enhanced robustness of these algorithms against noise.

***Hierarchical fast block-matching***
The hierarchical methods use the same image size but different block sizes at every hierarchical level for distance calculation. Some techniques employ different parameters and/or different algorithms at each level of hierarchy.

Recently conducted research includes:

- **Multigrid subsampling** of 8x8, 4x4, 2x2, 1x1 blocks on the luminance data for a H.263 coder was applied by [Panus 97].
- **HPDS** (Hierarchical Partial Distortion Search) is similar to the previous one and was proposed by [Kwan 97]. It is based on determining the candidate motion vectors (CMV) on a 8:1 subsampled image first, and then to refine the best CMV on a 4:1 subsampled image and so on, until the original resolution is reached. This method is based on the rationale that if a motion vector is the optimum in the full distortion measure, there is a high probability that it will be one of the motion vectors with minimal values in the partial distortion measure. A speed-up of 6 to 20 is gained compared to full-search.

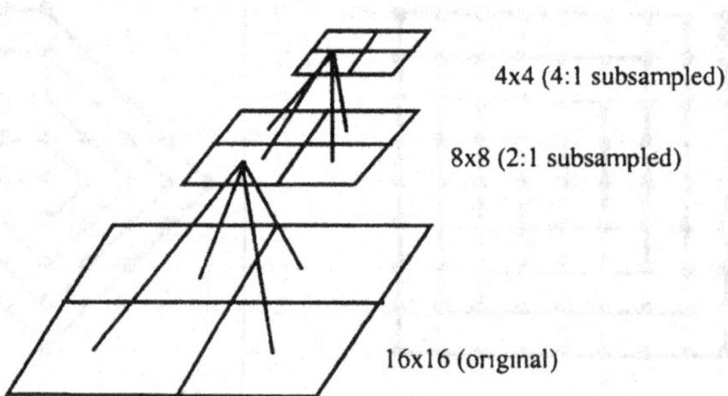


*Figure 2.18: Hierarchical motion estimation (here: mean pyramid, [Nam 95])*

***Multiresolution fast block-matching***
The multiresolution methods use different image resolutions with a smaller image size at a coarser level (i.e. of a pyramid form). Among several possibilities to gain multi-resolution representations some are listed here.

- **Gaussian pyramid** which employs calculation intensive filters as used in [Shi 97] together with hierarchy-dependent threshold settings.
- **Reduced pel resolution** [LeeC 96] which uses only a reduced number of bits of the pel for the coarse search, and the full number of bits only for the final search (cf. 2.4.2, DPC).
- **Mean pyramid** [Nam 95] which offers the advantage of reduced computational complexity and robustness against noise.
- **MADM** (Mean Absolute Difference of the Means) criterion, (cf. 2.4.2 MADM), is hierarchically applied by [Chun 94] p117.
- **BSPA** (Block Sum Pyramid Algorithm), [CHLee 97] which employs the hierarchical feature-matching criteria **SEA** (Successive Elimination Algorithm) described in 2.4.6.

***Note on VLSI-implementation***
From a implementational point of view a general drawback of hierarchical and multi-resolution fast block-matching algorithms are the increased memory requirements to store the image at several resolutions. Therefore, these algorithms are not described in detail here.

## 2.4.5 Pel decimation techniques

Pel decimation (also called pel subsampling) reduces the number of pels taken into account for the distance calculation at every search position and, in doing so, reduces the computational complexity for the distance criterion, as well as the memory bandwidth. Among the possibilities to reduce the number of pels per block, subsampling by the ratio of e.g. 2:1 or 4:1 [Liu 93:1] or simple filter algorithms like pel averaging etc. [Cheng 96], are often used. A theoretical analysis of pel subsampling techniques is given in [Tek 95].

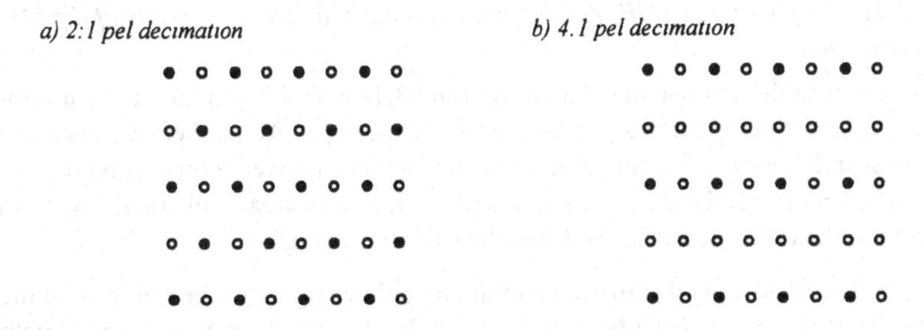


*Figure 2.19: a) 2:1, b) 4:1 pel decimation: only the dark marked pel positions are used.*

A widely known algorithm for fast motion estimation using 4:1 pel subsampling is described in [Liu 93:1] section II, where an alternating pel subsampling pattern is used, which makes it possible to take all pels of the search area successively into account. The pattern-matching scheme for the four alternating subsampling patterns is depicted in Fig. 2.20. The patterns of Fig. 2.21 a ... d depict the subsampling pattern, which has to be used at every search position for the alternating scheme in Fig. 2.20 a. Fig. 2.20 b depicts the other possible subsampling patterns.

The minimum MAD is calculated at every search position, using all patterns and resulting in four MAD loops ($MAD_A$ ... $MAD_D$) and four candidate MV ($CMV_a$ ... $CMV_d$). For these four motion vector candidates block-matching without subsampling is performed and the best result gives the final MV.

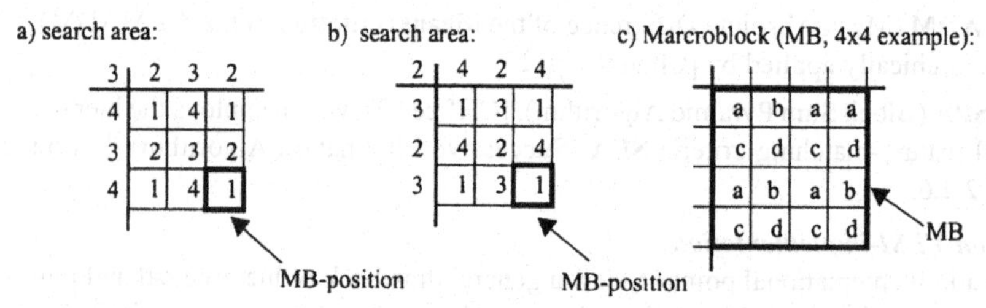


*Figure 2.20: a,b) Two different alternating schedules for pel subsampling patterns over the search area, c) pel decimation patterns of a macro-block used to calculate the distance criterion*

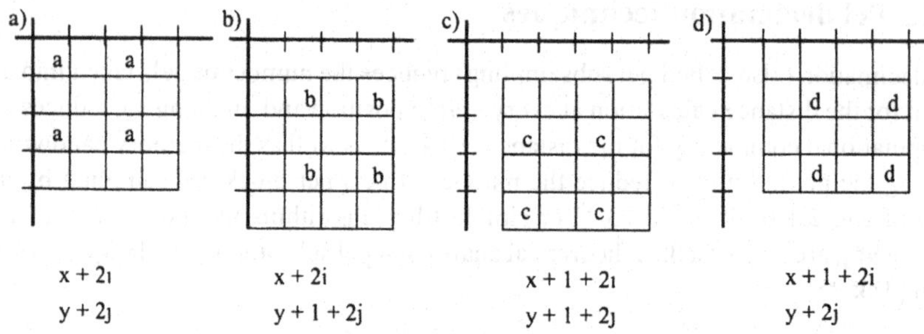


*Figure 2.21: Macro-block (MB) pel decimation pattern in detail for shifts of the MB in the search area*

[Yu 97] enhanced the scheme developed by [Liu 93:1] with 4:1 and 16:1 pel subsampling and a hierarchical search algorithm with a flexible (2 - 4) number of stages and a variable search area at different stages, which results in a lower computational complexity and memory bandwidth. The parameterizable search area of the final stage can be adapted to the network traffic (and therefore the frame rate).

[Cheng 96] described a pel decimation method by pel averaging of the four pels at the locations (2x, 2y), (2x + 1, 2y), (2x, 2y+1), (2x+1, 2y+1), which can be performed with low computational overhead and reduces the storage by performing motion estimation row by row, storing only the necessary averaged pixels in several line buffers. This pel averaging technique is combined with search area subsampling and a final full-search with a +/- 1 pel search area around the four best motion vector candidates without subsampling, which is reported to enhance the error robustness of the motion estimation algorithm.

***Adaptive Pel Decimation***
An adaptive pel decimation technique was reported in [Chan 96], with a 4:1 pel decimation being performed by subsampling in the first step. In the second step a scene-adaptive pel selection of the pel outside of this 4:1 subsampling raster is performed to gain MV candidates by a threshold-based absolute difference criteria between the candidate pel and the raster pels. In regions with higher activity more pels are selected as MV candidates. However, the irregular data flow of this adaptive algorithm seems not suitable for a (parallel) VLSI-based implementation. A gradient-based pel decimation scheme was proposed by [Tao 98], based on the observation that the expected prediction error of a pel is proportional to the gradient (defined there) of the pel.

***Note on VLSI-implementation***
Pel Decimation by subsampling techniques reduces the computational complexity and memory bandwidth of the fast motion estimation algorithms. However, especially for adaptive pel decimation schemes, this reduced computational complexity has to be paid for by irregular pel accessing schemes in the frame buffer, which may increase the hardware cost for the calculation of the pel addresses and the conflict-free parallel pel access. Hierarchical pel decimation algorithms require storage for the decimated pictures. This storage can be reduced considerably by performing the motion estimation row-by-row storing only on the necessary pels of the decimated blocks in several line buffers.

## 2.4.6 Feature-matching

The class of feature-matching algorithms can be seen as matching on meta information extracted from the current block and search area pels, Fig. 2.22.

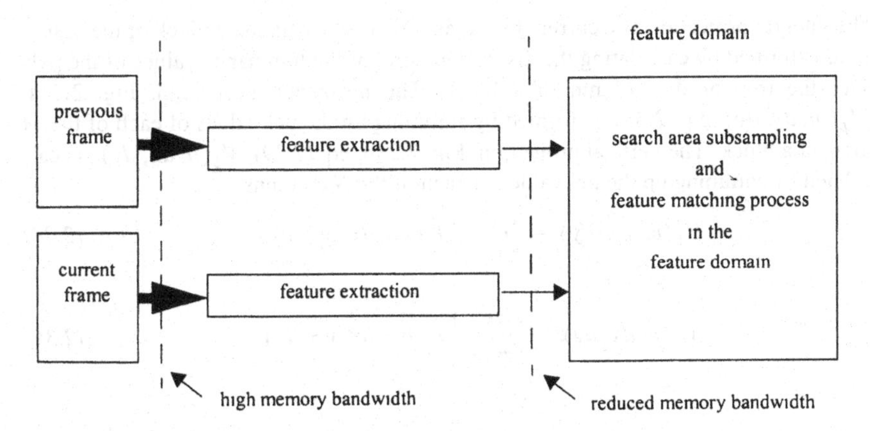


*Figure 2.22: Feature-matching algorithm*

In most of the feature-matching algorithms, the meta information has a higher entropy than the original information, thus reducing the arithmetic computational load, as well as the memory bandwidth within the matching process. The feature extraction is performed e.g. by morphological filters, reduced pel resolution, or projection methods, which are explained in more detail here. In the feature domain basically the same search area subsampling strategies as mentioned above can be applied, full-search and the three step search are most commonly used. Feature extraction can basically be performed a) within one step for the complete search area, which requires additional feature domain memory but enables pipelining of these two steps or b) subsequentially for every new search position, which requires no feature memory. Some distance criteria, which are in use for feature-matching, have already been described in section 2.4.2, some new matching criteria have been developed specially within emerging feature-matching algorithms.

#### 2.4.6.1 Integral Projection-Matching (IPM)

For this feature-matching algorithm integral projections were employed to extract the features [Cain 92], [Kim 92], [Ogura 95], [Sauer 96], and [Pan 96]. The basic principle of projection-matching is to calculate a one-dimensional cost function based on row and column sums, rather than a two-dimensional cost function, which reduces the computational power significantly. Projection-matching can be combined with other block-matching algorithms (e.g. pel decimation or search area subsampling) to reduce the number of operations further, e.g. [Kim 92], and [Sauer 96].

[Huang 81] showed that for simple translational motion, the information on the axes in the Fourier-transform domain is sufficient to estimate motion between two images. The computing of the horizontal and vertical frequency information is equivalent to discrete approximations of integral projections at these two orientations, according to the projection slice theorem. Seeking the minimum total squared error along the two axes in the Fourier-transform domain, is equivalent to finding the best match in integral projections among various spatial displacements.

The integral projection of a current block, as well as of a candidate block of the search area, is formed by calculating the discrete integral of the luminance values of the pels over the row or the column of a block. The horizontal projection, Fig. 2.23a, $H_k(m, dx, dy)$, eq. (2.38), is formed by summing up the pel values of each of the $M$ rows of a block. The vertical projection, Fig. 2.23b, eq. (2.39), $V_k(n, dx, dy)$, is calculated by summing up the pel values of each of the $N$ columns.

$$H_k(m, dx, dy) = \sum_{x=0}^{N-1} I_k(x + dx, m + dy) \tag{2.38}$$

$$V_k(n, dx, dy) = \sum_{y=0}^{M-1} I_k(n + dx, y + dy) \tag{2.39}$$

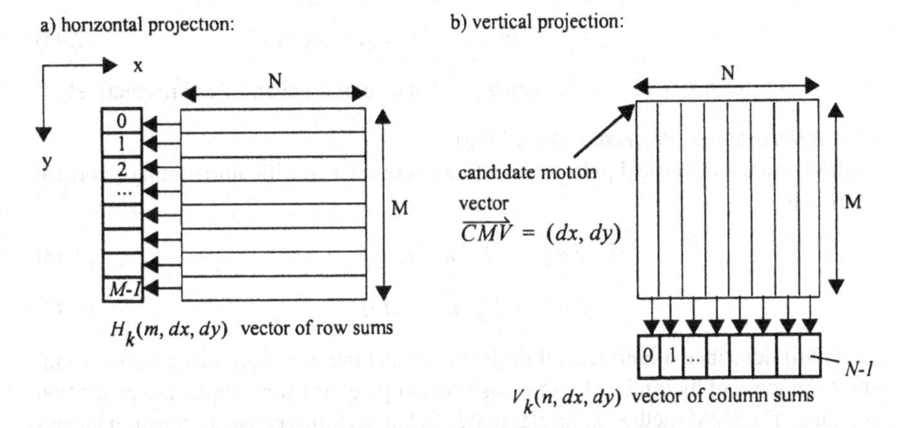


*Figure 2.23: Integral projection-matching: a) horizontal, and b) vertical projection*

For every search position *(dx, dy)* the MAE cost functions $D_{horizontal}()$ and $D_{vertical}()$, eq. (2.40), eq. (2.41), are calculated for a block size of M x N, with ($MV_x$, $MV_y$) being the resulting motion vector for the best match in the x- and y-direction:

$$D_{horizontal}(dx, dy) = \sum_{m=0}^{M-1} \left| H_k(m, 0, 0) - H_{k-1}(m, dx, dy) \right| \tag{2.40}$$

$$D_{vertical}(dx, dy) = \sum_{n=0}^{N-1} \left| V_k(n, 0, 0) - V_{k-1}(n, dx, dy) \right| \tag{2.41}$$

$$MV_y = \min_{(dx, dy) \in R^2} D_{horizontal}(dx, dy) \tag{2.42}$$

$$MV_x = \min_{(dx, dy) \in R^2} D_{vertical}(dx, dy) \tag{2.43}$$

Note that $D_{horizontal}()$ or $D_{vertical}()$ can be efficiently calculated in parallel by using a sliding window approach in which (after an initialization phase) at every step a new row (column) can be added to the current result, after subtracting the row (column) which falls out of the sliding window (cf. e.g. [Kom 93] p22).

***PTSS (Projection Three Step Search)***

[Kim 92] describes a projection matching scheme which combines the above described method with the TSS (Three Step Search) scheme, which is denoted later by [Sauer 96] as PTSS (Projection Three Step Search) and shows robustness of this scheme against Gaussian noise. PTSS performs 2:1 pel subsampling for the $H_k(m, dx, dy)$ and $V_k(m, dx, dy)$ calculation that can be described as following:

1. Calculate $H_k(m, dx, dy)$ and $V_k(m, dx, dy)$ for the current block.
2. Calculate $H_{k-1}(m, dx, dy)$ and $V_{k-1}(m, dx, dy)$ for the 9 subblocks SP1... SP9 of the reference frame, which are located at 4 pel distance. Note that already calculated results can be reused, where SP1 is located at the center of the current block.
3. Calculate the cost function *D()* and the candidate motion vector $(CMV_x, CMV_y)$ as denoted in eq. (2.44) and (2.45).

$$D(dx, dy) = D_{horizontal}(dx, dy) + D_{vertical}(dx, dy) \tag{2.44}$$

$$(CMV_x, CMV_y) = \min_{(dx, dy) \in R^2} D(dx, dy) \tag{2.45}$$

4. Perform steps 2. and 3. with a pel distance of 2 pel and 1 pel respectively.

***HPM (Hierarchical Projection-Matching)***

The horizontal and vertical pel sum is defined here for a specific horizontal or vertical sum vector $i$ :

$$H_{i,k}(m) = H_k(m, dx, dy) \tag{2.46}$$

$$V_{i,k}(m) = V_k(m, dx, dy) \tag{2.47}$$

[Sauer 96] described a hierarchical projection-matching with decreasing search radii, where the ideas of hierarchical motion estimation [Bier 86] were applied to projection matching. The HPM method is similar to PTSS, but performs exhaustive search in contrast to subsampling of the motion displacement space.

It only calculates one integral projection at every hierarchy level instead of 9. The procedure is:

1. Calculate $H_{1,k}(m)$ and $V_{1,k}(m)$ of a 2N x 2N block, centered at the N x N current block and calculate $H_{1,k-1}(m)$ and $V_{1,k-1}(m)$ for the (2N+2s1) x (2N+2s1) reference block, which is scaled for NxN block size. The cost function according to (2.40) ... (2.42) determines together with the motion vectors the center of the search area for step 2.

2. Calculate and scale $H_{2,k-1}(m)$ and $V_{2,k-1}(m)$ for the (2N+2s2) x (2N+2s2) reference block, and determine the center of the search area of step 3 similar to step 1.

3. Calculate and scale $H_{2,k-1}(m)$ and $V_{2,k-1}(m)$ for the (2N+2s3) x (2N+2s3) reference block.

The selection of s1=4, s2=2, and s3=1 allows a search distance of 7 pels for each coordinate. Subpel MV calculation is regarded to be of low complexity, as the interpolation of projections of one dimension is equivalent to projections after interpolation [Sauer 96].

As the theory of integral projections would require images of infinite extent, [Sauer 96] introduced a weighting of the horizontal and vertical pel sum $H_{i,k}(m)$ and $V_{i,k}(m)$ to reduce boundary effects. $H_{i,k}(m)$ and $V_{i,k}(m)$ are weighted according to the PDF (probability density function) of the motion vectors, which results in smaller weights at the block boundary compared to the block center.

This weighting can be done without multiplication, by allowing only weights with fractions of $2^n$. To further reduce the computational complexity of HPM, [Sauer 96] exploited spatial MV correlations with motion field subsampling techniques similar to [Liu 93:1], cf. section 2.4.7.

***Sp-Tp method for Integral Projection-Matching***
To meet VLSI implementation constraints [Ogura 95] described the Sp-Tp projection-matching algorithm and a VLSI-implementation thereof. The basic idea of this scheme is to reduce the projection vector $H_k(m, dx, dy)$ of length M to a projection vector $H^{Sp}{}_k(m, dx, dy)$ with length M/S and the vector $V_k(m, dx, dy)$ of length N to $V_k^{Tp}(m, dx, dy)$ with length N/S by summing up the data of adjacent rows and columns. S (and T) is the size of the summarized adjacent horizontal (vertical) projection data. This Sp-Tp method reduces the computational complexity for the distance calculation of the projections by decreasing the PSNR by about -0.15 dB for the 2p-2p method (Fig. 2.24) and -0.5 dB for the 4p-4p method, while reducing the matching data by the factors 2 and 4 respectively.

***Two Stage Projection-Matching***
A two-stage block-matching algorithm using projection-matching and a VLSI architecture is described in [Pan 96]. The first step performs exhaustive search with integral projection-matching for all candidate blocks of the search area, where the candidate MV is determined based on the 1-D MAD criterion in eq. (2.44) and (2.45). The second step performs a conventional block-matching with the 2-D MAD criterion in the 3x3 pel search area centered at the candidate MV of step 1.

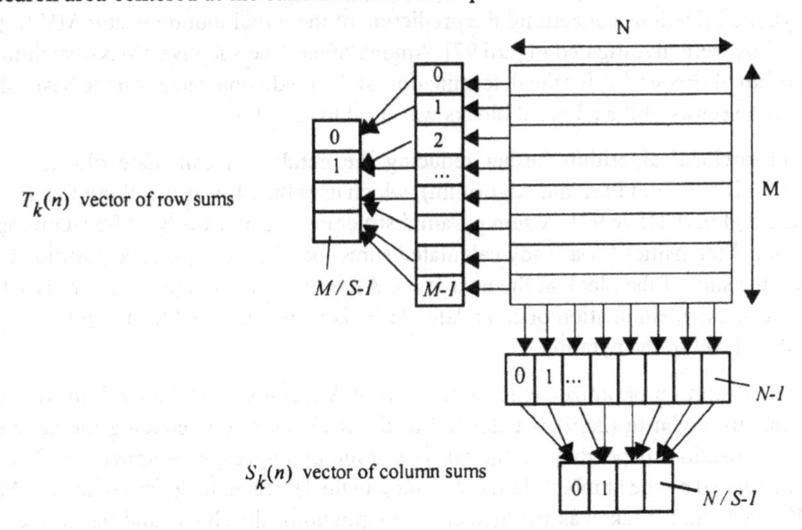


*Figure 2.24: 2p-2p integral projection-matching*

### 2.4.6.2 Other Feature-Matching Algorithms

***SEA (Successive Elimination Algorithm)***
[WLi 95] proposed a successive elimination algorithm for motion estimation, where successively MV positions of the search area are eliminated and thus the number of matching evaluations is reduced. The SEA algorithm is based on the triangle inequality, where:

$$\left|\left|I_k(m, n)\right| - \left|I_{k-1}(m + dx, n + dy)\right|\right| \le \left|I_k(m, n) - I_{k-1}(m + dx, n + dy)\right| \qquad (2.48)$$

From eq. (2.48) the inequality on which the SEA algorithm is based can be derived [WLi 95] eq. (2.49):

$$R - MAD(i,j) \leq M(dx, dy) \leq R + MAD(i,j)) \qquad (2.49)$$

with MAD (i, j) being the initial matching candidate block with the motion vector MV (i, j) and M (dx, dy) is the sum norm of the matching candidate block with motion vector MV (dx, dy):

$$M(dx, dy) = \sum_{m=x}^{x+N-1} \sum_{n=y}^{y+N-1} I_{k-1}(m+dx, n+dy) \qquad (2.50)$$

and R representing the sum norm of the reference block:

$$R = \sum_{m=x}^{x+N-1} \sum_{n=y}^{y+N-1} I_k(m, n) \qquad (2.51)$$

For the SEA algorithm only those blocks are further investigated which fulfill eq. (2.49) using a matching criterion like the MAD. A fast algorithm for calculation of the sum norm based on the reuse of already calculated row and column results was given in [WLi 95].

As the results of this algorithm are highly dependent on the initial motion vector MV (i, j), modifications concerning the prediction of the initial motion vector MV (i, j) to the SEA were investigated in [Oli 97]. Among others, the adaptive block-matching algorithm of [Feng 95a] for the determination of the prediction vector on the basis of the motion vectors of the adjacent blocks was used in [Oli 95].

A hierarchical algorithm further reducing the number of candidate blocks, called BSPA (Block Sum Pyramid Algorithm), which uses the SEA as matching feature was presented in [CHLee 97]. A sum pyramid structure of a block is built by summing up the 4 pel intensities (or already calculated sums) of a block's quadtree partitioning to gain the sum of the block at the next level. After the sum pyramid structure is set up, the successive elimination of candidate blocks is then performed from top level to the bottom level of the pyramid.

A rate/distortion-optimized extension of the SEA was given in [Cob 97], in which the inequality given in (2.49) is extended in the R/D sense, for reducing the candidate search positions depending on the rate constraint by Lagrange minimization. The rate is calculated by the number of bits according to the Huffman table for coding the MVs. MV prediction is taken as the median of the previous, the above, and the above right MVs. The remaining motion vectors are searched using a spiral search scheme around the zero motion vector at positions which satisfy the R/D extended inequality relation.

***BBM (Binary Block-matching)***

BBM (Binary Block-Matching) was proposed together with a VLSI architecture in [Nat 96], [Nat 97] and similar in [Miz 96]. BBM is defined as:

$$BBM(dx, dy) = \sum_{m=x}^{x+N-1} \sum_{n=y}^{y+N-1} \hat{I}_k(m, n) \otimes \hat{I}_{k-1}(m+dx, n+dy) \qquad (2.52)$$

in which $\otimes$ denotes the exclusive-or operation and $F_K(i, j)$ is a low-complexity preprocessing filter for binary edge extraction of the search window and the reference block.

$$\hat{I}_k(m,n) = \begin{cases} 1 & if \quad I_k(m,n) \geq F_K(I_k(m,n)) \\ 0 & otherwise \end{cases} \tag{2.53}$$

$$(MV_x, MV_y) = min_{(dx,dy) \in R^2} BBM(dx, dy) \tag{2.54}$$

In [Nat 96], [Nat 97] the filter $F_K(i, j)$ is defined as a 17x17 convolution filter with the kernel K:

$$K(i,j) = \begin{cases} 1/25 & if \quad i,j \in [1, 4, 8, 12, 16] \\ 0 & otherwise \end{cases} \tag{2.55}$$

This binary block-matching method can be combined with one of the different existing search strategies. A similar, but hierarchical feature-matching technique to determine the candidate MV for subsequent MAE block-matching was described in [Zhong 96]. The authors extract binary edge information for 4x4 blocks of the MB and perform a hierarchically edge pattern-matching process with motion field subsampling similar to [Liu 93:1] on these data. A similar hierarchical motion estimation with very low computational complexity, using a binary pyramid with 3-scale tilings was presented by [Song 98]. This algorithm is reported to result in a PSNR drop of about 1 dB compared to full-search.

[Le 98] reduced the luminance component of the search area and of the current block to a 1-bit binary representation using morphological filters. The morphological external gradient and the morphological internal gradient are used as morphological filters. The motion search is performed in the 1-bit feature space using an XOR-operator in a massive parallel SIMD (single instruction multiple data) architecture, built on computational RAM (i.e. DRAM/SRAM with embedded computational logic).

***BPM (Bit-Plane-Matching)***

In [Feng 95], [Feng 95a] BPM (Bit-Plane-Matching) was employed as the first step of a search range adaptive hierarchical scheme to determine candidate motion vectors, which are evaluated in a further step using the MAD criterion. The bit-plane $\delta(x, y)$ of the search-window is formed by comparing each pixel against the mean pixel value (*mean*) of the search window and equally for the reference block:

$$\delta(x,y) = \begin{cases} 0 & if \quad I_k(x,y) \leq mean \\ 1 & otherwise \end{cases} \tag{2.56}$$

Full-search is employed on the derived bit-plane. Only those locations, where the number of miss-matches is less than a predetermined threshold value, are further evaluated by using the MAD criterion.

***BFM (Block Feature Matching)***

The block feature-matching (BFM) criterion was introduced by [Luo 97], which allows an efficient VLSI implementation. The BFM is calculated as:

$$BFM(dx, dy) = |Mean_k(x,y) - Mean_{k-1}(x+dx, y+dy)| + \lambda\beta(dx, dy) \tag{2.57}$$

$$(MV, MV) = \min_{(dx, dy) \in R^2} \quad BFM(dx, dy) \tag{2.58}$$

with the Mean calculation being depicted as:

$$Mean_k(x, y) = \frac{1}{N^2}\sum_{m=x}^{x+N-1}\sum_{n=y}^{y+N-1} I_k(m, n) \tag{2.59}$$

Hamming distance:

$$\beta(dx, dy) = \sum_{m=x}^{x+N-1}\sum_{n=y}^{y+N-1} |N_k(m, n) - N_{k-1}(m + dx, n + dy)| \tag{2.60}$$

Binary sign vector:

$$N_k(m, n) = f(I_k(m, n) - Mean_k(x, y)) \tag{2.61}$$

Phase:

$$f(x) = \begin{cases} 1 & if \quad x \geq 0 \\ 0 & otherwise \end{cases} \tag{2.62}$$

NxN Bits Binary Sign Vector:

$$\vec{N}_k(x, y) = [N_k(x, y) + N_k(x+1, y) + \ldots + N_k(x+N-1, y+N-1)] \tag{2.63}$$

The weighting factor $\lambda$ was determined experimentally by the authors and is set to 2. The BFM matching criteria consists of the two parallel calculated features (these are compared for the current and the reference block), which are taken as cost function:

1. The mean, eq. (2.59), of a NxN block, which is the amplitude matching for the smoothness of the correlations among the subblocks [Xlee 96].
2. The Hamming distance, eq. (2.60), of the binary sign vectors of a NxN block, which is a phase-template matching method to evaluate the roughness of matching differences within a block. The Hamming distance can be efficiently calculated by a XOR operation and a look-up table.

A similar, but hierarchical approach with threshold levels determined by the quantizer step-size, was used by [XLee 96] and named STF (Sign Truncated Feature) matching. In this approach the Hamming distance of the binary sign vector of different hierarchical quadtree levels is minimized in order to gain the best motion vector.

***Others***

A family of fast block-matching algorithms employing various local features like adaptive pel decimation, block mean, variance, Cauchy-Schwartz inequality, and the central moment, was proposed by [Tao 98], where a computation reduction of up to 80% was reported for a specific implementation.

***Note on VLSI implementation***

In general, VLSI implementation advantages of feature-matching algorithms are the reduced amount of data fetching (i.e. memory bandwidth) and the reduced amount of computations, which is highly promising. However, in a typical HW/SW-partitioning scheme, the feature-matching criteria have to be implemented using dedicated VLSI, due to their high throughput. Therefore, they cannot by regarded as an algorithm modification (because of the different search area subsampling schemes), but have to be known beforehand.

## 2.4.7 Predictive Motion Estimation

To reduce the computational effort of motion estimation, a prediction of motion vectors is usually performed to gain an initial guess of the next motion vector. For predictive motion estimation the spatial and/or temporal method can be used.

### 2.4.7.1 Temporal prediction

Temporal prediction of motion vectors is based on the assumption of a temporal homogeneity of the motion field. The motion vector of a specific MB of the current frame is predicted from the MV of the previous frame at the same location.

### 2.4.7.2 Spatial prediction

Temporal prediction of motion vectors is based on the assumption of a spatial homogeneity of the motion field. The motion vector of a MB of the current frame is predicted from the MVs of the neighboring MBs of the same frame. The spatial motion prediction within the MPEG-1/MPEG-2/MPEG-4/H.26X standards have already been depicted in 2.2.4. For spatial MV prediction, taking into account the MVs of the neighborhood blocks, there are basically two methods [Lee 97], [Kos 97]:

- **Linear prediction** which uses the mean or the weighted mean, and
- **Non-linear prediction** which is based e.g. on the median calculation.

Examples to reduce the computational effort for motion estimation using spatial MV prediction are given here:

***Adaptive search area***

[Feng 95a] exploited the motion vector correlation to adapt the search area of a full-search method to the motion field (cf. the section on adaptive search area subsampling).

***Motion field subsampling***

[Liu 93:1] described in section III a technique with low complexity for subsampled motion field estimation. The first step includes the search for a subsampled MV-field by estimation of the MVs of a fraction of blocks by any motion estimation technique (blocks B, C, D, E in Fig. 2.25a) and the interpolation of the motion vector for e.g. block A, by the MV with the smallest SAD value of the four surrounding blocks.

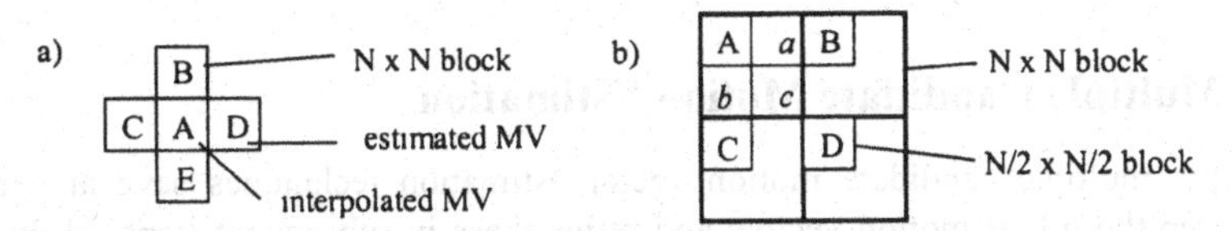


*Figure 2.25: a) Motion field subsampling [Liu 93:1], b) subblock motion field subsampling [Liu 93:1]*

A similar technique was proposed by [Song 98a], in which the MVs of four causally adjacent blocks are used as candidate MVs representing continuous motion, and regularly sampled MVs within the whole search area are used as candidate MVs representing complex/random motion. These candidate MVs are used as the center of a small search area. This proved to be of some benefit for large search areas.

[Liu 93:1] described in section IV a subblock motion field estimation technique. The N x N blocks are subdivided in N/2 x N/2 blocks, with the MV of the subblocks A, B, C, and D being determined by motion estimation, whereas the MV of *a, b, c* (in Fig. 2.25b) are determined by interpolation. For example the MV of block *a* is taken from the vertical neighbor blocks A or B, depending on which of the MVs gives the better match (MAE criteria) for *a*. For subblock *b* the MV from vector A or C is assigned and for *c* either the vector of A, B, C, or D is assigned, depending on which of these vectors gives the better match using the MAE criteria.

In [Xu 97] a linear prediction model based on the mean biased motion vector's minimum of the four causal neighbor MVs was presented to accelerate the motion estimation.

#### 2.4.7.3 Combined temporal and spatial prediction

Besides separate exploitation of the above mentioned prediction techniques, research was performed on combination of both techniques:

**SBMA** (Stochastic Block-Matching Algorithm) was presented by [Skim 95], [Skim 96] and (as an extended version for hierarchical motion estimation) in [Chali 97], exploiting the spatio-temporal correlations for fast motion estimation. SBMA is based on the three step search algorithm and basically determines the variable search window size by the spatial correlation and the shiftable search center using the temporal correlation. A speed-up factor of 70-150 is gained in comparison to full-search block-matching.

[Zheng 97] exploited the spatio-temporal correlation by minimizing a new criterion, DCCP (Distance from the true motion vector to the Closest Checking Point).

***Note on VLSI implementation***
For VLSI implementation the spatial prediction scheme requires only a few additional memory locations to store the motion vectors of the neighborhood macro-blocks. However, temporal MV prediction has the disadvantage, that memory space for all of the MVs (eventually also for the 1/2 pel MVs) of the previous frames is required. The calculation of the predictive MV is usually of low complexity and is subject to be implemented by a control processor.

### 2.4.8 Multiple Candidate Motion Estimation

Generally, multiple candidate motion vector estimation techniques save at every search step the *n* best motion vectors and refine these in subsequent steps. Multiple candidate MVs offer the advantage of being more robust to the local minimum problem.

### 2.4.9 Rate/Distortion-optimized fast motion estimation

Rate/distortion (R/D)-optimized motion estimation algorithms have already been discussed in section 2.3. An emerging research area is the rate/distortion-optimization for *fast* motion estimation algorithms. A R/D-optimization of the MV search and the residual coding by approximation with parametric functions was presented in [Chu 96]. R/D-optimized SEA (Successive Elimination Algorithm, cf. 2.4.6), was presented by [Cob 97] and R/D-optimized fast motion estimation for OBMC (Overlapped Block Motion Compensation) was presented by [Su 97].

### 2.4.10 Complexity/Distortion-optimized motion estimation

Similar to rate/distortion-optimized motion estimation, complexity/distortion-optimized (C/D-optimized) motion estimation algorithms have been recently proposed in literature. One proposal was to stop a search area subsampling process (e.g. a gradient-based or spiral search) at a certain point of confidence for the motion vector (e.g. [Zhe 97], [Cote 97]).

[Leng 98] proposed a complexity/distortion-optimized motion estimation by Lagrange optimization, eq. (2.64). In eq. (2.64) *C(dx, dy)* is the estimated complexity (here the estimated number of absolute differences to be summed up to stop the SAD calculation by summation truncation) needed to calculate the distortion measure for a specific candidate motion vector. Additionally [Leng 98] proposed a joint Lagrange optimization of the candidate motion vector selection and local refinement search for this SAD estimation technique (cf. 2.4.2).

$$J(dx, dy) = D(dx, dy) + \lambda \cdot C(dx, dy) \qquad (2.64)$$

## 2.5 BEYOND STANDARD-CONFORMANT MOTION ESTIMATION

### 2.5.1 Introduction

In this section motion estimation algorithms are described, which act as straightforward extensions of the currently standardized motion estimation schemes. Some of the basic ideas which have been investigated in terms of efficiency and visual quality within the context of the MPEG-4 standardization process are described as follows.

### 2.5.2 VBSME: Variable block size motion estimation

The basic scheme of variable block size motion estimation is to iteratively split up a block into smaller blocks depending on the quality of the matching. For each of the subblocks this procedure is repeated until either the smallest block size is reached, or the distance criterion is smaller than a threshold value. The merits of variable block size motion estimation (VBSME) for video sequence coding were investigated by numerous authors [Puri 87], [Chan 90], [Kim 93], [Lee 95], [Li 96], [Gis 96]. An overview of this topic was given by [Duf 95] and [Schu 96]. Usually the authors investigated a quadtree partitioning of the image in blocks of size 2x2, 4x4, 8x8, 16x16, and sometimes 32x32 along the motion boundaries. Variable block size algorithms were also successfully applied to image coding, e.g. [Stro 91] and as a combined scheme for inter- and intra-coding of video sequences [Macq 95], [Cal 97].

Compared to fixed block size ME, VBSME provides better estimation of small and irregular motion fields and allows a better adaptation of motion boundaries to object boundaries, which results in a reduced number of bits required for coding the prediction errors. However, supporting different block sizes requires more bits in the bit stream for signalling the size of the blocks and for encoding an increased number of motion vectors for smaller block sizes. Large blocks offer the advantage of producing a reduced number of motion vectors (and therefore a reduced bit rate), but may result in less accurate motion compensation and therefore higher prediction error, since a large block may contain two or more objects at different motion speed and motion direction. ME with small block sizes may be also easily interfered by random noise.

Standardized video compression schemes like H.263 and MPEG-4 generically support motion estimation with 16x16 block size. The advanced prediction mode extends the use of the 16x16 block size to the alternative use of four 8x8 blocks, determined by a simple SAD criterion (cf. 2.2.6). The SAD-calculation for VBSME at a specific blocksize $N$ is defined as depicted in eq. (2.2), where the previous restricted $N \in [8, 16]$ is now extended to a wider range:

$$N \in \aleph \tag{2.65}$$

whereas in the most cases

$$N \in [2, 4, 8, 16, 32] \tag{2.66}$$

is used.

***Rate/Distortion-optimization***

The optimum trade-off between block size (and therefore the number of blocks and coded vectors, which directly affects the bit rate) and distortion can be determined by Lagrange optimization, cf. section 2.3, [Berg 71] (fundamental R/D-theory), [Su 91], [Sul 94], [Lee 95], [Kim 95a], [Kim 96], [Gis 96], [HBi 96] (modified BFOS tree pruning), [Schu 96] (overview), and [Schu 97] (Lagrange based bit rate allocation among segmentation, motion and residual error).

Depending on the distance criterion (e.g. SAD) of the block comparison, the blocks are iteratively split up into smaller blocks until the smallest block size is reached (i.e. quadtree partitioning) or the SAD is within a predefined distortion level or the bit budget of the rate-control algorithm is empty.

### 2.5.3 Segment-matching

Segment-matching (Fig. 2.27), [Cal 97], employed a partial quadtree scheme, which allowed to represent the best combination (in a rate/distortion sense) of subblocks (cf. Fig. 2.27, Fig. 2.26) with one single motion vector.

This technique combines 1.) the advantage of VBSME to describe inhomogeneous motion fields by small blocks with 2.) the advantage of reduced motion vector coding costs, as for a cluster of subblocks (segment shape, Fig. 2.26) only one single motion vector is required instead of one motion vector for each single block.

The $segment_k(m,n)$ is one of 16 possible segment shapes depicted in Fig. 2.26, specifying which one of the four quadrants has to be described further by partial quadtrees of the next smaller block size level. The cluster of the dark quadrants are described by a single motion vector (Fig. 2.27). A white quadrant in Fig. 2.26 indicates, that this quadrant is to be described further by smaller quadtrees.

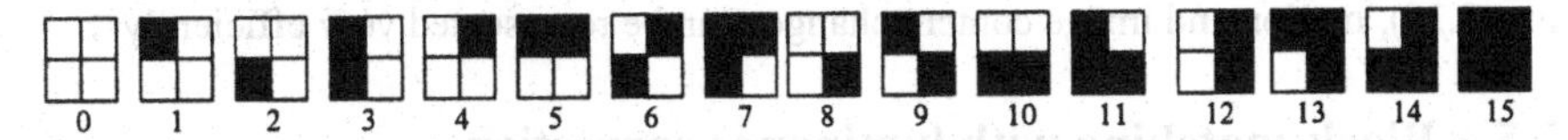


*Figure 2.26: Segment shapes ($segment_k$) with indices*

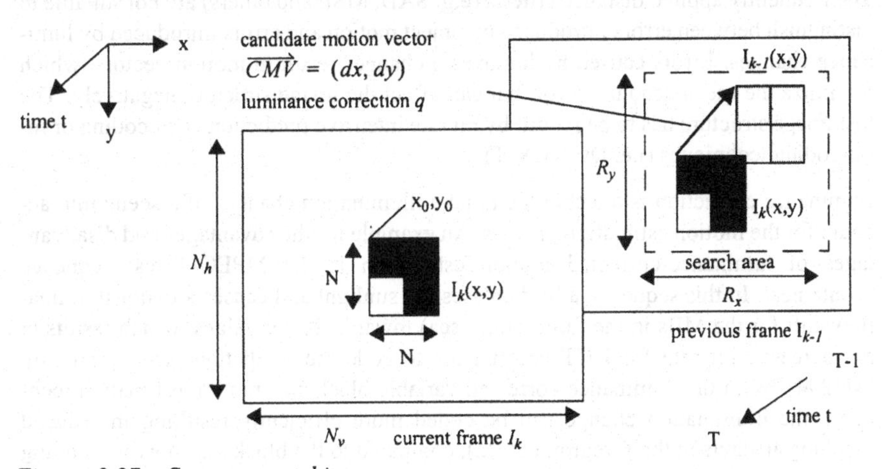


*Figure 2.27: Segment-matching*

| Level | Block-size (CIF: Y) |
|---|---|
| 0 | 32x32 |
| 1 | 16x16 |
| 2 | 8x8 |
| 3 | 4x4 |

*Figure 2.28: Partial quadtree partitioning*

The SAD calculation of a specific segment with index $k$ is defined as:

$$SAD(k, dx, dy) = \qquad (2.67)$$
$$\sum_{m=x}^{x+N-1}\sum_{n=y}^{y+N-1}|I_k(m,n) - I_{k-1}(m+dx, n+dy)| \cdot (segment_k(m,n) == 1)$$

E.g. for $k = 7$ $segment_k(m,n)$ is denoted as:

$$segment_k(m,n) = \qquad (2.68)$$
$$\begin{cases} 1 & if \quad ((0 < m < 15) \wedge (0 < n < 7)) \vee ((0 < m < 7) \wedge (7 < n < 15)) \\ 0 & otherwise \end{cases}$$

By iterative application on every hierarchical quadtree level (e.g. for $N = [4, 8, 16, 32]$, Fig. 2.28), motion and image content changes can be represented very efficiently.

### 2.5.4 Block-matching with luminance correction

The frequently applied distance criteria (e.g. SAD, MSE and others) are not suitable to distinguish between errors introduced by object motion and errors introduced by luminance changes. Errors caused by luminance changes result in motion vectors, which minimize the mean absolute error, but can affect the image structure negatively. The lost image structure has to be rebuilt by bit rate intensive prediction error coding or intra-coding techniques (i.e. DCT/IDCT).

Luminance correction is a technique to take illumination changes of a scene into account for the motion estimation process. An example for the advantages and disadvantages of luminance-corrected motion estimation is the MPEG-4 test sequence "container". In this sequence a bird crosses the sunlight and causes a major illumination shift for the MBs in the foreground (sea) towards darker values, which results in an increased bit rate for DCT coefficients to code the prediction error, [Har 96], [M1294]. With this luminance-corrected variable block size motion estimation technique the illumination change can be coded more efficiently resulting in reduced blocking artifacts in the foreground (sea), compared to the block-based hybrid coding concept as used within MPEG-1, MPEG-2 and MPEG-4. The visual quality is significantly better. Unfortunately, the overall PSNR for this specific sequence degrades, as the horizontal motion of the containerboat is estimated not homogeneously resulting in a so called "rubber-band" effect.

[Gilge 90] describes illumination correction for the MSE distance criterion, which was applied only on image blocks, where the brightness difference between the original frame $I_k(x, y)$ and the previous frame $I_{k-1}(x, y)$ exceeded a specified threshold. Illumination correction in space and time was described by [Nega 89] for optical flow. Pel-recursive motion estimation with illumination variation was studied by [Nico 95] and [Hampson 96].

The algorithm depicted here tries to reduce the SAD (or MAE) by changing the base luminance of the picture block under comparison with every search position of a block-matching unit. This results in reduced energy of the prediction error signal. However, luminance correction can also compensate real motion, requiring a co-working of luminance correction and motion estimation instead of a simple pipelining of these systems. Each pixel requires scaling by a multiplicative factor (the luminance correction coefficient), which changes temporarily and spatially. For small block sizes brightness variations can be considered to be very small, leading to an additive illumination correction coefficient with sufficient accuracy [Gilge 90].

Eq. (2.69) describes block-matching based on the SAD distance criterion, with an additive luminance correction coefficient $q$ which forms the basis for further investigations. Here the luminance correction coefficient $q$ is calculated as the quantized difference of the mean of the $I_k(x, y)$ pixels' intensity of the macro-block in frame $k$ and the mean of the $I_{k-1}(x, y)$ pixels' intensity of the macro-block in frame $k-1$. This algorithm tries to reduce the SAD by modifying the SAD with an illumination correction coefficient $q$, cf. eq. (2.69). Here the $q$-values are quantized to multiples of 4 and are in the range of [-252, 252], eq. (2.72).

$$SAD(k, dx, dy) = \sum_{m=x}^{x+N-1} \sum_{n=y}^{y+N-1} |I_k(m, n) + q(k, m, n, dx, dy) - I_{k-1}(m+dx, n+dy)| \tag{2.69}$$

$$q(k, m, n, dx, dy) = QUANT(mean(I_k(m, n)) - mean(I_{k-1}(m+dx, n+dy))) \tag{2.70}$$

## 2.5.5 Entropy constrained segment-matching with luminance correction

Based upon the previously described segment-matching and luminance correction motion estimation techniques, an iterative entropy-constrained variable block size motion estimation algorithm with luminance correction was described in detail by [Cal 97].

This technique was proposed by [M 0553] for the emerging MPEG-4 standard and was cross-checked by "core experiments" [Har 96], [M 1291], [M1294]. This advanced motion estimation technique results in a very low prediction error, which offers the advantage of decoding at very low complexity, as no intra-decoding technique (IDCT) has to be applied.

$$SAD(k, dx, dy) = \sum_{m=x}^{x+N-1} \sum_{n=y}^{y+N-1} |I_k(m, n) + q(k, m, n, dx, dy) - I_{k-1}(m+dx, n+dy)| \cdot (segment_k(m, n) == 1) \tag{2.71}$$

$$q(k, m, n, dx, dy) = QUANT(mean((I_k(m, n)) \cdot (segment_k(m, n) == 1)) - mean((I_{k-1}(m+dx, n+dy)) \cdot (segment_k(m, n) == 1))) \tag{2.72}$$

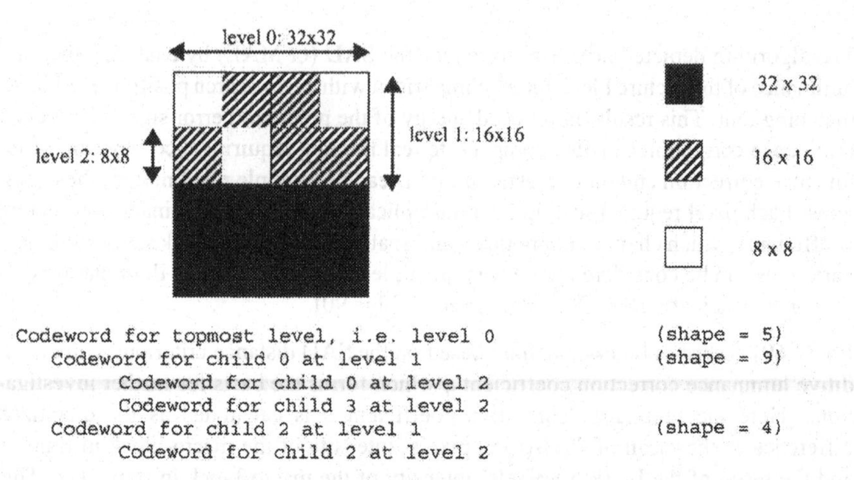


*Figure 2.29: Quadtree based bit stream description of a 32x32 luminance (Y) block*

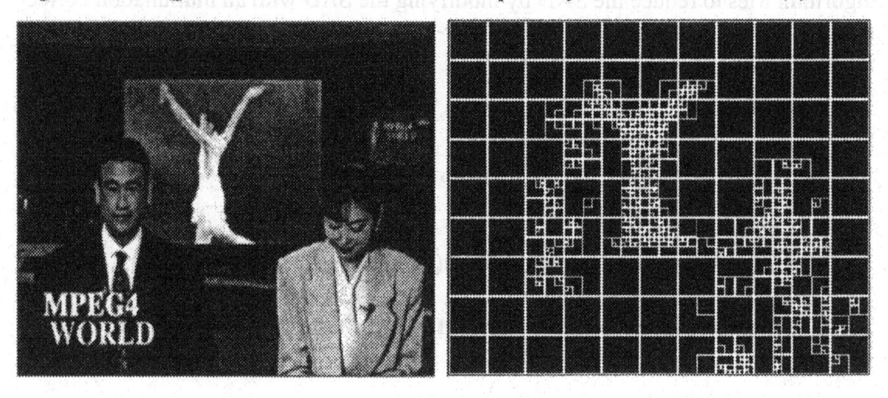


*Figure 2.30: Segment-matching with luminance correction (note the L-shapes on the right)*

In the bit stream each of the luminance macro-blocks (CIF: 32x32) is described on all levels (32x32, 16x16, 8x8, 4x4) with the motion vectors for the best segment shape, i.e. the best corresponding segment shape index $k$ and the corresponding luminance correction coefficient $q$. An example for a 32x32 block partitioning and the related bit stream description is given in Fig. 2.29. For level 0 (32x32 macroblocks) no luminance correction is used. Chrominance components are also processed by segment-matching with luminance correction. They are described accordingly in the bit stream, but with half resolution and half the block size of every level (16x16, 8x8 and 4x4). To reduce block artifacts OBMC (Overlapped Block Motion Compensation) is performed.

The decoder is based on motion compensation only, using the motion vectors on the subblock clusters and the $q$ coefficients. Fig. 2.30 depicts a single frame of the MPEG-4 sequence "news" at CIF resolution and the block partitioning of the Y (luminance) component. For areas with complex motion (e.g. the dancer in the background) smaller block sizes are preferred. Note the "L" shaped subblock clusters in Fig. 2.30. This algorithm was not optimized to find the true motion, but rather to find the best video sequence description based on segment-matching and $q$ correction.

# 2.6 SUMMARY

In this chapter an overview of MPEG-4 motion estimation has been given. It has been followed by an overview and classification of fast motion estimation algorithms. The fast motion estimation algorithms can be classified into low complexity distance criteria, search area subsampling with fixed search area and adaptive search area, hierarchical/multiresolution algorithms, pel decimation, different prediction techniques and feature-matching.

## 2.6.1 Evaluation of fast motion estimation algorithms for MPEG-4

***Distance criteria***

The various distance criteria with reduced implementation complexity, are basically considered not to impede MPEG-4 motion estimation using arbitrary-shaped objects, as every single pel of the VO (video object) is taken into consideration when calculating the distance criterion. The CCF (Cross Correlation Function) and the MAE (Mean Average Error) which show high computational complexity, are therefore not of interest. The SAD (Sum of Absolute Differences), MME (Minimum Maximum Error), RB-MAD (Reduced Bit Mean Absolute Difference), and the DPC (Different Pixel Count) were selected for further investigations, as these algorithms offer some advantages for VLSI implementation. MADM (Mean Absolute Difference) is not selected, as MADM is applied in a hierarchical scheme, requiring additional memory to save the results of the different layers.

***Search area subsampling***

For search area subsampling, search area adaptive- and hierarchical-methods seem to be more suitable than fixed search area methods for MPEG-4 motion estimation. Fixed search area methods may suffer from the fact that for small objects a considerable amount of search points may lie outside the video object. Therefore, only a small number of search points would provide significant data for the distance criterion, which might result in reduced quality. However, for hierarchical search area subsampling schemes (like the Three Step Search, TSS), the search area positions of the fine search lie within the VO (Video Object), as the best result of the coarse search is always inside the VO. For this reason, and because of being often used as reference, TSS was selected for further investigations.

***Hierarchical methods***

Hierarchical methods basically work with the arbitrarily-shaped objects of MPEG-4, however, for small video objects similar restrictions apply as discussed before, but with less impact, as the pels used for the distance criterion are gained more uniformly over the whole search area. However, for VLSI implementation hierarchical methods require additional memory to store the image data of several resolutions, or to store the results of the search in several resolutions. As a hierarchical method the mean pyramid was selected for further investigation, because of the additional feature of allegedly being robust against noise.

***Pel decimated motion estimation***
For pel decimation, the pels are also taken uniformly from the whole search area. Therefore, pel decimation is suited for MPEG-4 motion estimation, as already explained for the distance criteria. As pel decimation techniques offer the additional advantage of high regularity and are therefore suitable for VLSI implementation, the 4:1 subsampling technique was selected. 4:1 *alternate* pel subsampling was selected also, as it takes all pels of the search area into account successively. Adaptive pel decimation algorithms might provide some advantages within a MPEG-4 framework, but as they depict a very irregular data flow, which impedes VLSI-implementation, they were not selected for further research within the scope of this book.

***Prediction techniques***
Prediction techniques are regarded to show benefits solely for MPEG-4 motion estimation, in case the VO depicts a significant size. As for the small session size of the envisaged MPEG-4 video communicator only small VOs can be used, the advantage of motion vector prediction beyond the MPEG-4 method is regarded to be questionable and prediction techniques were therefore not selected for further investigation.

***Feature-Matching***
The algorithms of the feature-matching class differ considerably from one another and for this reason, it is impossible to provide a common evaluation. The integral projection-matching (IPM) algorithm uses, in it's adapted representation for MPEG-4, all pels of a VO to calculate the row and column sums, and is therefore basically suitable for MPEG-4 motion estimation. The SEA (Successive Elimination Algorithm) is regarded not to be directly implementable for MPEG-4, as every pel of a block is required to fulfill the inequality equation, eq. (2.48). BBM (Binary Block Matching) is regarded to be suitable for MPEG-4. The integral projection-matching algorithm has been selected for further investigations, as with the sum calculation of row and columns some error robustness is introduced. However, at this point it is not clear, how the IPM algorithm can be supported advantageously by a VLSI architecture.

***Rate/Distortion- and Complexity/Distortion-optimized motion estimation***
Rate/distortion and complexity/distortion optimization can be basically applied to MPEG-4 motion estimation without any restrictions.

### 2.6.2 Methods beyond MPEG-4

Emerging motion estimation algorithms had been discussed and investigated within the MPEG-4 standardization. The algorithms presented here were of the luminance-corrected variable block size motion estimation and segment-matching type, and showed the advantage of visual quality improvement for a selected number of sequences. However, these algorithms depict a significantly higher computational complexity than the currently used block-matching technique based on fixed block sizes.

# 3

# COMPLEXITY ANALYSIS METHODOLOGY

## 3.1 MOTIVATION

There are different means to measure the computational complexity of algorithms. For fast motion estimation algorithms, most of the complexity analysis results presented in literature are based on the average number of search points per macro-block. However, with this simple method of using the number of search points, the computational and the memory bandwidth requirements of the entire algorithm (which includes e.g. pel addressing, pel access, decision calculations, filtering, etc.) are not taken into account.

In the course of this book a new, automatic complexity analysis method is developed which is not only applicable for motion estimation algorithms. This method is based on a time-dependent analysis of the number of RISC-like instructions required. These results are grouped into 1.) arithmetic instructions, 2.) control instructions and 3.) memory access instructions and their associated data type (integer/floating-point and wordlength).

## 3.2 PROCESSOR BASED COMPLEXITY METRICS

In this section complexity metrics based on instruction level statistics are discussed. These complexity metrics can be used in conjunction with a visual quality metric to jointly evaluate fast motion estimation algorithms. However, these metrics are also applicable for other algorithms and applications.

### 3.2.1 General parameters of implementation specific complexity

Establishing rigid metrics for complexity analysis is considered to be a difficult task. Nevertheless, efforts have been made to quantify the computational and silicon requirements of video algorithms. Within the MPEG-4 standardization context, the task arose to compare several algorithms with the same functionality in terms of computational complexity and visual quality and to select the better algorithm as a standard. Within the context of this work, an easily to apply complexity metric was required to compare several fast motion estimation algorithms in terms of their computational complexity for a) processor based implementations, b) algorithm specific ASIC implementations, or c) a combination of both. A list of general parameters to determine the implementation-specific complexity for estimating the resulting cost for processor based implementations or algorithm-specific VLSI is given here. An overview of quantitative and qualitative complexity-metrics is given in Fig. 3.1.

#### 3.2.1.1 Memory

The memory size and the required memory bandwidth are considered to be the dominant factor for the implementation costs of a MPEG / ITU-T compliant video coding system. Video coding and especially motion estimation is *memory bound*, as high amounts of data have to be processed by relatively simple arithmetic operations. The memory bandwidth determines the speed and type (DRAM/SRAM) of the memory modules, and therefore the clock rate, the power consumption, and the cost of the system.

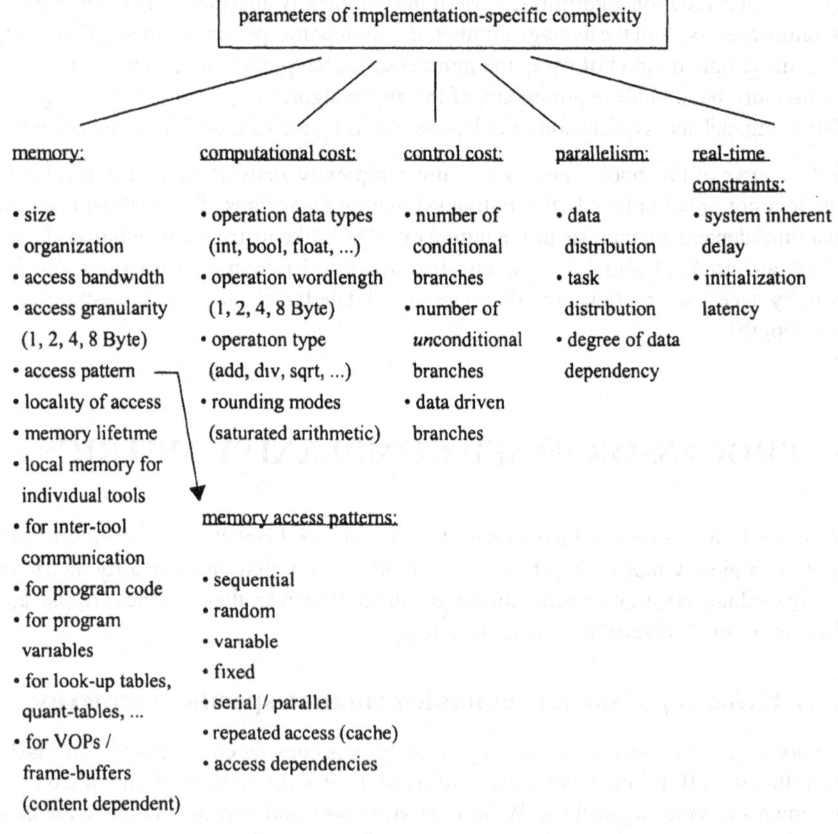


*Figure 3.1: Parameters of implementation-specific complexity*

The memory related cost of data transfer is a function of the memory size, memory type, and the memory access bandwidth. As the memory access bandwidth is directly proportional to the power consumption of most of the low-power RAMs [Itoh 95], the evaluation of memory bandwidth requirements is of high importance (cf. 3.2.6).

For external memory, the cost of memory is determined by the non-linear function of the physical size of the memory, as the memory is fabricated with predefined sizes of DRAM/SRAM/ROM-chips (e.g. 4 MByte physical RAM have to be provided in case of 2.1 MByte are required). The width of the memory access limits the possibility to use standard memories (in case 8-, 16-, 32-, 64-bit memory access granularity are used). Higher memory bus sizes increase the packaging costs by higher pin-counts and require more chip size or wider internal buses.

The locality of memory access is determined here by the lifetime of data structures of individual video tools (i.e. memory lifetime) as well as for inter-tool communication and determines e.g. the size of caches/buffers in processor/ASIC-based solutions.

Memory access patterns can be differenciated in sequential, random, fixed, or variable access patterns, whereas sequential memory access allows prefetching of data. This results in lower hardware costs than random memory accesses do. Fixed memory access patterns allow simpler hardwired AGUs (Address Generation Units) than variable memory access patterns (cf. also the section on AGUs, chapter 6). Repeated memory accesses to the same data units allow caching of data. Parallel memory access offers advantages towards sequential memory access in terms of throughput. Memory access dependencies can occur, when the memory location of the second data unit is determined, based upon the result of the calculations on the first data unit. If memory access dependencies exist, data prefetching is often not applicable.

Memory is also required for program code (RAM/ROM), constant program variables (ROM), static program variables (RAM), and scratch RAM for non-static program variables. Content-independent data like quantization tables, lookup-tables for DCT/IDCT can be also located in ROM. Besides the content-independent data in memory, there also exists content-dependent memory, which consists mainly of memory for the VOPs (for MPEG-4) or as frame-buffers (otherwise). The efficient memory allocation and management for multiple VOPs is considered a non-trivial task.

#### 3.2.1.2 Computational (arithmetic) complexity

The computational (arithmetic) complexity is defined as the actual processing of the data usage. The floating-point data types comprise a significant higher computational complexity than integer or boolean datatypes for most ASIC- and processor-based implementations.

Some types of arithmetic operations like additions and subtractions are less costly in terms of VLSI-area or processing time than others, like e.g. divisions, square roots and transcendental functions. Typically, current processors use data path sizes of 8, 16, 32, 64 bit. Using other wordlengths than these results in a) inefficient use of data paths, b) inefficient use of memories, and c) additional hardware or software to perform saturation at e.g. the 12th bit in a 16-bit data path. Saturated arithmetic is typical for some tools within MPEG video. Some rounding modes are harder (and therefore slower) than others, depending on the kind of operation. Rounding is particularly expensive if the rounding point is fixed, which occurs if fractional values of non-standard width are mapped on arithmetic data paths of standard width.

#### 3.2.1.3 Control costs

Algorithms, which consist of a high number of (conditional) branches impose restrictions on VLSI implementations in terms of irregular data-flow and are therefore difficult to parallelize. Conditional branches inside loops are more costly than other branches, because these take the possibility for loop-unrolling and parallelization of a loop. Conditional branches inside loops result in pipeline stalls, whereas current processors try to reduce this cost by branch prediction tables. Data driven branches depend e.g. on the content of the scene for the encoder or on the bit stream (at the decoder) and can significantly affect processor and VLSI-implementation cost.

#### 3.2.1.4 I/O-bandwidth (of single functions, of function blocks, global)

I/O-bandwidth can be differenciated in burst data access and singular data access. Burst data access offers significant advantages in case of direct memory access (DMA) compared to single memory word access. Within MPEG-4 new constraints arise, as several VOPs may be accessed at the same time, resulting in the demand for an efficient I/O-scheduling mechanism.

#### 3.2.1.5 Ability to exploit or introduce inherent parallelism or pipelining

Data parallelism of operations with small wordlength ("packet arithmetic") can be exploited advantageously by a SIMD (Single Instruction Multiple Data) model, which is supported by several processor architectures on the market. Task parallelism can be gained by pipelining of video coding tools, which work subsequently on the same data set. Deep nested loops without conditional branches in the loop kernel are good candidates for exploitation of parallelism or pipelining.

#### 3.2.1.6 Real-time and scheduling constraints

Real-time constraints define an upper bound for the system inherent delay. To meet real-time constraints, the exploitation of parallelism is required for some algorithms. B-VOPs in MPEG-4 (PB-frames in H.263) introduce additional implementation- and processing-independent delay.

#### 3.2.1.7 Complexity Scalability

The complexity scalability of an algorithm is often useful to scale down computational demands of an algorithm while limiting the loss in visual quality. Generally algorithms with inherent complexity scalability are preferred.

### 3.2.2 Theoretical worst case evaluation

Some algorithms depict variable complexity over time. For these algorithms the Theoretical Worst Case (TWC) is defined here.

***Definition:***
The Theoretical Worst Case (TWC) of complexity is the maximum number of instructions a program can consume.

***TWC calculation:***
The calculation of the TWC of the complexity is performed in two parts:

1. preselection of the most complex flow graphs of the program, and
2. calculation and selection of the program flow path with TWC complexity.

The preselection of the most complex flow graphs of a program is done by spanning up a data flow graph (DFG) for the program. The root corresponds to the beginning of the program and a node corresponds to a choice of different branches. A branch corresponds to a list of consecutively executed instructions without conditional jumps or branches to other code. A leaf corresponds to the end of a possible flow path of the program. The extraction of a DFG from the program code can be performed manually or by automated tools using a top-down approach.

The calculation and selection of the program flow path with TWC complexity is achieved by traveling from the leaves to the root (bottom-up). A subpath is built starting from the leaves with one common node. Each subpath is weighted by the computational complexity. Once all subpaths leading to a common node are evaluated in terms of complexity, their computational complexity is compared and the subpath with the maximum complexity is chosen. Then the next subpath group leading to the closest common new node is compared with the existing subpath until the root is reached.

However, for evaluation of fast motion estimation algorithms within the MPEG-4 context the TWC complexity has not beed applied, because low-power MPEG-4 encoding systems can be designed more economically by taking into account average complexity demands than worst case complexity demands (cf. chapter 1).

### 3.2.3 Instruction level complexity analysis

For complexity analysis of video compression algorithms, the counting of RISC-like operations (arithmetic, control, and memory access) and their data types (integer or floating-point, wordlength) is performed within this book by instrumentation of a C/C++ implementation of the algorithms.

***Definitions:***
A processor architecture supports $n$ instructions, with the instruction set $I$ being defined for a specific processor with an instruction set of a total of $n$ instruction types:

$$I = [insntype_0, insntype_1, insntype_2, \ldots, insntype_{n-1}]^T \quad (3.1)$$

The number of executions of each instruction type is defined as:

$$C = [cnt_0, cnt_1, cnt_2, \ldots, cnt_{n-1}]^T \quad (3.2)$$

A *weight* for each single instruction type is defined by:

$$W = [weight_0, weight_1, weight_2, ..., weight_{n-1}]^T \quad (3.3)$$

The $weight_0$ can be, e.g. the number of cycles required to execute instruction $insntype_0$, normalized by power consumption or processor clocks required for execution, etc. A mix of instructions executed on a processor for a specified function $func_i$ is defined as:

$$InsnMix = C \cdot I \quad (3.4)$$

Generally a weighted instruction mix is reflected by:

$$InsnMix = C \cdot I \cdot W \quad (3.5)$$

### 3.2.4 Instruction classification

The arithmetic operations are classified as: multiplications (MUL), additions (ADD), subtractions (SUB), shift operations (SHIFT), and divisions (DIV). As the estimation of the memory bandwidth is essential for the design of low-power mobile video devices, memory read and memory write operations (data-type: 8-, 16-, 32-, 64-, 128-bit) are also analyzed in detail.

#### 3.2.4.1 WNAO (Weighted Number of Arithmetic Operations)

***Related work***

From the complexity analysis of half-rate speech coders within the European GSM ("Group Speciale Mobile") standardization a complexity-metric has been derived, which was used by [Dall 94], [Marca 94], and [Oli 97]:

$$Complexity(C) = \quad (3.6)$$

$$\frac{DynamicMemory(DM)}{5} + \frac{StaticMemory(SM)}{20} + ArithmeticOperations(O)$$

in which $DM$ is the number of dynamic memory positions (RAM), $SM$ is the number of static memory positions (ROM) and $O$ is the weighted sum of arithmetic operations. $O$ is also referred to as WNAO (Weighted Number of Arithmetic Operations), [Oli 97]. The amount of program memory is not included in this measure, since it depends on the actual processor architecture (variable length instructions result often in tighter code than fixed length instructions or VLIW, Very Long Instruction Word, instructions). This complexity-metric was applied manually by the insertion of counters into the source code representation of the algorithms.

The weights defined here were chosen with a specific DSP processor implementation in mind, which has e.g. access to a MAC (multiply-accumulate) unit. However, to use this method for a particular processor, the weights have to be specially defined for this specific target processor. Within the context of this work, no specific target processor was defined. Therefore, all weights were chosen to be equal to one.

The weighted number of arithmetic operations is calculated as follows:

| Operation | Example | Weight |
|---|---|---|
| Additions | a + b = c | 1 |
| Multiplications | a * b = c | 1 |
| Multiply-additions | a * b + c = d | 1 |
| Data moves | float, int | 1 |
| Logical | shift, modulo | 1 |
| Divisions | a / b = c | 18 |
| Square roots | sqrt() | 25 |
| Transcendental | sin(), log() | 25 |
| Arithmetic test | if, if then else | 2 |
| Add-compare-select | Viterbi decoding | 6 |

*Figure 3.2: Weights for arithmetic operations [Dall 94], [Marca 94]*

### 3.2.5 Definition of a generic complexity and quality metric

Within the context of this book there is an algorithm named "*org*" which is of high complexity, but provides good visual results (e.g. the full-search motion estimation algorithm). This algorithm will be compared with another algorithm "*fast*" which is of reduced complexity and delay, but, however results in degraded visual quality. In the previous chapter numerous fast motion estimation algorithms were presented. Therefore, as a next step, a methodology is developed to compare these algorithms for a given application in terms of visual quality, complexity and delay.

***Definition of a FOM (Figure of Merit):***
A figure of merit (FOM) is defined as follows

$$FOM = \alpha \cdot Q - \beta \cdot O - \gamma \cdot D \tag{3.7}$$

where:

$$Q = quality_{org} - quality_{fast}\,[dB] \tag{3.8}$$

$$O = \frac{complexity_{org}}{complexity_{fast}} \tag{3.9}$$

$$D = delay_{org} - delay_{fast}\,[ms] \tag{3.10}$$

The $delay_{fast}$ may further be constrained e.g. by real-time conditions by

$$delay_{fast} < delay_{max} \tag{3.11}$$

and $\alpha$, $\beta$ and $\gamma$ are application-dependent weights, for which Fig. 3.3 lists a table of examples. However, these weights have to be chosen for a specific application.

Note that $Q$ must not be lower than $-\frac{\alpha}{2}$ dB, $O$ should not exceed $1/\beta$, and $D$ should not be larger than $1/\gamma$ ms.

| Application | visual quality<br>weight: $\alpha$ | complexity<br>weight: $\beta$ | delay<br>weight: $\gamma$ |
|---|---|---|---|
| high quality broadcast TV, off-line coding | high quality required | no rigid complexity constraints | delay is of lower priority |
| real-time videophone | consumer quality<br>$\alpha = 2$ (quality < 1dB) | low cost<br>-> low complexity<br>$\beta = 1/4$ (complexity < 4) | delay < 25ms (real-time):<br>$\gamma = 1/25$ (delay < 25 ms) |

*Figure 3.3: Weights for some application examples*

***Definition of a complexity metric:***
In the following, the complexity metric is defined.

$$O = \tag{3.12}$$
$$W_{arith} \cdot C_{arith} \cdot I_{arith} + W_{control} \cdot C_{control} \cdot I_{control} +$$
$$W_{mbr} \cdot MBr + W_{mbw} \cdot MBw + W_{sm} \cdot SM + W_{sps} \cdot SPS + W_{dps} \cdot DPS$$

This metric consists of:

- $C_{arithm}$: the number of arithmetic operations, $I_{arithm}$
- $C_{control}$: the number of control operations, $I_{control}$
- MBr: memory bandwidth (8/16/32 bit)
- MBw: memory bandwidth (8/16/32 bit)
- SM: the static memory size used
- SPS: the static program size
- DPS: the dynamic program size

and their respective weights. The weights depend on the target application. The problem now is, first to find a relationship between the parameters of the complexity metric listed above and the parameters needed to assess the cost of a hardware or a software implementation (cost and maximum delay). The second problem is to find the weight each parameter has in this relationship. However, the solution depends on application constraints.

***Definition of a quality metric:***
For the visual quality metric the sum of the squared PSNR difference of all sequences is used to characterize a fast motion estimation algorithm. The PSNR difference is calculated by the subtraction of the result of the fast motion estimation algorithm from the rate/distortion-curve of the full-search method (cf. appendix for details).

### 3.2.6 Instruction level memory power profiling

Memory access bandwidth determines mainly the power consumption in most modern low-power RAMs, and not the clock rate of the RAM [Itoh 95]. With accurate power models the memory power consumption can be directly calculated from the memory bandwidth data.

These data can be obtained from an instruction level profiler as developed within this book. The memory power consumption is defined as:

$$Power_{memory} = Power_{static} + (f(MBr) + f(MBw)) \cdot C_0 \cdot V^2_{dd} \qquad (3.13)$$

with $Power_{static}$ being the power consumption of the memory without any data access (standby-mode), $C_0$ being the capacitive load, and $V_{dd}$ representing the supply voltage. As accurate memory power models are difficult to obtain and technology-dependent, this work discusses only memory data transfers. However, it always has to be kept in mind that the memory bandwidth numbers given in this work are directly related to memory power consumption.

### 3.2.7 Instruction level arithmetic power profiling

The low-power requirements of mobile systems make it necessary to analyze and minimize the power consumption of all components of the system. With the increasing computational power of embedded core processors, the system depends increasingly on functionalities implemented by processor instructions than by silicon gates. For power constrained HW/SW-partitioning the need arises therefore to evaluate the power consumption for a given program on a processor.

There is a number of power analysis tools on transistor and gate level. However, determining the power requirements of a given program for a specific processor with these tools is very time consuming. It also requires knowledge of the architecture and the low-level circuit details of the processor, which often cannot be obtained. Therefore, active research is performed on gaining the power consumption for a particular instruction of a processor. The power consumption values of single instructions are combined with the execution frequencies of the respective processor instructions. The power consumption of single machine instructions can be gained by measurement on a physical existing device, or by estimation from low-level design models. For example, software power optimization techniques for a given processor were proposed in [Tiw 96], and a power profiler (PPROF) for a programmable microprocessor with sixteen instructions was described in [Meh 96].

The results of an instruction level profiler - as developed in this book - can form the basis for power profiling by means of an appropriate power model. This power model can be a simple weight for each single instruction or an appropriate weight for each type of instruction class. A more advanced power model [Tiw 96] also considers inter-instruction effects like pipeline stalls and cache misses.

The power consumption can be calculated from the instruction level profiling data as:

$$Power_{arithmetic} = W \cdot C \cdot I \cdot C_0 \cdot V^2_{dd} \qquad (3.14)$$

However, as the values of these power weights depend significantly on the VLSI technology and the type of the processor (some processors depict similar power requirements for every single instruction, others diverge significantly), no research has been conducted in this direction, within the context of a particular processor architecture.

It was observed by [Nacht 98] that the IDCT, as the computationally most dominant tool in the H.263 decoder standard, depicts a power consumption in the range of about two magnitudes lower than the power consumption of the (unoptimized) memory access of this IDCT module. Therefore, profiling and optimization of memory accesses is given here high priority for low power video architectures.

## 3.3 METHODS FOR INSTRUCTION LEVEL COMPLEXITY ANALYSIS

### 3.3.1 Introduction

Instruction level profiling provides the number and type of processor instructions, which a program executes during runtime. These data give information on computational, control and memory access costs and can be used for complexity evaluation, as well as for performance-tuning of programs and algorithms. The constraints for an instrumentation tool are regarded as following: no restrictions for the programmer, no source code modifications, as well as high portability between different computer architectures, operating systems (OS) and compilers. Further aspects for the assessment of instrumentation tools and methods are the consideration of library functions, moderate slow-down of the instrumented application and a low amount of trace data.

There are several tools for instruction level tracing and simulation available. A comprehensive overview was given e.g. in [Pie 95] or [Pardo]. However, the existence of a portable tool for instruction level profiling and memory bandwidth analysis is not known so far. As heterogeneous environments are common and existing tools are not very portable, it is regarded to be important to achieve portability. This new contribution, iprof, concentrates on providing a (GNU platform-) portable tool for gathering instruction usage statistics and memory bandwidth data between user defined events of a program with a minimum of program slow-down. An overview of some principles to solve this task is given in 3.3.3, and some of the available tools are summarized in section 3.3.3.3. The design and implementation of iprof is described in section 3.4. Application of this complexity analysis method on the visual part of the emerging MPEG-4 multimedia standard is described in the next chapter.

### 3.3.2 Overview: Collection of runtime information

The evaluation of computational and control costs can be carried out by software implementation of the algorithms on standard microprocessors. In this case data are traced during program execution to gain information on instruction usage, memory access, etc. A full instruction trace provides the exact temporal sequence of instructions and their operands. These data, however, are usually huge and difficult to gather and provide no information on instruction usage for single program functions. In contrast, instruction level profiling does not save the full instruction trace but provides instruction usage statistics on several levels of detail: from function level down to program statement level; however, the full temporal sequence of the instructions is not preserved.

Fig. 3.4 depicts a classification of methods for gathering runtime information.

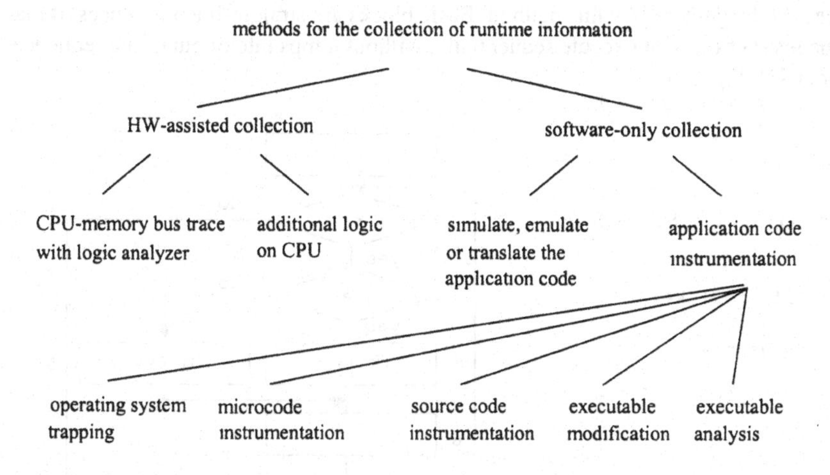


*Figure 3.4: Methods for collection of runtime information*

Runtime data gathering can be classified as hardware-assisted and software-only collection schemes.

### 3.3.2.1 Hardware-assisted collection schemes

Hardware-assisted collection schemes usually trace the activity of the CPU memory bus with an external logic analyzer or employ additional logic for processor embedded performance monitoring. Additional logic is used for on-chip counters, which increment at specific runtime events (e.g. mispredicts, total cycles, cache misses, pipeline status, various instruction classifications) as implemented with the DEC Alpha processor [Bha 96], or provide specific signals for execution tracing, like the Intel Pentium [Be 95].

### 3.3.2.2 Software-assisted collection schemes

Software-only trace collection methods avoid adding hardware to the monitored system, resulting in inexpensive and an easier portable data collection. The first approach is to simulate, emulate, or translate the code to be instrumented, and the second is the instrumentation of the code. Emulation of the application code means running this code through a program (simulator) which executes each instruction of the application code in the same manner it would run on the target architecture. This technique is mainly used in profiling new (physically not existing) architectures before VLSI fabrication [Pie 95], [Tremb 95]. Due to the complex modeling of the CPU behavior (instruction fetching, decoding, pipelining, ...) this approach is considered to be more complex and slower than code instrumentation.

Software-only runtime instrumentation can be based on application code instrumentation (at several stages of the compilation process), microcode instrumentation, or can be obtained by operating system means. Code instrumentation tools usually work on rewriting the program to be instrumented to gain runtime information during execution. Typically information on basic block and instruction usage, as well as address traces can be gathered by this method. Basic blocks are straight-line sequences of machine instructions that execute sequentially, without jumps into or out of the sequence, [Larus 93], Fig. 3.6.

```
for (i=0; i < 5; i++)
{
        if (i>2) k++;

        else m++;

}
```

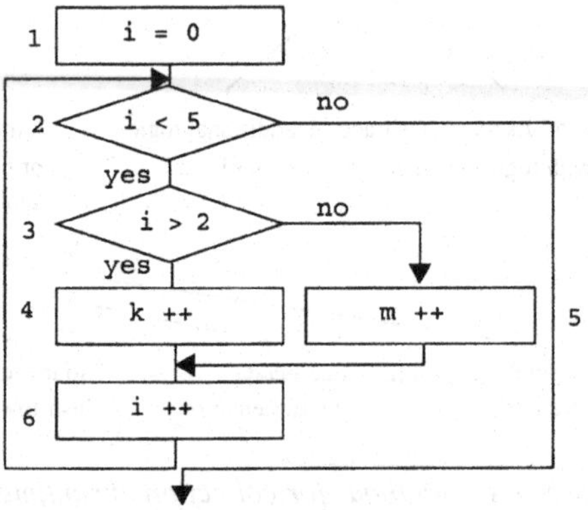

*Figure 3.5: Example program code and control flow graph consisting of 6 basic blocks*

The instrumentation code is inserted directly into source code, during compilation, object linking or directly into executable with minimal overhead. The logical program behavior and hardware environment remains the same. Regular jumps into trace routines are used to collect and save the desired data. This approach offers very restricted portability between different architectures and operating systems (OS). The efficiency of instrumentation through operating system modifications (OS traps) is limited by the overhead and the speed of the OS primitives. Accessibility of OS primitives could be difficult in proprietary operating systems and there is also another disadvantage, which is the lack of portability.

Generally, software-only instrumentation tools are less expensive and provide better portability than hardware-assisted tools. Compared to operating system trapping (no portability), code instrumentation tools provide a viable solution with (restricted) portability. However, code instrumentation tools are restricted by the type of information collected and impose an additional load on the system.

Fig. 3.6 depicts the stages in which instrumentation is applicable, which are

1. during the source compilation process,
2. during object linking, and
3. late executable modification.

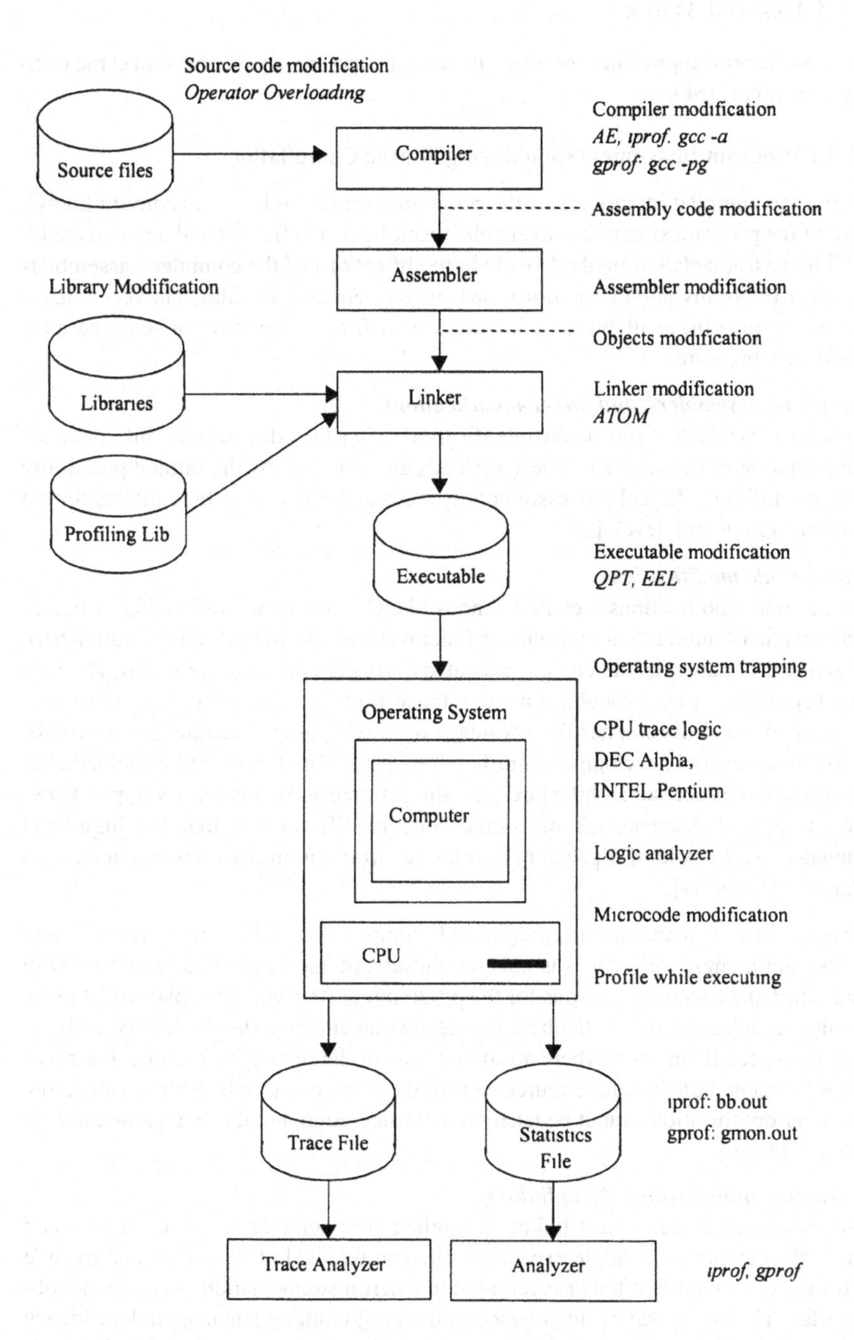


*Figure 3.6: Stages during compilation and program execution in which instrumentation is possible*

### 3.3.3 Related Work

There are several approaches for program instrumentation. An overview on some techniques is presented here.

#### 3.3.3.1 Program Instrumentation during Source Compilation

During the compilation process of the program's source code, the maximum knowledge of the program structure is available, resulting in an efficient and detailed analysis. The instrumentation methods include modifications of the compiler-, assembler- or linker-programs and/or modifications of the source code files. However, these methods require the availability of the source code files, which may not be the case for proprietary programs.

***Compiler-, assembler-, and linker-modifications***
Compiler-, assembler- and linker-modifications offer the advantage of automatic instrumentation of the program. These methods are restricted by the limited portability between different OS and processor architectures and often cause high efforts for the instrumentation tool developers.

***Source Code modifications***
Source code modifications include methods like C++ operator overloading or manual/automatical introduction of counting functions into the source code to summarize the number of particular statements, executed during the run of the program. The purpose behind is to gain a rough estimate of the arithmetic complexity of an algorithm. Source code modification has the advantage of causing only moderate difficulties for instrumentation tool development. Furthermore, portability is provided through the independence of processor architecture, operating system and compiler developer. However, a general disadvantage of source code modification is that the high-level language source code of a program is often far from the instructions executed by a computer [Larus 94].

Measurement of instruction usage (control, computation and memory access statements) and memory references is not straightforward and can only be achieved with high effort and severe restrictions for the programmer (this will be explained lateron). Another problem is the difficulty of code instrumentation should library calls be present, especially in cases where no source code of the library is available. For accurate instrumentation the whole source code of the libraries has to be instrumented, too. Compiler optimization cannot be taken into consideration, but this is regarded to be a minor constraint.

***Instrumentation during Object linking***
Instrumentation during object linking is applied after compilation of the objects and before the creation of a single executable file and removal of relocation and module information. A modified linker is required to insert instrumentation code into the object files. The advantage of this method is the programming language independency and unrestricted source code development. However, the source code has to be available to create the object files and a sophisticated linker has to be developed.

### 3.3.3.2 Program Instrumentation by Executable Modification

Instrumentation of the executable is a technique offered by several tools. This is the most difficult approach, because due to the absence of source code information, no program structure is available and a huge amount of architectural- and system-specific details has to be considered. The tool's tasks are to locate the `.text` section (contains the program code) of the executable (Fig. 3.7) and to recognize the basic blocks. Then it has to disassemble and instrument the code by inserting additional machine instructions (without affecting the logical behavior of the program). This expanded code requires the translation of all addresses of the text segment and the modification of the jump table, as well as the relocation of the old and new code while rebuilding the executable. During execution of this modified program, usually a trace of basic block usage and memory access is generated. Most of the tools employ compiler knowledge or code heuristics to gain program structure information and overcome performance and reliability restrictions. Some tools are able to instrument stripped binaries (without symbol table), others require the compiler to include additional symbol table information (usually produced with option "`-g`" during compilation).

An instrumentation tool based on executable modification has to recognize different executable representations (in UNIX world: COFF, BSD a.out, ELF, ...). Most executables consist of an operating system reserved section, a `.text` section (which contains the program code), a `.data` section and a `.bss` (block starting symbol) section, which allocates space for uninitialized data. An example of a typical executable containing basic blocks, is depicted in Fig. 3.7. The usual procedure is as follows: The `.text` section is disassembled and split into basic blocks which are instrumented by insertion of additional code. Code insertion increases the size of the basic blocks which now have to be relocated, the new addresses of the basic blocks being stored in a relocation table. The instrumentation tool has to deal with the restriction that some compilers do not put instructions but data in the code segment. These data objects cannot be moved to other addresses, because the code may reference this data with direct jumps or this data may be shared by multiple processes [Con 95]. Restrictions for executable disassembling are given by the different instruction set architectures of the processors: As a RISC instruction set usually employs 32-bit aligned machine instruction format [Hen 90], the disassembly and code insertion is straightforward. Variable length instruction set architectures (as employed for example in x86 architectures) require more effort to prevent the disassembler to step out of alignment and disassemble non-instruction bytes. Some tools require the symbol table to be included in the binary to assist in disassembly.

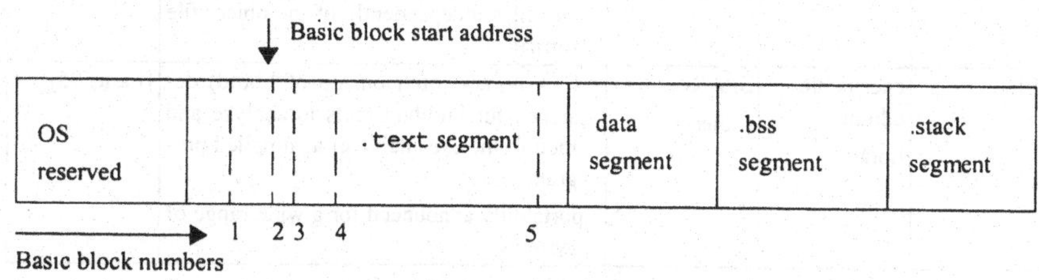


*Figure 3.7: Executable with basic block's start addresses*

### 3.3.3.3 Overview and Description of some Tools for Program Instrumentation

| Tool | Method | Platform | Comments | Reference |
|---|---|---|---|---|
| **Instrumentation during Source Compilation** | | | | |
| AE | abstract executions, compiler modification | MIPS | • a very small trace file is generated, as well as an application specific regeneration program, which allows to expand the full trace file from the small trace file<br>• based on gcc internal RTL (Register Transfer Language) representation | [Larus 90] |
| iprof | basic block statistics, executable analysis | all GNU supported platforms | • portable to all GNU platforms<br>• instruction classification<br>• memory access statistics<br>• analysis between user definable events during the program run | this book |
| **Instrumentation during Object Linking** | | | | |
| ATOM | instrumentation library | DEC Alpha | • provides library routines to insert instrumentation procedure calls before or after any program, procedure, basic block or instruction | [Sri 94] |
| **Executable Modification** | | | | |
| QPT | executable modification | SPARC, MIPS | • similar to AE<br>• produces compact address and basic block traces, which can be regenerated for later simulations | [Larus 93]<br>[Larus 94] |
| Fast Breakpoints | debugger breakpoints | SPARC | • employs the process control mechanisms of interactive debugger breakpoints<br>• dynamically patches a running executable to jump into the instrumentation code | [Kess 90] |
| pixie | executable modification | MIPS | • basic block statistics,<br>• traces | [Smith 91] |
| IDtrace | executable modification | x86 | • basic block statistics,<br>• traces | [Pie 95] |
| **Libraries for Executable Reading and Modification** | | | | |
| libbfd | binary file descriptor library | all GNU supported | • reduces the expenditure of executable format recognizing<br>• package which allows applications to employ the same C-routines to operate on object files independently of the object file format | [Cha 91] |
| EEL | executable editing library | SPARC (others) | • C++ library (built on top of libbfd) designed for building tools to analyze and modify the executable of a compiled program<br>• portability announced for a wide range of systems | [Larus 95] |

*Figure 3.8: Overview of some instruction level profiling tools*

***Advantages***
An advantage of the executable modification is the source code independence, which is beneficial for the instrumentation of third-party programs, for which no source code is available. The independence of the underlying programming language results in unrestricted usage of assembler or any higher language in the source files. No special or modified compiler is necessary to enable the employment of different object files. These can be produced by different compilers and linked together to one executable binary. No modification of the compiler-link process or the setting of special compiler flags is required. Since static library code is included in the executable, full instrumentation of static linked libraries is possible. No source recompilation is necessary for repeated instrumentation runs, which results in fast turnaround times. Code optimization of a compiler can also be investigated.

***Disadvantages***
However, there are some disadvantages: The portability of the instrumentation tool is restricted, because executable modification is highly dependent on the processor architecture, the operating system and the executable representation. Further disadvantages are the high complexity of the instrumentation tool and the restrictions imposed by peculiarities of the processor's instruction set architecture (e.g. variable length instructions). A disadvantage are reliability restrictions, as executable modifications may be based on invalid assumptions of the compiler technology used, leading the instrumentation tool to an insertion of invalid code in the executable.

A selection of tools and methodologies is described in the following.

***Operator Overloading***
The operator overloading feature of C++ is regarded as a possible solution to implement simple source code instrumentation. C++ permits redefinition of the meaning of operators such as +, -, *, & for any class [Strou 91], [Bak 91]. In other words, the meaning of operators can be overloaded. A class is a new abstract data type. Such overloaded operators offer the possibility to define new operations on this data. This technique can be used for instrumentation within each arithmetic operation. Information like the count of arithmetic operations (total or per function), data type (float or integer), precision of the data type (number of bits affected with each operation) can be gathered. Special tracing is possible, for example for division (division by 4 can be implemented less costly with a shift operation than e.g. division by 7). For application of this method, all basic data types like `int`, `double`, etc. have to be redefined to `instrum_int`, `instrum_double`, etc, because operator overloading is not applicable to basic data types. This results in a huge amount of manual editing of the source files. Derived data types have to be derived from `instrum_int`, `instrum_double` now. The data collected by overloading can be summarized just before program completion.

Apart from instrumentation of arithmetic operators, functions such as `new()` and `delete()` or `malloc()` and `free()` can be replaced by instrumented copies of themselves, like `instrum_new()`, `instrum_delete()`, to obtain a rough value of total memory usage. This requires that every memory slice used by constants, variables, etc. has to be allocated by the `instrum_new()` function.

A disadvantage are the underestimated values of memory used. This is mainly due to the fact that the overhead of function calls (stack size) is not taken into account. A more exact memory usage statistics can be gathered only by special memory access methods and requires extensive editing of the source files. A main disadvantage of the operator overloading technique is the use of pointers in the source code, which cannot be overloaded to count memory transfers and therefore require therefore significant manual modifications of the source code by introducing non-pointer data structures.

Overloading of arithmetic operations introduces high computational overhead by a function call with every single arithmetic command resulting in significant slow-down of the application runtime. One of the prerequisites for this technique is that the source code has to be written in C++ to allow introduction of classes for overloading and compilation with a C++ compiler. Operator overloading is not directly applicable to arithmetic commands used in library functions, meaning that the complete library source code (e.g. `stdio.h`, `math.h`, `string.h`, `stdlib.h` in C or `iostream.h`, `iomanip.h`,... in C++) has to be modified and recompiled.

***AE (Abstract Execution)***

Larus and Ball [Larus 90] developed AE (Abstract Execution), which is based on a modified GNU C [GNU CC] compiler. AE is available under GNU public license (GPL). As program tracing usually generates a huge amount of trace data, the goal for AE was to generate a very small trace file which could be saved and reused for several simulation runs. The compiler also produces (besides the executable) an application-specific regeneration program (i.e. an abstract version of the application program), which allows expansion of the full trace from the produced (small) trace file.

AE relies heavily on the gcc internal RTL (Register Transfer Language) representation, which is a code similar to the instruction set of a simple load/store architecture. This load/store architecture has an infinite number of registers and an orthogonal instruction set. The gcc parser translates a program into RTL and the compiler modifies the RTL description with every optimization step. AE analyses the RTL code sequence to find basic block boundaries, loops and instructions that have to be instrumented. In addition AE builds the complete control flow graph of the program and inserts the tracing code before the instructions. AE also counts the number of machine instructions produced by gcc for an RTL instruction. The AE concept is able to provide very detailed information, which cannot be obtained by executable tracing.

***ATOM***

ATOM [Sri 94] is a tool-building system for DEC Alpha and OSF/1 which enables the development of customized instrumentation and analysis tools. ATOM provides library routines to insert procedure calls before or after any program, procedure, basic block or instruction. ATOM was built using OM, a link time program modification system. OM's input consists of a collection of object files (which may be produced by C, Fortran or other language compilers) and libraries that make up a complete program. It builds a symbolic intermediate representation on which instrumentation and optimization is applied, and finally outputs an executable.

***QPT***

QPT [Larus 93], [Larus 94] is similar to abstract execution (AE), [Larus 90], aiming at the production of compact address and basic block traces, which can be regenerated for later simulations. AE instrumentation was implemented in the C Compiler, where QPT instruments the executable on a SPARC- or MIPS-based UNIX computer with `a.out` executable format. QPT builds a control flow graph (CFG) from the executables symbol table and puts instrumentation on the edges of the CFG.

***EEL (Executable Editing Library)***

EEL (Executable Editing Library), [Larus 95], is a C++ library designed for building tools to analyze and modify the executable of a compiled program. EEL provides classes for machine-independent executable editing. Portability is announced across a wide range of systems and currently SPARC architectures are supported. The system- and architecture-specific components of EEL that manipulate the executable files use the GNU libbfd library. Symbol table information is required in the executable file for accurate program analysis, but this information is sometimes incomplete or misleading. EEL refines the symbol table by analyzing the application program to find data tables, hidden routines, as well as multiple entry points and builds an internal RTL (Register Transfer Language) description of the program. EEL constructs the control flow graph (CFG) of each routine and finds subroutine calls in case the executable has no symbol table. The CFG is an architectural-independent approach of representing control flow on basic block level. A tool using EEL is able to add foreign code before or after almost any instruction without affecting local control flow and does not need to be aware of architectural details such as delayed branches. EEL internal instructions are abstractions of RISC-like machine instructions. Instructions are divided into functional categories and provide operations to inquire about semantics. These categories are implemented as C++ classes: memory references (load and store), control transfers (calls, returns, system calls, jumps and branches), computation (add, mul, ...) and invalid (i.e. data). New instruction categories can be built by employing inheritance from more than one class. The earlier developed QPT was rewritten using the EEL class library, resulting in QPT2. QPT2 can be modified to trace the number and type of instructions used during program execution.

***Fast Breakpoints***

A similar approach for executable instrumentation is to employ the process control mechanisms of interactive breakpoints used by debuggers. Interactive breakpoints have (in a typical UNIX environment) the disadvantage of severe performance degradation introduced by two context switches per interaction. To avoid this slow breakpoint mechanisms, an about 1000 times faster non-interactive breakpoint scheme was introduced by [Kess 90], which dynamically patches a running executable to jump to the instrumentation code. As fast breakpoints modify the machine language representation of the code without recompiling, executables produced by any compiler with any source code language can be instrumented, but implementation is highly processor-architecture dependent. Problems (among others) are that the `.text` segment of the executable, which is usually allowed to be read only during program execution must be writeable and the branch instructions may not have sufficient range to transfer control from the old branch address to the new breakpoint code.

***Pixie***

Pixie was developed by [Smith 91] as a full execution trace generation tool for the MIPS processor architectures based on basic block statistics. [Con 95] reports the instrumented executable to be 4 to 5 times slower, when tracing both instruction and data references. The pixie tool can work on executables generated by different compilers and uses a heuristic to decide on basic block boundaries. To circumvent the runtime overhead of pixie, a successor called nixie had been developed, which is, however, limited in its number of instrumentable applications.

***GNU[1] C Compiler Options***

The GNU C compiler offers several options to output data gained during the compilation process. These data can be processed further by other tools. The gcc [GNU CC] debugging option "-dletters" generates a debugging dump during compilation. The option "-dr" generates a dump after RTL generation to the file "file.c.rtl". This RTL file can provide input for an external instrumentation tool, but the complexity of the RTL syntax inhibits the straightforward use for an instrumentation tool.

The GNU CC [GNU CC] debugging option "-pg" comprises extra code in the executable, to output profile information suitable for the analysis program gprof (see section on GNU gprof). The profiling information is sampled at periodic times during program execution and is written to the file "gmon.out". gmon.out mainly contains sampled execution times and the duration (in ms) of each function call. The option "-pg" has to be used during compilation (and linking) of each source file for which data has to be gathered.

The GNU CC [GNU CC] debugging option "-a" generates extra code to write profile information for basic blocks. During program execution it is traced how many times each block is run, as well as the basic block's start address, and the function name containing this basic block. These data are written to the file "bb.out" by calling of the function exit(). If option "-g" is used, the line number and filename of the basic block's start are also saved. Option "-a" is used by the instruction profiler iprof, which will be described later.

***GNU libc_p.a***

There is a profiling C library, usually located in `/usr/lib/libc_p.a` on UNIX machines, which can be used by writing "-lc_p" instead of the usual "-lc" linker call argument [Fen 93]. Together with gprof, this option supplies information about the number of calls for standard library functions, such as read and open.

***GNU libbfd***

The GNU BFD (Binary File Descriptor) library [Cha 91], libbfd, reduces the expenditure of executable format recognition. BFD has been developed as a package which allows applications to employ the same C-routines and to operate on object files independently of the object file format. GNU libbfd is part of the GNU gdb and GNU binutils distributions, providing the advantage of portability and distribution in terms of the GPL (GNU Public License). BFD is split up into two parts: first the front end,

---

1 GNU is an acronym for: Gnu is Not Unix

which provides the interface to the user and calls back end routines, and second, the back end for interfacing to object files. BFD subroutines automatically determine the format (ELF, COFF, a.out, ...) of the input object file. Then they build a descriptor in memory with pointers to BFD routines that will be used to access and process elements of the object files data structures. A typical application is the conversion of symbol tables. The advantage is that instrumentation tools built on GNU libbfd are independent of the operating system flavour and object file formats.

***GNU gprof***

The GNU project profiler gprof provides profiling at function level and was built using the GNU libbfd library. GNU gprof provides the advantage of portability and distribution in terms of the GPL (GNU Public License) and was described in [Grah 82], [Grah 83] and [Fen 93]. The GNU CC [GNU CC] debugging option "-pg" is required during compilation for embedding extra code in the executable to save periodically sampled profile information during program execution. GNU gprof postprocesses this saved profile information together with the executable. The GNU profiler delivers a call graph of the profiled program, as well as function call statistics (number of calls and calling function) and function execution duration in milliseconds.

At the time gprof had been developed, most computers/OS only supported timers in which the resolution had not been fine enough to measure the execution times of functions, or the overhead accessing of the timers had been too high. To solve this problem, the status of the program execution is sampled at regular time intervalls by noting the location of the program counter (PC). Many of such samples are collected, and from their distribution across the addresses of the executable's `.text` segment, the distribution of time spent in each routine is calculated. This method is advantageous for highly structured programs with numerous small routines, but cannot give profiling information on statements below function level. With this statistical sampling method, the gathered execution times of the functions may have some quantization error.

### 3.3.4 Summary

In this section several tools and methodologies for program instrumentation were reviewed. However, none of them offers portability to different processor architectures, operating systems, and compilers. Therefore, in the next section the development of a portable instruction level profiler will be described.

## 3.4 A NEW TOOL AND METHODOLOGY FOR COMPLEXITY ANALYSIS: IPROF

### 3.4.1 Introduction

This section presents a new portable instruction level profiling tool: iprof [iprof]. This tool allows to produce an instruction usage statistics of a program or of user-specificable parts of a program. Portability is achieved by reusing functionalities of the GNU C/C++ compiler (gcc/g++) and the GNU disassembler. Basic block execution counts are obtained by instrumentation of the program under test with the compiler option "-a". The basic block execution count facility of gcc was extended to reset and write basic block statistics at user-specified events during the program run, enabling event-driven analyses of parts of the program under test.

The iprof tool is built upon the GNU disassembler for portability reasons, and uses basic block counts to allocate counts to individual instructions. The output processor of iprof also groups instructions, such as load/store, arithmetic, control and memory access by data type. The advantages of this tool are: portability among GNU-supported platforms, no required modifications of source code or object files, no restrictions for the programmer, only little slow-down of program execution (about 5%) allowing for the profiling of interactive applications, flexible trace scope (from function level to single lines of source code), high reliability and a smaller amount of trace data to be saved.

### 3.4.2 Goals for the development of a complexity analysis tool

The constraints for an instruction level profiling tool are regarded as:

- No restrictions for the programmer,
- portability between different computer architectures, operating systems and compilers,
- no source code modifications, as well as
- event-dependent instruction level profiling.

For program instrumentation there are several tools and methods, but most show deficits in terms of portability or the ability to gather instruction level data. Therefore, a new portable tool for instruction level profiling and instruction classification was developed, which is described in the following section.

### 3.4.3 Description of iprof

iprof is able to gain statistics of the complete program, as well as of single source code files, functions and daughter functions or basic blocks (Fig. 3.7) with low effort. In contrast to other tools, this tool has not been written from scratch as functions from the GNU C/C++ compiler and the GNU (dis-) assembler have been reused. This offers the advantage, that iprof is portable with less effort to all GNU-supported architectures, platforms, and operating systems.

In short, the principle of iprof is first of all, to obtain basic block execution counts by running the program under test, after compiling with "gcc -a" (or "g++ -a"). Then the disassembler is used to assign instructions to basic blocks and hence to allocate counts to individual instructions. The output also summarizes instruction counts for classes of instructions, such as memory loads/stores, arithmetic and control instructions.

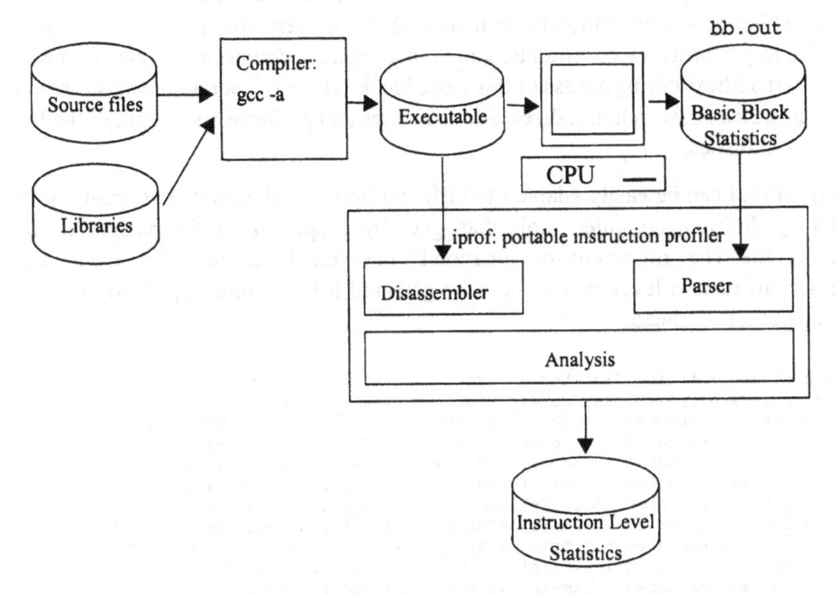


*Figure 3.9: Functionality of the portable instruction level profiler: iprof*

Different executable representations (in UNIX world e.g.: COFF, BSD a.out, ELF, ...) exist. Most executables consist of an operating system reserved section, a `.text` section (which contains the program code), and a `.data` section and a `.bss` (block starting symbol) section, which allocates space for uninitialized data. An example of a typical executable is depicted in Fig. 3.7.

To circumvent the limitations and unreliabilities of *executable modification*, iprof employs *executable analysis*. A program is subdivided into basic blocks (BBs) by the compiler. Basic blocks are straight-line sequences of machine instructions that execute sequentially, without jumps into or out of the sequence. Fig. 3.5 depicts a short sample C program with the control flow graph containing 6 basic blocks and Fig. 3.9 an outline of the functionality of iprof. With every control instruction (jump, branch, call, ...) the current basic block is left and another is entered. The temporal sequence of instructions is provided for straight code between jumps and branches with small amount of trace data.

***Instrumentation of the program***

As described before, there are several methods to instrument the executable in order to gain basic block statistics. The option "`-a`" of the GNU C/C++ compiler *[GNU CC]* was found to be a portable and reliable method for this critical task. This option is used to instrument the executable in order to save statistics on basic block usage after program completion. The option "`-a`" generates extra code which is included in the executable, to write profile information for basic blocks. During program execution, the number of times each basic block is called, the basic block start address and the function name, containing this basic block, is traced and written to the file "`bb.out`" by calling the function `exit()` (or `__bb_write_file()` as described later) at program completion (Fig. 3.10). The option "-a" of the GNU compiler is implemented by a machine-dependent part, written in assembler for every target machine architecture due to performance constraints, which increments a counter for a specific basic block start address during access of this basic block. There is a portable part written in C in `libgcc2.c`, which initializes data structures and performs file output of the basic block statistics.

The iprof tool can be easily adapted to different basic block statistics' formats (produced by different compilers/tools than gcc) by adaptation of the parser for the `bb.out`-file. The program instrumentation by basic block statistics offers the advantage that instruction level statistics can be gathered for an arbitrary program by low computational overhead.

```
Basic block profiling finished on Sun May 10 19:49:08 1998
File /home/seql/p9/work/shorty/shorty.d, 9 basic blocks
    Block #1: executed 1 time(s) address= 0x10824 function= main
    Block #2: executed 6 time(s) address= 0x10840 function= main
    Block #3: executed 1 time(s) address= 0x10860 function= main
    Block #4: executed 5 time(s) address= 0x10878 function= main
    Block #5: executed 2 time(s) address= 0x10898 function= main
    Block #6: executed 3 time(s) address= 0x108c0 function= main
    Block #7: executed 5 time(s) address= 0x108e0 function= main
    Block #8: executed 1 time(s) address= 0x10908 function= main
    Block #9: executed 1 time(s) address= 0x10924 function= main
```

*Figure 3.10: Example basic block statistics' file of the C program of Fig. 3.5 (SPARC)*

Additionally, the compiler option "-g" provides program's source code file names and source line numbers of BB's entries in the statistics file (bb.out, Fig. 3.10). For each basic block the machine instructions are contained in the executable's .text section (Fig. 3.7). The basic block statistics' file contains the number of calls for every basic block and the start address of the corresponding instructions of these basic blocks in the executable file. The analysis program iprof jumps into each of the executable's basic blocks, decodes and counts the machine instructions of every basic block and multiplies for each instruction the number of calls for this basic block. A summing up over all basic blocks gives the total usage count of a particular instruction. iprof consists of a parser, a disassembler and an analysis section. The parser section of iprof reads through the basic block statistics' file produced during program execution and puts this data in the internal basic block data structures.

***Parser***
As the basic block statistics file bb.out has an uniform format, which is independent of the platform, architecture, and operating system, a parser could be created to extract the number of calls for each basic block.

***Disassembler***
To provide high portability, the GNU disassembler is included in iprof to access and decode the instructions of the executable. As the format (but not the machine instructions) of the disassembled data comprises a uniform format for different architectures, it is possible to extract the name and the operands of each instruction, independently

of the underlying architecture. A modified dump of the executable's `.text` segment is provided as an additional output by iprof, which depicts the number of executions of each instruction ("`calls`" in Fig. 3.11) and the corresponding basic block numbers ("`bb#`" in Fig. 3.11).

The GNU disassembler is based on the GNU BFD (Binary File Descriptor) library [Cha 91], libbfd, which reduces the expenditure of executable format recognizing already described before.

```
Disassembly of section text:
                                                                              added by
address      bb# calls        label    insn        arguments                  gcc - a
..
00010804       -                -      <main> save   %sp, -128, %sp
00010808       -                -      <main+4> sethi   %hi(0x21000), %o0
0001080c       -                -      <main+8> ld  [ %o0 + 0x48 ], %o1! 00021048 <force_to_data>
00010810       -                -      <main+c> tst  %o1
00010814       -                -      <main+10> bne  00010824 <LPY0>
00010818       -                -      <main+14> add  %o0, 0x48, %o0
0001081c       -                -      <main+18> call  00010cd4 <__bb_init_func>
00010820       -                -      <main+1c> nop
00010824       0               (1)     <LPY0> sethi  %hi(0x21000), %g1
00010828       0               (1)     <LPY0+4> ld  [ %g1 + 0xa8 ], %g2! 000210a8 <.LLPBX2>
0001082c       0               (1)     <LPY0+8> inc  %g2
00010830       0               (1)     <LPY0+c> st  %g2, [ %g1 + 0xa8 ]
00010834       0                1      <LPY0+10> clr  [ %fp + -28 ]
00010838       0                1      <LPY0+14> clr  [ %fp + -24 ]
0001083c       0                1      <LPY0+18> clr  [ %fp + -20 ]
00010840       1               (6)     <LPY0+1c> sethi  %hi(0x21000), %g1
00010844       1               (6)     <LPY0+20> ld  [ %g1 + 0xac ], %g2! 000210ac <.LLPBX2+4>
00010848       1               (6)     <LPY0+24> inc  %g2
0001084c       1               (6)     <LPY0+28> st  %g2, [ %g1 + 0xac ]
00010850       1                6      <LPY0+2c> ld  [ %fp + -20 ], %o0
00010854       1                6      <LPY0+30> cmp  %o0, 4
00010858       1                6      <LPY0+34> ble  00010878 <LPY0+54>
0001085c       1                6      <LPY0+38> nop
00010860       2               (1)     <LPY0+3c> sethi  %hi(0x21000), %g1
00010864       2               (1)     <LPY0+40> ld  [ %g1 + 0xb0 ], %g2! 000210b0 <.LLPBX2+8>
00010868       2               (1)     <LPY0+44> inc  %g2
0001086c       2               (1)     <LPY0+48> st  %g2, [ %g1 + 0xb0 ]
00010870       2                1      <LPY0+4c> b  00010908 <LPY0+e4>
00010874       2                1      <LPY0+50> nop
...
```

*Figure 3.11: iprof disassembled executable with basic block statistics generated by iprof (SPARC)*

After disassembling the code of the `.text` segment, the total usage count of each instruction and the grouping of instructions into instruction classes is printed. Fig. 3.11 shows the disassembled executable. The code at the addresses 0x10804 - 0x10830 is inserted by the option "`-a`" of the GNU compiler, which performs mainly the initialization of basic block statistics (call to `<__bb_init_func>`) and the counting of the basic block execution times. Simple tracing of flat instruction usage per source file, per function and per basic block can be gathered by, e.g., passing the name of the function to be traced to iprof. No recompilation of the application program source code is necessary for the evaluation of different functions or program sections. The analysis section also summarizes the number of calls of every function used.

***Analysis***

```
Instruction Usage Statistics: 0
----------------------------
Executed Instructions in Basic Blocks:
    opcode    times executed
  0 ld                              21 (19.44444 %)
  1 nop                             20 (18.51852 %)
  2 cmp                             11 (10.18519 %)
  3 ble                             11 (10.18519 %)
  4 mov                             10 (9.25926 %)
  5 add                             10 (9.25926 %)
  6 st                              10 (9.25926 %)
  7 b                                9 (8.33333 %)
  8 clr                              4 (3.70370 %)
  9 ret                              1 (0.92593 %)
 10 restore                          1 (0.92593 %)

Total:                             108 (100%)
```

*Figure 3.12: Instruction usage statistics output for Fig. 3.5*

### 3.4.3.1 Event-Driven and Call Graph Profiling by GNU gcc modifications

***Event Driven Profiling***

As the basic block statistics, as originally provided by the GNU compiler option "-a", is only written during program exit, the gcc compiler has been modified to enable a start and stop of instrumentation at arbitrary events in the C/C++ source code of the instrumented program. This allows to reset and write new basic block statistics data with just a single function call in the source code of the instrumented program at a defined event (which can also be conditionally determined). For this purpose, two new commands for source code instrumentation have been implemented in the gcc compiler (`libgcc2.c`), which can be treated like other function calls:

`__bb_set_zero();`

Action: All internal basic block counters of the instrumented program are set to zero to prepare for new accounting.

`__bb_write_file();`

Action: Writes the basic block statistics since their last set to zero to a new basic block statistics file.

All machine instructions executed by the processor between these two function calls are accounted, regardless of their occuring in one or several daughter functions or recursive function calls. Over the runtime of the instrumented executable, several basic block statistics files are produced with every call of `__bb_write_file()` and can be evaluated by iprof to produce a time-dependent instruction usage statistics. These compiler modifications do not effect code generation, but also allow call graph instruction level profiling, by simple inserting these commands at the beginning and the end of the function to be analyzed (Fig. 3.13).

This feature has been employed to generate time-dependent instruction usage analysis and call graph instruction usage analysis (cf. [Grah 82], [Ball 96]) of specific parts of the program (e.g. instructions in single or multiple functions and all their daughter functions called only by *these* parent functions, and not by other parent functions of the daughters earlier or later on).

```
int function_to_be_analyzed(){
      int i, j, k;
      __bb_set_zero(); /* set statistics counters to zero */
      ...
      daughter_function_1();
      ...
      daughter_function_2();
      ...
      __bb_write_file(); /* write statistics to file */
      return i;
}
```

*Figure 3.13: Example: instrumentation of a function with daughter functions*

### 3.4.3.2 Instruction Classification and Memory Access Statistics

To provide additional gross performance metrics (as discussed before) of the instrumented program, the analysis section of iprof classifies the processor's instructions in terms of: memory reference instructions (load, store), control instructions (call, return, jump), computation instructions (add,mul,sub, ...), float instructions and type conversion instructions. For each instruction and instruction class a counter was implemented, which increments with the instruction's occurrence and its matching of one of the classes mentioned before. Instructions embedded in libraries can only be summarized, if the libraries are compiled with "gcc -a" and static linking of these libraries is performed. However, instruction classification is architecture-dependent, currently instruction classification is provided for SPARC and x86 architectures.

***SPARC architectures:***
The arithmetic instructions are classified and summarized further:

- MUL: smul, fmuld, fmuls, .umul, .mul, ...
- ADD: add, addx, faddd, fadds, ...
- SUB: subx, sub, fsubs, ...
- SHIFT: sll, sra, srl, ...
- DIV: sdiv, fdivd, fdivs, .udiv, .div, ...

In Fig. 3.14 an example for the instruction classification and in Fig. 3.15 an example for the memory access statistics for the example program of Fig. 3.5.

Statistics of the counts and destinations of the "call"< instruction are also provided, giving the number of call times of a particular function. As the SPARC is a load/store-architecture, memory accesses and memory access granularity (8-, 16-, 32-, 64-, 128-bit) are analyzed with low effort. There are load (copies a byte, halfword, word, or doubleword from memory) and store (copies to memory) instructions, which are summarized to gain the memory access bandwidth and granularity. However, these results depend strongly on the employed compiler optimization and data types used in the instrumented C/C++ program.

```
Instruction Grouping:
---------------------

Arithmetic Instructions:
    add                                 10 (9.25926 %)
 Subtotal:                              10 (9.25926 %)
    call .umul:                          0 (0.00000 %)
    call .mul:                           0 (0.00000 %)
    call .udiv:                          0 (0.00000 %)
    call .div:                           0 (0.00000 %)
 Subtotal + calls:                      10 (9.25926 %)

 MUL:
 Subtotal:                               0 (0.00000 %)
    call .umul:                          0 (0.00000 %)
    call .mul:                           0 (0.00000 %)
 Subtotal + calls:                       0 (0.00000 %)
 ADD:
    add                                 10 (9.25926 %)
 Subtotal:                              10 (9.25926 %)
 SUB:
 Subtotal:                               0 (0.00000 %)
 SHIFT:
 Subtotal:                               0 (0.00000 %)
 DIV:
 Subtotal:                               0 (0.00000 %)
    call .udiv:                          0 (0.00000 %)
    call .div:                           0 (0.00000 %)
 Subtotal + calls:                       0 (0.00000 %)
Jump, Test and Compare Instructions:
    b                                    9 (8.33333 %)
    ble                                 11 (10.18519 %)
    cmp                                 11 (10.18519 %)
    restore                              1 (0.92593 %)
    ret                                  1 (0.92593 %)
 Subtotal:                              33 (30.55556 %)

Load/Store Instructions:
    ld                                  21 (19.44444 %)
    mov                                 10 (9.25926 %)
    st                                  10 (9.25926 %)
 Subtotal:                              41 (37.96296 %)

Type Conversion Instructions:
 Subtotal:                               0 (0.00000 %)

Float Instructions:
 Subtotal:                               0 (0.00000 %)

Other Instructions:
    clr                                  4 (3.70370 %)
    nop                                 20 (18.51852 %)
 Subtotal:                              24 (22.22222 %)

Total Instructions used:                     108 (100 %)
                              =      0.00010800 Million
```

*Figure 3.14: Instruction grouping for the example program of Fig. 3.5*

```
Memory Access Statistics:
-------------------------
    instruction  accesses (% of loaded/stored Bytes)

Load from Memory:
   Byte           ( 8 bit)
 Subtotal:                         0 (0.00000 %)
   Halfword       (16 bit)
 Subtotal:                         0 (0.00000 %)
   Word           (32 bit)
    ld                            21 (67.74194 %)
 Subtotal:                        21 (67.74194 %)
   Double Word ( 64 bit)
 Subtotal:                         0 (0.00000 %)
   Quad Word     (128 bit)
 Subtotal:                         0 (0.00000 %)
 Loaded Data from Memory total: 84 Bytes   (67.74194 %)
                                 =        0.00008011 MBytes

Store to Memory:
   Byte           ( 8 bit)
 Subtotal:                         0 (0.00000 %)
   Halfword       (16 bit)
 Subtotal:                         0 (0.00000 %)
   Word           (32 bit)
    st                            10 (32.25806 %)
 Subtotal:                        10 (32.25806 %)
   Double Word ( 64 bit)
 Subtotal:                         0 (0.00000 %)
   Quad Word     (128 bit)
 Subtotal:                         0 (0.00000 %)

 Stored Data to Memory total: 40 Bytes   (32.25806 %)
                                 =        0.00003815 MBytes

Total Memory Bandwidth Consumption: 124 Bytes   (100 %)
                                 =        0.00011826 MBytes (100 %)
```

*Figure 3.15: Example for memory access statistics for example program of Fig. 3.5*

***x86 Architectures:***

For x86 processors an arithmetic and control instruction classification is performed. Additionally a classification in terms of protected mode instructions, stack instructions, string instructions and CPU instructions is provided. Gathering memory access statistics for x86 architectures requires more effort than for the SPARC (load/store) architecture and is currently not implemented.

### 3.4.3.3 Use of iprof

With the iprof tool the programmer is not impeded by any restrictions and no source code modifications are necessary. However, the source code of the program has to be available for compilation and the GNU compiler gcc or g++ has to be used. The instrumented program is fast, a slow-down of about 5% for an application program is typical, and for repeated runs no source recompilation is necessary. As iprof employs static executable analysis instead of the sometimes critical modification of object files, this tool proved to be reliable on several platforms and operating systems.

#### 3.4.3.4 Limitations

There are some instructions of the executable's `.text` segment, with no basic block statistics being available, which are not included in the instruction profile. This is the case for the small amount of instructions for program initialization and program closing, as well as the instrumentation code added by "gcc -a" (which is usually of no interest and is marked with parentheses within each basic block). When executing iprof with option "-s" (this gives an annotated disassembly of the executable) these lines of instructions are marked with a "-" for not-summed up instruction counts. Library functions, which were not instrumented are also marked by a "-". In some cases of very accurate instruction usage statistics demands, instrumentation of system libraries may be also required. However, system library calls usually share only a small part of the total computation time. The instruction statistics is summarized iprof-internally, by using long data types enabling extended profiling runs. Self-modifying code can not be profiled, because iprof is based on static executable analysis.

#### 3.4.3.5 Portability

The instruction level profiler iprof is portable for all GNU platforms with low effort. This tool contains, besides the disassembler (which can be compiled for different targets), very little system specific code: for instruction classification and for detection of basic block bounds. Usually the end of a basic block is determined by the start address of a new basic block. There are rare cases (e.g. the end of a program) when a basic block is finished and no new basic block starts immediately. To detect these cases, iprof tests the occurrence of statements which indicate the end of a subprogram. These statements are architecture-dependent and need architecture specific code, which is, however, of neglectable size: one line of source code in iprof for x86 architectures (instruction-group: `ret`) and two lines for SPARC (`ret, restore`). Instruction classification depends heavily on the instruction set architecture of the target platform and, for this reason, cannot be portable.

### 3.4.4 Design Alternatives for iprof

For the design of iprof there had been several possible alternatives. For example, for the analysis of the computation costs of each basic block several alternatives would have been conceivable, which are discussed as following: BB analysis of the source code, BB analysis of the assembler file, function call analysis in assembler file, RTL (Register Transfer Language) analysis and BB analysis of the executable (which was finally chosen).

***Basic block analysis of the source code***
BB analysis of the source code files means to parse the C/C++ statements of a source file's basic block section and to assign each statement to an abstract instruction or an architecture neutral performance metric. An advantage is the independence of machine representation, which enhances portability. However, disadvantages are the limited precision (source code representation is far from the actual instructions executed), the treatment of library calls (a library call is not a separate basic block) and the fact that

a C/C++ interpreter has to be built or modified for this application which results in high expenditures. Another problem is the detection of the end of basic block boundaries, especially in the case of several basic blocks in one source code line. With this method only a very general complexity assessment and no instruction level statistics can be gathered.

***Assembler file analysis***
A different approach which had been considered, was to analyze the assembler files, which can be saved during compilation with the GNU CC option "-save-temps" and parsed to gain a complexity assessment. Unfortunately, gcc/g++ provides no means to identify the assignment of the basic blocks to the produced assembler code. Function call analysis of the program would have also been possible to gather instruction level statistics. The GNU profiler "gprof" ([Grah 82], [Grah 83], [Fen 93]) realizes profiling at function call level which allows assignment of function calls to assembler code. Functions usually contain branches (basic blocks do not), and there is no way to calculate the executed instruction count on basic block level.

***RTL (Register Transfer Language) analysis***
The RTL (Register Transfer Language) is a gcc/g++ compiler internal instruction description of each basic block, which can be saved to files. The option "-dr" generates a dump after RTL generation to the file "`file.c.rtl`". This RTL file can provide input for an external instrumentation tool, but due to the complexity of the RTL syntax, this point was not further pursued. The RTL description can be parsed and assigned to an abstract performance measure. Unfortunately, the RTL syntax cannot be easily accessed, and R. M. Stallmann stated in [GNU MAN]: "GNU CC was designed to use RTL internally only. A correct RTL for a given program is very dependent on the particular target machine. And the RTL does not contain all the information about the program".

### 3.4.5 Assessment of the iprof approach

The machine instruction counts and the memory bandwidth delivered by the described approach using iprof, are exactly those, which are executed by the instrumented code. The memory bandwidth resulting from this data is the total memory bandwidth, which is usually buffered by a cache hierarchy on common processor based architectures. Usually a program also depends on library functions (e.g. `libc`) which are not instrumented automatically (but can be instrumented by recompilation of the libraries). However, the total time spent in library calls is very short (2.2 % of the total time for the instrumented MPEG-4 verification model decoder) and most of the library calls (e.g. `printf`) are of minor interest from the complexity analysis point of view. Machine code written by a human being, which is used for some libraries cannot be instrumented by this approach. Other sources for bias of the complexity result is the C/C++ description of the analyzed algorithms, which is often optimized in readability, portability, and modularity instead of speed.

# 3.5 SUMMARY

In the first section of this chapter, parameters for implementation-specific complexity were discussed. Then existing methods to perform instruction level complexity analysis were reviewed and a new portable tool for instruction level complexity analysis was presented: `iprof`. Compared to other tools, this tool depicts, besides portability, the capability of memory bandwidth analysis, which is most important to analyze data-driven algorithms used for video compression.

# 4

# COMPLEXITY ANALYSIS OF MPEG-4 VISUAL

## 4.1 MOTIVATION

A complexity analysis of the video part of the emerging ISO/IEC MPEG-4 standard was performed as a basis for HW/SW-partitioning for VLSI implementation of a portable MPEG-4 terminal. While the computational complexity of previously standardized video coding schemes was predictable for I-, P- and B-frames over time, the support of arbitrarily-shaped visual objects as well as various coding options within MPEG-4 introduce now content- (and therefore time-) dependent computational requirements with significant variance. In this chapter the results of a time-dependent complexity analysis of the encoding and decoding process of a binary shape coded VO (video object) and the comparison with a rectangular shaped VO is given for the complete codec, as well as for the single tools of the encoding and decoding process. It is shown that the average MB complexity per arbitrarily-shaped P-VOP (video object plane) depicts significant variation over time for the encoder and minor variations for the decoder.

Memory bandwidth requirements by video coding algorithms are one of the key considerations in the implementation of either general purpose or dedicated processors. The study of memory requirements (on-chip as well as off-chip) is one of the key points leading to energy efficient designs, because the memory power consumption is directly related to the memory bandwidth. Off-chip memory usually requires large amounts of energy to drive high-capacitance off-chip bus loads, whereas on-chip memories can be accessed faster and with lower energy budget. Therefore, the trade-off on on-chip or off-chip memories is controlled by the memory bandwidth requirements.

Neither MIPS or MOPS can measure secondary performance issues like memory usage and power consumption. This is a severe limitation because execution time means little if memory requirements exceed system design constraints. This chapter reveals the memory bandwidth of the MPEG-4 video standard in total, as well as for the single tools.

# 4.2 INTRODUCTION

The video part of the emerging MPEG-4 (Moving Pictures Experts Group) standard [MPEG 4], [Sik 97a], [Sik 97b] advances, compared to previous video compression standards, in the support of arbitrarily-shaped video objects. This enables new applications based on e.g. object manipulation. The shape of a video object is referred to as an alpha-plane which is generated by techniques (e.g. by the blue screen technique or by segmentation) that are not covered by the MPEG-4 standard. The alpha-plane can be of the binary type (a pixel is translucent or not) or it can be a grey scale alpha-plane, where each pixel's intensity is weighted by a value between 0 and 255, allowing partially translucent objects. The presented analysis is based on the Committee Draft (CD) of MPEG-4 version 1 [MPEG4], which was delivered in November 1997. The CD included the following coding tools for natural video sequences: I-VOP coding, P-VOP and B-VOP prediction, interlace coding, binary shape coding, rectangular and object based temporal scalability, rectangular spatial scalability, error resilience for rectangular- and arbitrarily-shaped objects, static sprites (basic sprites / low latency sprites), 12-bit video, as well as rectangular- and object-based texture coding.

Previously standardized MPEG [MPEG 1], [MPEG 2] and ITU-T (International Telecommunications Union) [H.261], [H.263] video communication algorithms were block-based and characterized by analytically determinable computational and memory access requirements. This enables a more or less straightforward system design for the worst case event. However, for VLSI implementation of the video part of the MPEG-4 standard the evolution from block-based video to arbitrarily-shaped visual objects (of variable number and size within time) limits the applicability of the analytically methods of computational power estimation employed so far for block-based video: block-based video consisted of a predictable complexity for I-, B- and P-frames over time. Arbitrarily-shaped video objects (VOs), as well as various coding options within MPEG-4 for different VOs, now introduce a content-dependent complexity with significant variance, which results in the following demands:

- A time-dependent approach for automatic complexity analysis in terms of arithmetic and control operations, as well as memory access bandwidth and granularity.
- Definition of complexity-driven bounds for encoding/decoding of multiple VOs to guarantee real-time encoding and decoding, as well as a specified QoS (Quality of Service). Real-time operation is required to be achieved by a system designed for an average computational load rather than by providing the computational power for the rare worst case event of maximum computational load for every VO, which is not feasible in e.g. a mobile environment.

In section 4.2.2 the complexity analysis method applied within this work is explained. In sections 4.3 and 4.4 the results for the MPEG-4 visual encoder and decoder for binary-shaped video objects are given. Section 4.4.5 shows the results for the special case of a rectangular VO and section 4.5 summarizes this chapter.

### 4.2.1 Previous Work on Complexity Analysis of Video Compression Algorithms

There are already several studies on complexity analysis of previously standardized MPEG and ITU-T algorithms:

[Zhou 95] described a theoretical complexity analysis in terms of MOPS (Million Operations per Second) of a MPEG-1/2 style video decoder for 720x480 pel at 30 fps, for the purpose of employing the Ultra-SPARC visual instruction set for MPEG decoding. The basic operations of a video decoder are divided up by the authors into: 1.) bit stream parser and variable length decoder (12 MOPS), 2.) dequantization and inverse DCT (206.1 MOPS), 3.) motion compensation (50.8 MOPS), 4.) YUV to RGB conversion (360.3 MOPS) where one multiplication is counted (i.e. weighted) as 4 generic operations. Altogether, these authors calculated a complexity of 620-750 MOPS for the decoder.

[Bha 95] analytically calculated numbers of "RISC-like" operations for the ITU-T H.261 video compression standard for the case 30 fps and CIF, where the compression (with a fast 25 step motion estimation algorithm and a fast DCT algorithm) requires 968 MOPS and the decompression requires totally 198 MOPS. The definition of RISC-like operations (as given by [Bha 95]), which is similar to the RISC instructions of a real machine used in this book, is explained by an example: The evaluation of `r1 = a + b`, where `a` and `b` are numbers in memory, requires 3 RISC-like operations: two loads from memory and one addition, the result is saved in register `r1`. Modern superscalar RISC architectures are capable of executing more than one of these RISC instructions per clock cycle. In [Gut 92] the RISC processing requirements for ITU-T H.261 video compression and decompression at 30 fps are stated to be 1193 MOPS, where e.g. 51% of the computational power is used for motion estimation, 16.1% for IDCT, 9.2% for loop filtering and 6.2% for DCT. [Kap 98] performed a simulative analysis of the cache memory requirements for the H.263 video codec using the QPT (cf. previous chapter) trace generator and a cache simulator. The author analyzed H.263 video coding and decoding for several H.263 coding options (unrestricted motion vector mode, syntax-based arithmetic coding mode, advanced prediction mode, and PB-frames mode) and different cache sizes and strategies.

These complexity analysis studies are based mainly on theoretical considerations of arithmetic operations for block-based video. Control operations and the required memory access bandwidth are not taken into account. In this chapter the results of a simulative complexity analysis of the video part of the emerging MPEG-4 standard are presented, which was performed as a basis for HW/SW partitioning for the VLSI implementation of a mobile handheld multimedia terminal supporting MPEG-4. The focus of this chapter is on time-dependent complexity analysis of the encoding and decoding part of a binary shape coded arbitrarily-shaped VO, and the comparison with a rectangular-shaped VO. The content of this chapter focuses on the complexity analysis of a collection of MPEG-4 video tools, which are regarded as essential for low-power mobile communication. For these reasons, the MPEG-4 tools e.g. for sprites, 12-bit video, texture, error resilience and interlaced prediction are not analyzed here.

## 4.2.2 Experiment Conditions

### 4.2.2.1 Complexity Analysis Method using iprof

The MPEG-4 video verification model encoder and decoder software was instrumented and analyzed by the instruction level profiler iprof [iprof], which was developed by the author and explained in the previous chapter. The tool iprof is able to gain instruction usage statistics of the complete program, as well as of single source files and functions with low effort.

### 4.2.2.2 Test sequence parameters for complexity analysis

The complexity analysis was produced on a Sparc Ultra 2 using an early, unoptimized MPEG-4 video verification model C implementation [MomVM] compiled with the GNU gcc 2.7.2.3 compiler with optimization options (`-mv8 -O3`). A fully optimized implementation of an MPEG-4 video codec is expected to require a lower instruction count and memory bandwidth by about a factor of five to ten. The MPEG test sequence "weather" was used with the following parameters: QCIF, 10Hz (i.e. every 3rd frame), 10s length, 56 kbit/s, VM 5+ rate control, Q-Intra first: 10, Q-Inter first: 10, error resilience disabled, DC/AC-prediction enabled, H.263 quantizer, full-search motion estimation with a search range of [-16; +15.5] pel and a non-optimized DCT/IDCT implementation. The sequence was coded IPPPP... without B-VOPs. For the binary shape coding case only VO1 (woman) was used without coding the background. Fig. 4.1 depicts three VOPs of VO1 of the sequence "weather". The huge variations of the computational complexity depicted in this chapter are caused by the small movement of the woman's hand.

*Figure 4.1: VOP 0, VOP 195, and VOP 267 of the VO1 of sequence "weather"*

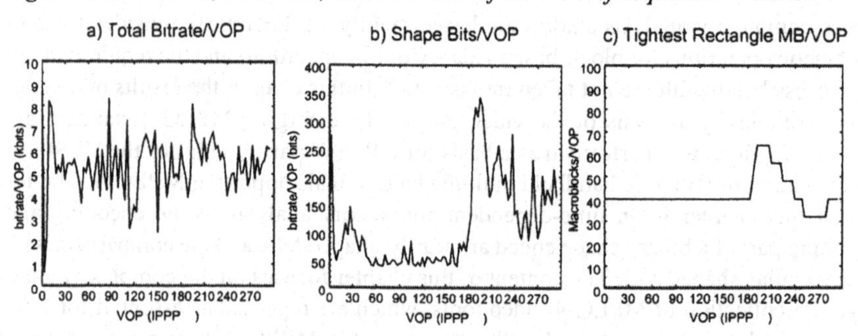


*Figure 4.2: a) Total bitrate, b) shape bits and c) the number of macroblocks of the tightest rectangle of the video object VO1 (woman)*

To find the minimum number of macro-blocks that contain the object (for high coding efficiency) the tightest rectangle is determined by the MPEG-4 video encoder (VOP formation), which contains all the MBs of an object. Fig. 4.2a gives the total (rate controlled) bit rate, Fig. 4.2b depicts the bit rate for the shape coding of binary shaped VO1 and Fig. 4.2c gives the number of MBs of the tightest rectangle over time.

# 4.3 MPEG-4 ENCODER: ARBITRARILY-SHAPED VIDEO OBJECTS

## 4.3.1 Total computational power requirements

Fig. 4.3 depicts the total computational power requirements of the MPEG-4 video verification model encoder using binary shape encoding. The instruction usage for the I-VOP is 49.87 million instructions and the I-VOP memory bandwidth is 61.68 MByte. Note the variability of the instruction usage and the memory access bandwidth over time.

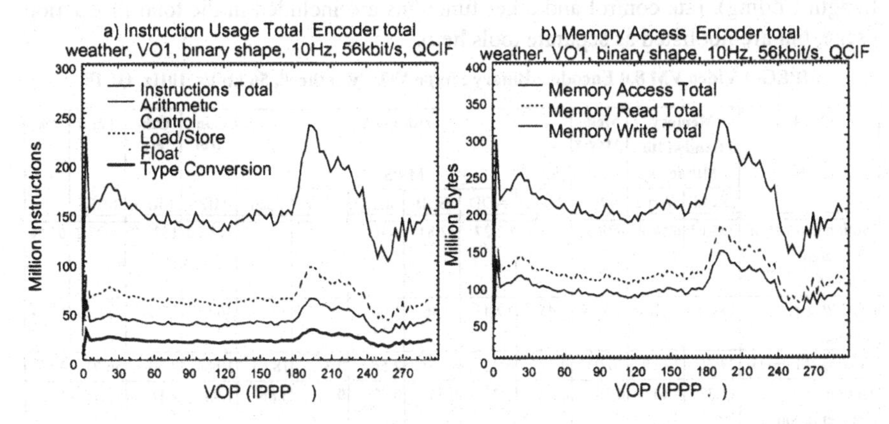


*Figure 4.3: MPEG-4 encoding of an arbitrarily-shaped video object: (a) total instruction usage, and (b) total memory access over time*

Note that the first P-VOP requires 226.63 million instructions and a memory bandwidth of 281.63 MByte, which is mainly caused by the motion estimation and the shape coding. Fig. 4.4a depicts the MB-normalized (division of the number of instructions per VOP by the number of coded MBs per VOP) instruction usage of the tightest rectangle of a VOP. Fig. 4.4b gives the probability density function (pdf) of the MB-normalized instruction counts per VOP (i.e. the pdf of the total instructions depicted in Fig. 4.4a), with the mean instruction usage/MB per VOP being 3.63 million instructions with a standard deviation of 0.45 million instructions.

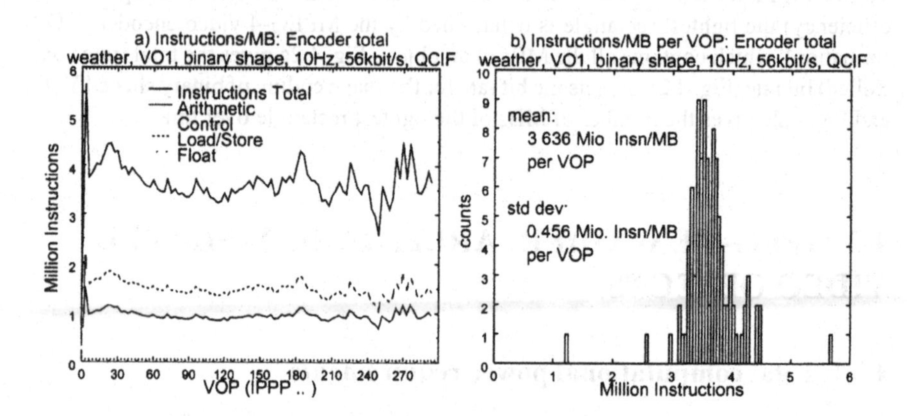


*Figure 4.4: MPEG-4 video encoder: (a) normalized instruction usage per coded MB and (b) the pdf of the total instructions in (a)*

Fig. 4.5 shows the average million instructions per second (MIPS) for this IPPP... sequence of 10s length. Note that file I/O, bit stream coding, RVLC (Reversal Variable Length Coding), rate control and other functions are included in the total instruction usage, but are not listed as separate tools here.

**MPEG-4 Video VM 8.0 Encoder: Binary Shape VO, "weather", 56 kbit/s, 10Hz, QCIF**

| Tool | Memory Bandwidth MByte / s | | Mem. Access | Arithmetic MIPS | | | | | | Con-trol | Oth-ers | Total | % |
|---|---|---|---|---|---|---|---|---|---|---|---|---|---|
| | Read | Write | MIPS | MUL | ADD | SUB | SHIFT | DIV | Total* | MIPS | MIPS | MIPS | |
| Motion Estimation full-search [-16,+15.5] | 783.68 | 624.40 | 438.38 | 1.63 | 62.27 | 32.81 | 4.40 | 0.12 | 241.57 | 168.70 | 153.69 | 1002.3 | 66 % |
| CAE-Shape Coding | 114.38 | 102.40 | 66.35 | 0.45 | 19.47 | 0.46 | 3.60 | 0 | 57.23 | 46.61 | 31.67 | 201.87 | 13 % |
| Padding (Motion) | 29.25 | 27.89 | 20.03 | 0.37 | 6.42 | 0 | 2 56 | 0.3 | 18.63 | 10.64 | 7.60 | 56 91 | 4 % |
| Motion Compensation | 11.33 | 6.61 | 6.39 | 0.12 | 2.75 | 0.57 | 0.92 | 0 | 5.86 | 1.94 | 1.18 | 15.38 | 1 % |
| IDCT | 16.48 | 13.88 | 0 | 0.45 | 2 18 | 0.48 | 0.30 | 0 | 5.16 | 1.94 | 1.68 | 14.35 | 0.9% |
| H263 Quantization | 7.32 | 5.57 | 3.74 | 0 | 0.23 | 0.11 | 0.12 | 0 11 | 1 93 | 2.18 | 2 10 | 9.97 | 0.7% |
| DCT | 8.75 | 6 51 | 3 75 | 0.45 | 1 53 | 0.48 | 0.30 | 0 | 3.48 | 0 88 | 0 98 | 9 11 | 0.6% |
| Inverse H263 Quantization | 3.73 | 3 30 | 1 86 | 0 | 0.46 | 0 | 0 | 0 | 1.34 | 1.34 | 0.85 | 5.38 | 0 3% |
| Reconstruction | 1.07 | 0.83 | 0 56 | 0 01 | 0.29 | 0 | 0 12 | 0 | 0.72 | 0 72 | 0.72 | 1.76 | 0 |
| Padding (Texture) | 0 81 | 0 63 | 0.44 | 0.02 | 0.07 | 0 | 0.15 | 0.03 | 0.58 | 0 266 | 0.371 | 1.67 | 0 |
| Prediction | 0 77 | 0.99 | 0.52 | 0.01 | 0.27 | 0 | 0.11 | 0 | 0 67 | 0.260 | 0 17 | 1.63 | 0 |
| others (file I/O, RVLC, ...) | (not instrumented separately) | | | | | | | | | | | 188 31 | 12 % |
| Total | 1096.2 | 892.9 | 615 2 (40 7%) | 4 1 (0 27%) | 114.6 (7 60%) | 36.0 (2 39 % | 16.4 (1 08%) | 0 9 (0 06%) | 392.3 (26 0%) | 274.7 (18 2%) | 226.3 (15 0%) | 1508.7 (100%) | 100 % |

*Figure 4.5: MPEG-4 encoder with binary shape VO: computational power requirements (*Total arithmetic MIPS includes also loop inc., etc.)*

Full-search motion estimation and content-based arithmetic binary shape encoding share the major part of the encoder complexity. These tools are analyzed in detail in the sections 4.3.2 and 4.3.3. Macro-block based padding shares also a significant part of the computational complexity and is necessary for the motion estimation and motion compensation on the current and reconstructed VOP (cf. 4.4.2 for detailed analysis).

For the encoder texture padding (low pass extrapolation, LPE padding) is also applied on the 8x8 INTRA blocks that have at least one transparent and one non-transparent pixel in its associated alpha information. LPE texture padding is performed before the DCT coding. The DCT/IDCT (Discrete Cosine Transformation and inverse DCT) is employed for I-VOP (intra-VOP) coding and P-VOP (predicted-VOP) prediction error coding. For quantization (and inverse quantization) of the DCT (and IDCT) transform coefficients the H.263 method is used here. Prediction and reconstruction of VOPs is similar to the block-based scheme. For VOP motion compensation the computational complexity of the encoder per VOP is close to the decoder (Fig. 4.14).

## 4.3.2 Motion Estimation

Full-search motion estimation (ME) with fcode 1 (search area: [-16, +15.5] pel), advanced prediction and unrestricted motion vector options are used here, sharing the major part of the computational power (Fig. 4.6).

The main operation is the sum of absolute difference calculation (SAD) on 16x16 MBs. For a low-power multimedia terminal, the ME computational complexity is subject to be reduced by fast ME algorithms or dedicated ME VLSI. However, full-search motion estimation was used for this complexity analysis. Development- and time-dependent complexity analysis of fast motion estimation algorithms supporting arbitrarily-shaped video objects is depicted in chapter 5.

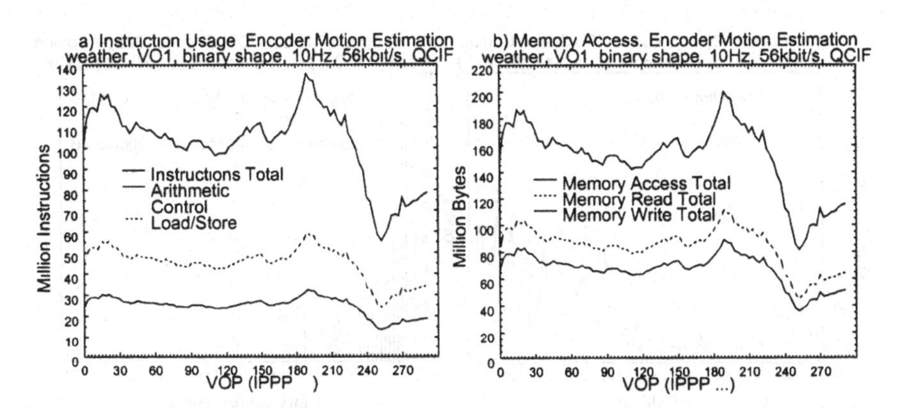


*Figure 4.6: MPEG-4 video encoder: full-search motion estimation using advanced prediction mode with search area: [-16, +15.5]*

### 4.3.3 CAE Binary Shape Coding

Modified CAE (Content-based Arithmetic Encoding) binary shape coding is used for coding the shape of binary alpha-planes (grey scale alpha-planes are encoded by a different, DCT-based method). The method adopted for the MPEG-4 verification model is able to encode a binary alpha plane in INTRA-mode for I-VOPs and in INTER-mode for P-VOPs. CAE binary shape coding consists of the following tools: shape motion estimation, shape motion compensation, shape border subsampling, size conversion (upsampling and downsampling for rate control purposes), mode decisions for BAB (Binary Alpha Block) coding, and arithmetic encoding of the BAB pixels.

Fig. 4.7 and Fig. 4.8 depicts the computational complexity of CAE-binary shape coding.

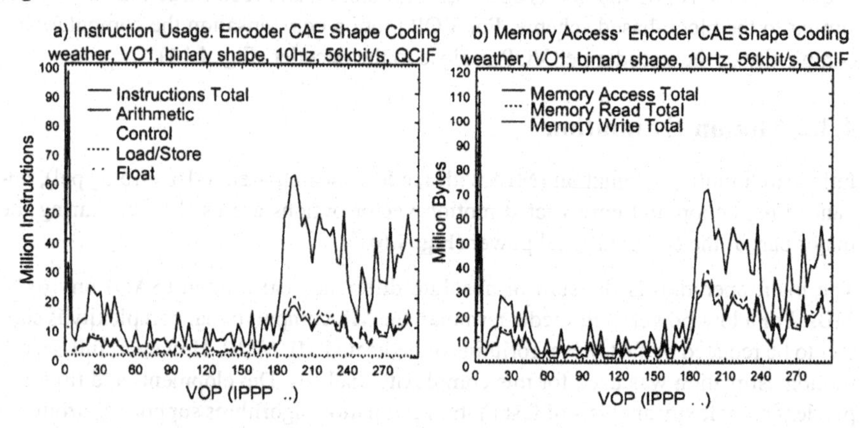


*Figure 4.7: MPEG-4 video encoder: CAE binary shape encoding*

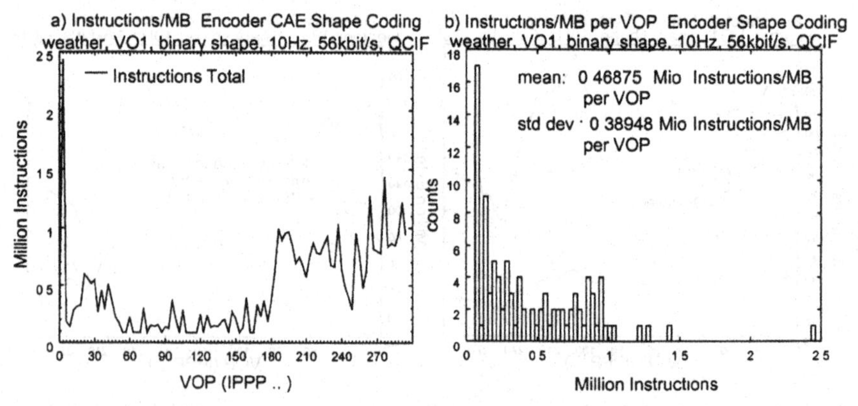


*Figure 4.8: MPEG-4 video encoder: CAE shape coding with MB-normalized complexity: a) over time; b) pdf*

The I-VOP binary shape coding in this example requires a computational complexity of 4.73 million instructions and a memory bandwidth of 4.50 MBytes. Note that for the encoding of the first P-VOP alpha-plane a significant higher computational complexity (97.47 million instructions) and memory bandwidth (105.75 MByte) is required than for shape-coding of an average P-VOP (20.18 million instructions per VOP and 21.67 MByte per P-VOP). This graph also shows that an increased computational power is required at major changes of the object's shape. These figures indicate that the number of coded MBs is with a mean of 0.46875 million instructions/MB per VOP (instructions per VOP were MB normalized) and a standard deviation of 0.389 million instructions/MB per VOP an inadequate estimate for the P-VOP shape coding complexity. A more refined complexity model, taking into account e.g. the number of boundary MBs and the BAB coding modes, is therefore required.

# 4.4 MPEG-4 DECODER: ARBITRARILY-SHAPED VIDEO OBJECTS

## 4.4.1 Computational Power Requirements

Fig. 4.9 depicts the total computational power requirements of the MPEG-4 video verification model decoder for the binary shape decoding. The instruction usage for the I-VOP with 40 MBs is 64.60 million instructions and the memory bandwidth for the I-VOP is 77.81 MByte.

Fig. 4.10a depicts the normalized instruction usage per coded MB of the tightest rectangle of a VOP. Fig. 4.10b gives the probability density function (pdf) of the instruction counts per MB for P-VOPS (the I-VOP instructions/MB are not shown), where the mean instruction usage per VOP is 0.588 million instructions/MB with a standard deviation of 0.109 million instructions.

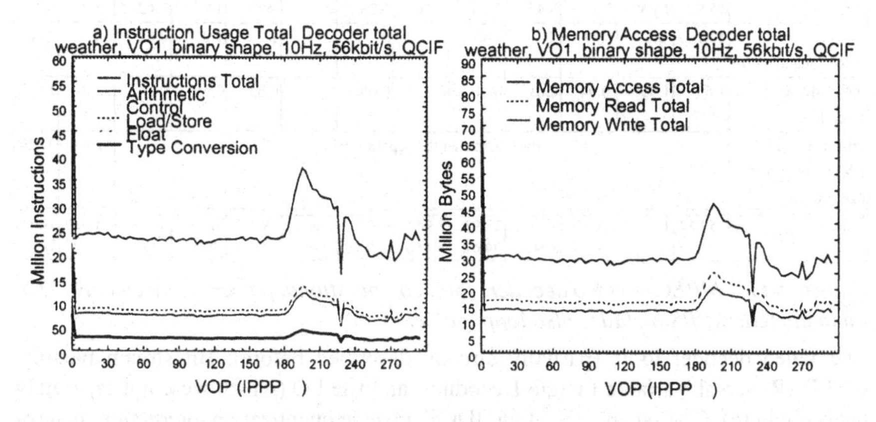


*Figure 4.9: MPEG-4 video decoder for binary-shaped objects: total computational requirements*

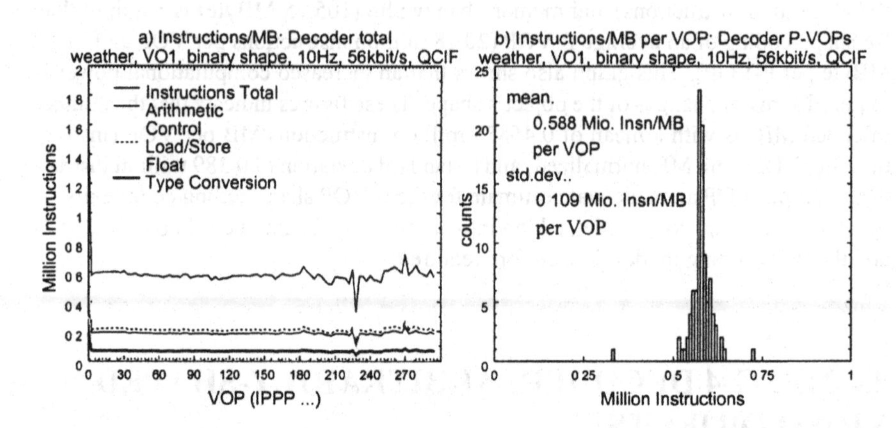


*Figure 4.10: MPEG-4 video decoder: (a) normalized instruction usage per coded MB and (b) pdf for the P-VOPs*

**MPEG-4 Video VM 8.0 Decoder: Binary Shape VO, "weather", 56 kbit/s, 10HZ, QCIF**

| Tool | Memory Bandwidth MByte / s | | Mem Access | Arithmetic MIPS | | | | | | Con-trol | Oth-ers | Total | % |
|---|---|---|---|---|---|---|---|---|---|---|---|---|---|
| | Read | Write | MIPS | MUL | ADD | SUB | SHIFT | DIV | Total* | MIPS | MIPS | MIPS | |
| Deringing Postfilter | 45.23 | 34.86 | 25.44 | 0.32 | 6 44 | 0.60 | 3.89 | 0 | 25 14 | 13 92 | 9 34 | 73 86 | 30 % |
| Padding (Motion) | 12.71 | 12 04 | 8 68 | 0.18 | 2.57 | 0 | 0.68 | 0 15 | 7.41 | 4 65 | 3.44 | 24.20 | 10 % |
| Motion Compensation | 12 55 | 8.13 | 7.18 | 0 12 | 2.91 | 0.57 | 0 92 | 0 | 6.34 | 2.58 | 1 50 | 17 62 | 7 % |
| Deblocking | 12.07 | 7 23 | 5.97 | 0 03 | 1 28 | 0.92 | 0 71 | 0 | 4 97 | 2.34 | 1.77 | 15 07 | 6 % |
| IDCT | 15.90 | 13 39 | 5 35 | 0.44 | 2.10 | 0 46 | 0 28 | 0 | 4 98 | 1.87 | 1.62 | 13 84 | 6 % |
| Inverse H263 Quantization | 3.61 | 3.19 | 1.79 | 0 | 0 44 | 0 | 0 | 0 | 1 29 | 1.27 | 0.82 | 5 20 | 2 % |
| CAE-Shape Decoding | 1 53 | 1.95 | 0.94 | 0.01 | 0.20 | 0 | 0 03 | 0 | 0.63 | 0.55 | 0 34 | 2.48 | 1 % |
| others (File I/O, RVLC, bit stream parsing, etc ) | (not instrumented separately here) | | | | | | | | | | | 92 62 | 37 % |
| Total | 159.19 | 132 32 | 88 95 (36 3%) | 1.51 (0 61%) | 25 93 (10 5%) | 2.94 (1 20%) | 9 18 (3 75%) | 0 16 (0 06%) | 78 79 (32 1%) | 45.96 (18 7%) | 31.18 (12 7%) | 244.89 100% | 100 % |

*Figure 4.11: MPEG-4 video decoder: total computational power requirements (*Total arithmetic MIPS includes also loop inc., etc.)*

The video decoder tools (partially already described before): Bit stream parsing, RVLD (Reversal Variable Length Decoding) and file I/O (which were not separately analyzed here), CAE-shape decoding, IDCT, inverse quantization, prediction, macro-block padding (4.4.2), VOP motion compensation (4.4.4), as well as the optional de-blocking and deringing postfilters (Fig. 4.12).

## 4.4.2 Macro-block-based padding

Marco-block based padding at the decoder is performed for VOP motion compensation and defines the values of luminance and chrominance samples outside the VOP for prediction of arbitrarily-shaped objects. Padding consists of the following parts: 1.) horizontal repetitive padding (each sample at the boundary of a VOP is replicated horizontally to the left and/or right direction, in order to fill the transparent region outside the VOP of a boundary MB), 2.) vertical repetitive padding (filling up the remaining unfilled transparent horizontal samples) and extended padding (filling up of exterior MBs immediately next to the boundary MBs by replicating the samples at the border of the boundary MBs). Padding consists basically of repetition of border pels. Fig. 4.12 depicts the VOP padding complexity of the decoder. I-VOP padding requires 3.49 million instructions and a memory bandwidth of 3.33 MBytes. The padding complexity is approximately proportional to the number of boundary MBs.

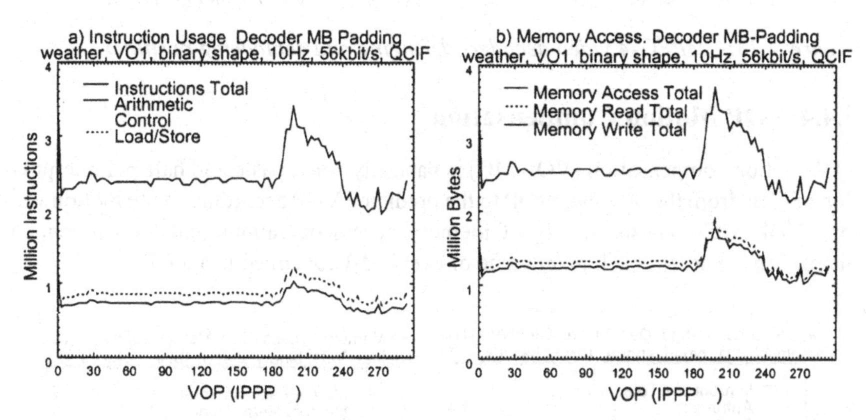


*Figure 4.12: MPEG-4 video decoder: macro-block padding for VOP motion compensation*

## 4.4.3 Postfilter: Deringing-filter and Deblocking-filter

Fig. 4.13 depicts the complexity of the optional (not normative) deringing and deblocking filter, which is roughly proportional to the number of coded MBs. For the I-VOP the deringing and deblocking filter requires a computational power of 25.32 million instructions and a memory bandwidth of 28.41 MByte. However, at frame 198 (where a high movement of the woman's hand occurs) a peak of computational power of 16.41 million instructions and 18.39 MBytes memory access is required.

For the deblocking postfilter, smoothing operations are performed along the 8x8 block edges at the decoder. The deringing filter comprises three parts: 1.) threshold determination by calculation of the maximum and minimum grey value within a block in the decoded image, 2.) binary index acquisition to determine the smoothing region, and 3.) adaptive smoothing by a 3x3 filter. The postfilter shows a high amount of memory accesses due to the filter access on every pel indicated by the binary index, but also shares a significant part of control operations for the threshold determination.

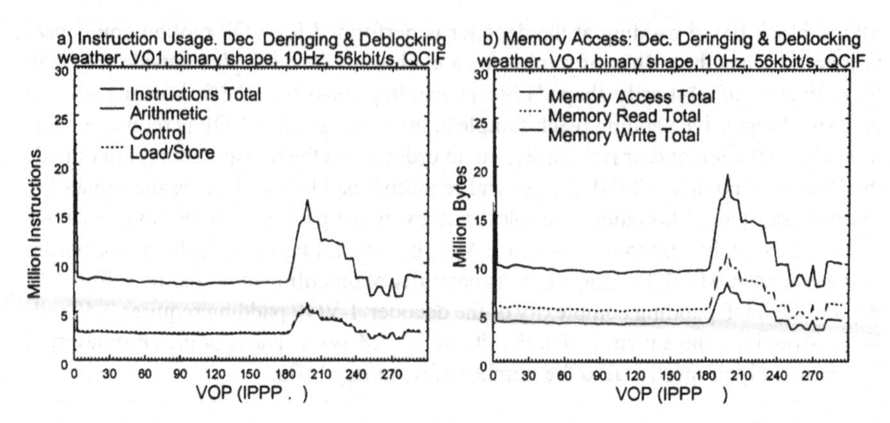


*Figure 4.13: MPEG-4 video decoder: deringing and deblocking postfilter*

## 4.4.4 VOP Motion Compensation

VOP motion compensation (VOP MC) is basically copy, paste and half-pel interpolation of MBs from the previous VOP to the predicted VOP according to the motion vectors. VOP MC consists mainly of memory access operations and loop increment instructions to access each pixel value of every MB contained in a VOP.

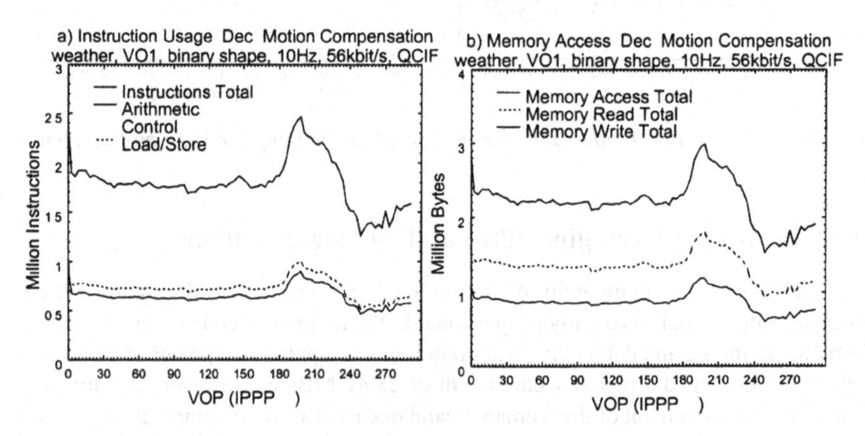


*Figure 4.14: MPEG-4 video decoder: VOP motion compensation*

## 4.4.5 MPEG-4: Rectangular-shaped video object (special case)

For comparison purposes, the complexity of encoding and decoding of a rectangular VO (the other conditions were the same as stated before) is given here. For QCIF there are 99 MBs for every VOP, where in the previous figures the number of MBs per VOP is time-dependent (cf. Fig. 4.2).

**MPEG-4 Video VM 8.0 Encoder & Decoder: Rectangular VO, "weather", 56 kbit/s, 10 Hz, QCIF**

| | Memory Bandwidth MByte / s | | Mem. Access | Arithmetic MIPS | | | | | | Con-trol | Others | Total |
|---|---|---|---|---|---|---|---|---|---|---|---|---|
| | Read | Write | MIPS | MUL | ADD | SUB | SHIFT | DIV | Total* | MIPS | MIPS | MIPS |
| Encoder Total (full-search ME) | 2543.1 | 2075.0 | 1426.8 (43 30%) | 8 97 (0 27%) | 227.39 (6 9%) | 111 36 (3 37%) | 27.76 (0 84%) | 27.76 (0 84%) | 838.79 (25 45%) | 532.02 (16 14%) | 497 27 (15 0%) | 3294 9 (100%) |
| Decoder Total | 257.71 | 209.03 | 141.92 (35 78%) | 2.35 (0 59%) | 38.44 (9 69%) | 5.5 (1 39%) | 17.92 (4 51%) | 0.01 (0 003%) | 129.51 (32 65%) | 75 23 (18 96%) | 49 95 (12 59%) | 396 63 (100%) |

*Figure 4.15: MPEG-4: Computational complexity for a rectangular VO (i.e. frame) of QCIF size (*Total arith. MIPS includes also loop inc., etc.)*

# 4.5 SUMMARY

A complexity analysis of the video part of the emerging MPEG-4 standard has been reported in this chapter. Where the computational complexity of previously standardized video coding schemes was predictable for I-, P- and B-frames, the support of arbitrarily-shaped visual objects within MPEG-4 now introduces content- (and therefore time-) dependent computational requirements with significant variance. During the MPEG-4 standardization process complexity analysis had been already performed at several stages to make sure that MPEG-4 can be implemented by cost-effective solutions.

In november 1997, MPEG-4 version-1 reached Committee Draft (CD) status [MPEG 4], meaning that the set of MPEG-4 version-1-tools had now been converged at a stabilized status and no major modifications were introduced until the International Standard (IS) stage in 1999. Complexity analysis of the MPEG-4 video verification model offers a reasonable value for other VLSI and processor implementors to support systems and applications based on MPEG-4.

For task scheduling purposes on a processor controlled VLSI implementation of MPEG-4, an approximation for the computational requirements for each VOP is required. The proposal, to take the number of MBs (macro-blocks) per VOP as a rough approximation for the expected computational power at the encoder or decoder for a specific VOP, is analyzed within in chapter.

It is shown that the complexity per MB for arbitrarily-shaped P-VOPs depicts significant variation over time for the encoder and smaller variations for the decoder. The encoder requires an average of 3.63 million instructions/MB with a standard deviation of 0.45 million instructions/MB for every VOPs. The decoder requires 0.588 million instructions/MB with a standard deviation of 0.109 million instructions for P-VOPs.

The computational complexity of the most computational demanding tool (beside the motion estimation), the CAE binary shape coding, is not directly proportional to the number of coded MBs. The complexity of macro-block based padding (which has high impact on the memory access bandwidth) is dependent on the number of boundary blocks and may be insufficiently approximated in some cases with the number of coded MBs.

Generally, it can be observed by the high amount of control operations, that VLSI implementations of MPEG-4 are expected to migrate from pure algorithm-specific solutions (as seen for MPEG-1 and MPEG-2) to processor supported systems. This means that dedicated, but processor controlled, accelerators would be employed for the computational intensive MPEG-4 tools, e.g. for motion estimation, shape coding, DCT/IDCT, macroblock padding and (post-) filters, depending on the video session size. The memory bandwidth variations of some of the tools indicate the requirement of a suitable memory hierarchy, with special emphasis on addressing and memory efficient storage of arbitrarily-shaped objects. The real-time scheduling of the encoding and decoding process of the MBs of several arbitrarily-sized video objects on a limited number of dedicated VLSI accelerator modules (one for each tool for a common case), memory hierarchy aspects and the development of fast object-based motion estimation algorithms are issues for future research.

# 5

# ANALYSIS OF FAST MOTION ESTIMATION ALGORITHMS

## 5.1 INTRODUCTION

A complexity and visual quality analysis of several fast motion estimation (ME) algorithms for the emerging MPEG-4 standard has been performed as a basis for HW/SW partitioning for the VLSI implementation of a portable multimedia terminal. While the computational complexity for the ME of previously standardized video coding schemes was predictable over time, the support of arbitrarily-shaped visual objects (VO), various coding options within MPEG-4, as well as content-dependent complexity (caused e.g. by summation truncation for SAD, cf. p30) now introduce content- (and therefore time-) dependent computational requirements, which cannot be determined analytically. Therefore, a new time-dependent complexity analysis method, based on statistical analysis of memory access bandwidth, arithmetic and control instruction counts utilized by a real processor, was developed (cf. chapter 3) and applied.

Fast ME algorithms (cf. chapter 2) can be classified into search area subsampling, pel decimation, feature-matching, adaptive hierarchical ME and simplified distance criteria. Several specific implementations of algorithms belonging to these classes are compared in terms of complexity and PSNR to ME algorithms for arbitrarily- and rectangular-shaped VOs. It is shown that the average macro-block (MB) computational complexity per arbitrarily-shaped P-VOP (Video Object Plane) depicts a significant variation over time for the different ME algorithms. These results indicate that theoretical estimations and the number of MBs per VOP are of limited applicability as approximation for computational complexity over time, which is required e.g. for average system load specification (in contrast to worst case specification), for real-time processor task scheduling, and for Quality of Service guarantees of several VOs.

## 5.2 ANALYSIS METHODOLOGY

### 5.2.1 Common conditions

The results were produced on a Sun Sparc Ultra 2 using the MPEG-4 video verification model, *[VM 8.0]*, C implementation [MomVM] compiled with the GNU gcc 2.7.2.3

compiler with optimization options (`-mv8 -O3`) and iprof 0.41 [iprof], [M3204], [M2863], [M1056], [M0921], [M0838]. All sequences are coded IPPPP... without using B-VOPs. The common conditions of the PSNR/bit rate are given here:

- advanced prediction: on (i.e. [-2, +2] pel search for 8x8 blocks around the best 16x16 $\overrightarrow{CMV}$ of a fast ME algorithm
- half-pel motion estimation: on
- disabled: rate control, error resilience, SADCT
- enabled: DC/AC-prediction
- deblocking filter: rectangular VO: off, arbitrarily-shaped VO: on
- combined motion/shape/texture coding
- h.263-quantization

For the visual quality metric the sum of the squared PSNR difference of all sequences is used to characterize one fast motion estimation algorithm. The PSNR difference is calculated by subtraction of the result of the fast motion estimation algorithm from the rate-distortion curve point of the full-search method (cf. annex).

# 5.3 FAST MOTION ESTIMATION FOR MPEG-4

There exist stringent requirements on low power consumption for battery powered real-time visual communications applications. For a typical real-time video application with p=16, f=30, $N_h$=352, and $N_v$=288 (CIF) for full-search motion estimation, a computational load of 9.34 billion integer arithmetic operations (with 8- and 16-bit data) is required per second, as well as a memory bandwidth of 6.22 billion 8-bit accesses per second. These numbers are not feasible using today's processor technology at low power constraints. Therefore, fast motion estimation algorithms with reduced computational complexity are considered here. The basic strategies are:

- Search area subsampling
- Pel-subsampling
- Hierarchical/multiresolution algorithms
- Feature-matching algorithms
- Low-complexity distance criteria
- R/D-optimized motion estimation

Beside the R/D-optimized motion estimation algorithm, the reduced computational complexity of fast motion estimation algorithms is paid for in terms of lower visual quality. Some of the simplified distance criteria offer the advantage of reduced VLSI area but have a higher computational load for processor implementation. The other fast motion estimation algorithms are more suitable for processor implementation, but show irregularities in data flow resulting in more difficult VLSI implementation.

### 5.3.1 Investigated Motion Estimation Algorithms

The algorithms given in this table were selected in section 2.6.1 and are investigated in this chapter.

| Name | Motion Estimation Algorithm | Literature |
|---|---|---|
| org | **Full-Search Motion Estimation**<br>Original MPEG-4 video verification model 8.02 code Search range. [-16, 15]. | [MomVM] |
| fs | **Full-Search Motion Estimation**<br>Modified MPEG-4 video verification model 8 02 code to serve as a flexible basis for several fast motion estimation algorithms with search range: [-16, 15]. | - |
| rd | **Rate/Distortion-optimized Full-Search Motion Estimation**<br>A cost function (similar to [Cob 97]) is calculated by multiplication of *MVBits(dx,dy)*, i.e. the number of bits for MV encoding after VLC (Variable Length Coding), by a predefined Lagrange multiplier $\lambda = 2Q_p$, [TMN 9] and addition of the calculated SAD for the MV candidate under consideration: $cost = SAD(dx,dy) + \lambda \cdot MVBits(dx,dy)$. The motion vector with the minimum cost is selected. Full-search is used with as search range of [-16, 15]. | [Cob 97] |
| liu | **4:1 Alternate Pel Subsampling**<br>The alternate pel subsampling method of is implemented for the full-search algorithm. | [Liu 93.1] |
| tss | **TSS: Three Step Search**<br>The original algorithm of Koga with a search range of [-8,7] is implemented. | [Koga 81] |
| cote | **5 Region Diamond Search**<br>A five-region diamond search was performed similar to the algorithm of [Cote 97] The differences are that no R/D-optimization was performed for the simulations here and the originally proposed semi-fixed length MV encoding was omitted because of its incompatibility with the MPEG-4 framework. | [Cote 97] |
| hier | **Hierarchical Motion Estimation**<br>In this algorithm a hierachical mean pyramid is used offering the advantage of reduced complexity and robustness against noise, as only a mean calculation of the blocks is employed. | [Nam 95] |
| null | **Zero Motion Vector**<br>This algorithm always uses the zero motion vector (i.e. the motion vector predictor) without any motion estimation. This technique is implemented mainly for test purposes | - |
| fsrbsad | **RBMAD & Full-Search**<br>The full-search [-16, +15] algorithm with the distance criteria RBMAD. (Reduced Bit Mean Average Difference) is used. RBMAD reduces the number of bits used in absolute difference calculation (i.e. bit truncation) and therefore reduces the VLSI area, power consumption and enables higher operating speed, as the difference unit and the adder can be of reduced wordlength | [Baek 96] |
| fsmme | **MME & Full-Search**<br>The full-search algorithm [-16, +15] with the distance criteria MME (Min/Max-Error) is used. | [Chen 95·1] |
| fspdc | **PDC & Full-Search**<br>Full-search algorithm [-16, +15] with the distance criteria PDC (Pixel Difference Criteria) is used | [Ghar 90] |
| fsdpdc | **DPC & Full-Search**<br>The full-search algorithm [-16, +15] with the distance criteria DPC (Different Pixel Count) is used | [LeeC 96] |
| fsbbm | **BBM & Full-Search**<br>Full-search algorithm [-16, +15] with the distance criteria BBM (Binary Block-Matching) is used | [Nat 97] |
| proj | **Integral Projection-Matching**<br>Integral Projection-Matching was used to find the candidate motion vectors in a full-search [-16, +15] scheme The SAD was applied to the candidate motion vectors in the final step | [Kim 92] |

*Figure 5.1: MPEG-4: Investigated full-search and fast motion estimation algorithms*

## 5.4 RESULTS: PSNR/BIT RATE

In this section the results of the simulations are presented. Fig. 5.2 and Fig. 5.3 depict the results for sequences with rectangular- and arbitrarily-shaped objects. Fig. 5.4 shows the PSNR of the reference full-search implementation within the MPEG-4 verification model. A detailed description of the test sequences is given in the annex.

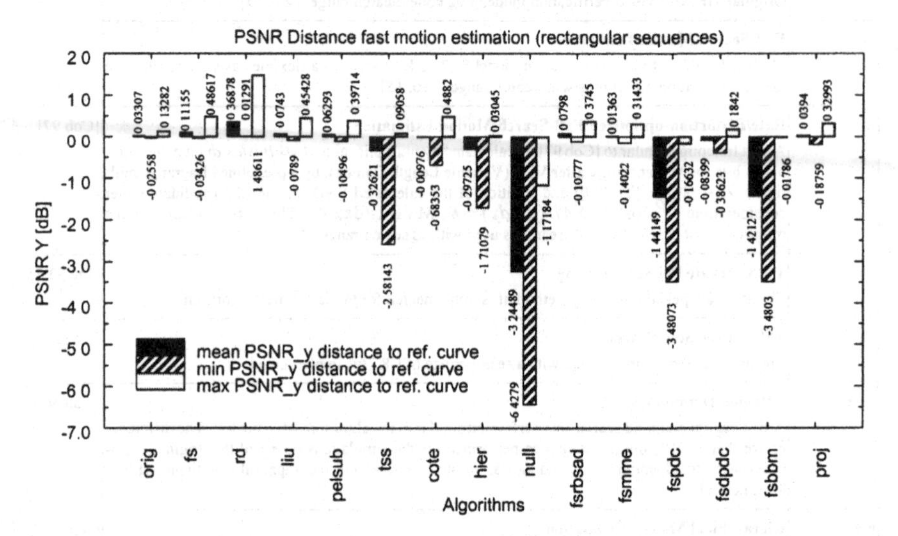


*Figure 5.2: Fast motion estimation within MPEG-4: PSNR distance (rectangular sequences)*

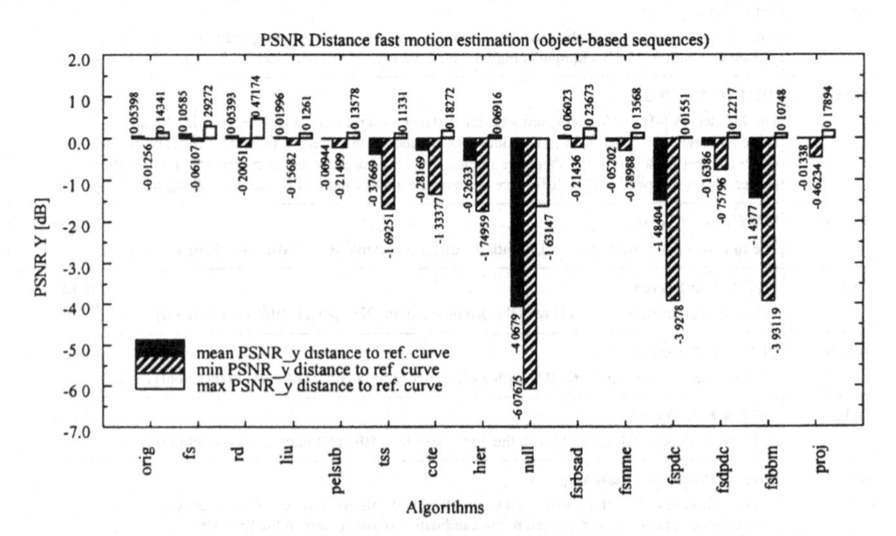


*Figure 5.3: Fast motion estimation within MPEG-4: PSNR distance (object-based sequences)*

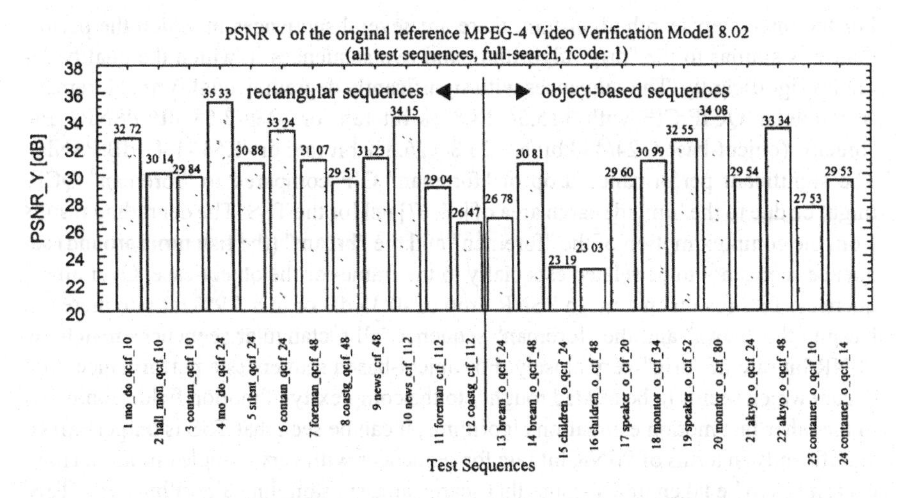


*Figure 5.4: PSNR (Y) results for the test sequences with full-search motion estimation and fcode: 1*

## 5.4.1 Rate/distortion-optimized full-search motion estimation (rd)

Rate/distortion-optimized full-search motion estimation is the only algorithm which results in a +0.33 dB mean PSNR improvement for frame-based sequences (equals to about 10 % bit rate saving at the same PSNR). For object-based sequences the gain is only +0.05 dB (or -3% bit rate). From the results it can be seen that the R/D-algorithm shows its best performance for frame-based sequences such as "foreman" and "coastguard" at 112 kbit/s, where a bit rate reduction of about 30% is achieved (or +1.49 dB improvement for foreman CIF and +0.89 dB for coastguard CIF). This is believed to be due to the high amount of motion in these sequences. However, it has to be taken into account that the R/D-optimized full-search motion estimation algorithm requires more computational power than the conventional full-search motion estimation algorithm.

## 5.4.2 4:1 Alternate pel subsampling (liu)

Alternate pel subsampling depicts very similar results as the original "orig" algorithm. In some cases a slight bit rate decrease/PSNR gain can be seen, where no extreme case of performance degradation (worst case: rect "news", QCIF 112 kbit/s, -0.08 dB) occurs. The rectangular "coastguard" sequence benefits highly from this method (+0.45 dB PSNR, -16% bit rate), which might be due to an aliasing effect from the subsampling on the water parts of the images. On the other hand, no significant performance differences for this algorithm occur in regard to rectangular- or object-based sequences. However, this algorithm offers only small gains in terms of complexity reduction, but this technique can be combined with other fast motion estimation algorithms.

### 5.4.3 Three step search (TSS)

For the three step search algorithm, there are several sequences in which the performance is similar to the "orig" algorithm, and two sequences in which the quality degrades significantly. The sequences with significantly degrading quality are "foreman" (rectangular QCIF/CIF with +15.5/+18.8 % bit rate or -0.9/-2.58 dB PSNR) and "bream" (object-based, 24/48 kbit/s, +23.8/+26.3% bit rate or -1.54/-1.47 dB PSNR). The significant performance drop of "foreman" CIF compared to "foreman" QCIF might be due to the limited search area of [-8,+7] pel for the TSS. The degrading results from the complex motion of the "foreman" and the "bream" (the fish turns around) sequence seems not to be related especially to the frame- or the object-case. Over all sequences, TSS results in a mean PSNR drop of -0.35 dB or in a 3.9% bit rate increase. Despite the "news" and the "foreman" sequence, all rectangular sequences match the PSNR/bit rate of "orig" very closely. For object-based sequences a performance drop occurs which seems to be related roughly to the complexity of motion field. Compared to the other fast motion estimation algorithms, it can be seen that TSS is outperformed significantly in terms of PSNR/bit rate for sequences with very complex motion. However, it has to be taken into account that search area subsampling algorithms, like TSS, offer a high capability of reducing the computational complexity compared to similar algorithms.

### 5.4.4 Five region diamond search (cote)

The five region diamond search algorithm, called "cote" algorithm here, also belongs to the type of algorithms with significantly reduced computational complexity. This algorithm depicts a mean of -0.14dB PSNR-loss for all sequences or an mean bit rate deviation of only -0.5%, while the object-based sequence "bream 24" (+20 % bitrate) deviates by -1.19 dB. Compared to the TSS algorithm, the "cote" method depicts a significant performance drop for object-based sequences with complex motion ("bream"), whereas the frame-based sequence with high complex motion "foreman" is handled very well (better than TSS) by "cote" (-17% bit rate/+0.14dB PSNR at 112 kbit/s). This is believed to be due to the limited search range of "cote" [-4, +4] pel compared to the [-8, +7] pel of TSS. The "cote" algorithm seems to be more advantageous for the compression of rectangular medium complex motion head and shoulders sequences like "mother and daughter" (+0.15dB at 10 kbit/s), "silent" (+0.49dB PSNR for 24 kbit/s), or "news" (+0.151 dB at 48 kbit/s) compared to the "orig" algorithm. For object-based sequences with simple motion the results of "cote" are comparable within a +/- 0.4 dB range to "orig".

### 5.4.5 Hierarchical motion estimation (hier)

The hierarchical motion estimation algorithm "hier" also offers significantly reduced computational complexity. The average PSNR-drop of "hier" on all sequences is 0.412 dB which means that a bit rate increase of 4.8% would be necessary to obtain the same PSNR value as "orig". The "hier" algorithm depicts a significant performance drop compared to "orig" and shows an inferior performance in some cases compared to

"TSS" and "cote". Object-based sequences with complex motion depict significant losses: "bream" (-1.76 dB PSNR /24% bit rate at 24 kbit/s), as well as frame-based sequences with high complex motion: "foreman" (-1.7dB PSNR/+10% bit rate at 112 kbit/s). Sequences with motion of low complexity also show some losses, especially if they are object-based.

### 5.4.6 Zero motion vector (null)

The zero motion vector method uses no motion estimation at all, and had only been implemented as a test-case in order to test the situation when the motion estimation is completely switched off. This could be the case in an ultra low-power mode. Only zero motion vectors were transmitted using this method. As expected, the results are inferior (especially for the high motion sequences as "foreman" and "bream", which were in the -6dB PSNR-range), but, as one can see, the low complex motion sequences were still within a -3 dB PSNR-range, some (like "silent") are in the range of a -1.17 dB PSNR-loss, and "hall monitor" only suffered a mere -1.2 PSNR loss.

### 5.4.7 Reduced bit mean average difference (RBMAD)

The RBMAD (fsrbsad) distance norm (which is combined with full-search) is mainly targeted on reduced VLSI implementation complexity, as a reduced number of bits for the SAD is employed. The RBMAD depicts a positive effect on the "coastguard" sequence at 112 kbit/s, where a PSNR gain of 0.375dB is obtained (or a -15% bit rate loss). This is believed to be due to a filtering effect by the masking of the LSB (least significant bits) of the SAD. The worst case sequence for RBMAD is object-based "monitor" at 80 kbit/s with a PSNR drop of -0.04dB. Taking into account the results of all sequences, the average PSNR difference to "orig" is within a +/-0.3 dB range, which means that most of the results are very similar to "orig".

### 5.4.8 Min/max criterion

The Min/max-norm (fsmme) results are similar to the results of RBMAD, with "coastguard" at 112k again benefitting with a +0.314 dB PSNR performance-gain. On average, in all sequences, the PSNR difference to "orig" lies within a +/-0.5 dB range, with most of the results being very similar to "orig".

### 5.4.9 Pixel difference criterion (PDC)

The PDC (fspdc) distance criterion depicts an average performance drop of -1.46 dB PSNR and a worst case performance drop of -3.928 dB PSNR. For example, the sequence "coastguard" QCIF requires a bit rate increase of 87% to reach the same PSNR as "orig". Therefore, this algorithm is not considered further.

### 5.4.10 Different pixel count (DPC)

The DPC (fsdpc) distance criterion depicts an average performance drop of -0.124dB

PSNR and a worst case performance drop of -0.65dB PSNR (object-based sequence "bream", QCIF 48k). Beside the sequence "bream", the other sequences are within a range of +0.184/-0.4 dB PSNR range or a bit rate range of +7.1 %/-13.8%. The sequence "coastguard" 112k benefits from this algorithm with a +0.184 dB PSNR performance gain (or a -13.8% bit rate decrease).

### 5.4.11 Binary block-matching (BBM)

The BBM (fsbbm) distance criterion depicts an average performance drop of -1.42 dB PSNR and a worst case performance drop of -3.72 dB PSNR for the sequence "bream" at QCIF 24k. For example, "coastguard" QCIF requires a bit rate increase of 86% to reach the same PSNR as "orig". Therefore, this algorithm is not considered further.

### 5.4.12 Integral projection-matching (proj)

The integral projection-matching "proj" algorithm depicts about the same mean average PSNR as the "orig" full-search algorithm. However, the PSNR ranges from -0.36 dB for sequence "bream" at QCIF 48k (equals to +8.8 % bit rate increase) to +0.33 dB PSNR (or a -14.8% bit rate saving) for sequence "coastguard" at CIF 112k. An interesting fact is that the integral projection-matching outperforms the TSS, "cote" and "hier" algorithms for the difficult sequence "foreman" with complex motion at CIF, whereas "proj" is able to achieve a PSNR-gain of +0.27 dB over the "orig" algorithm (or a bitrate saving of -12%).

## 5.5 RESULTS: COMPLEXITY

The computational complexity of the object-based sequence "akiyo", QCIF, 10kbit/s is depicted in Fig. 5.5 and Fig. 5.5 for various motion estimation algorithms.

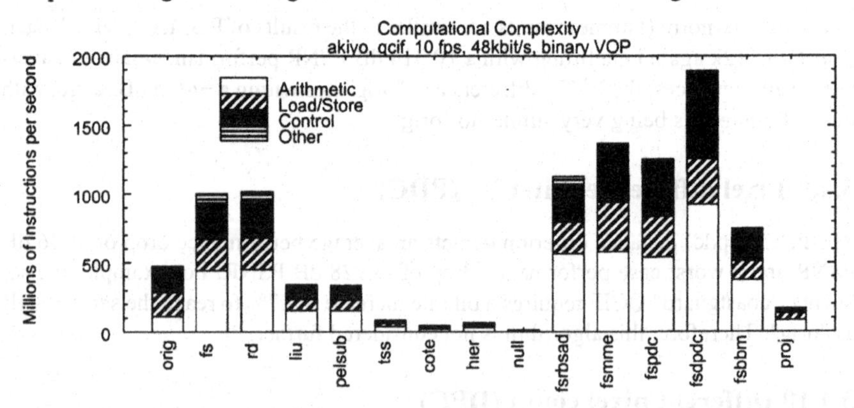


*Figure 5.5: Average computational complexity of several motion estimation algorithms within MPEG-4*

Note that the arithmetic complexity depicted here consists only of the types Mul, Add, Sub, Shift. Fig. 5.7 depicts the computational complexity over time for the full-search, the "cote", and the projection-matching algorithm using the same sequence. Note that all types of instructions are weighted equally.

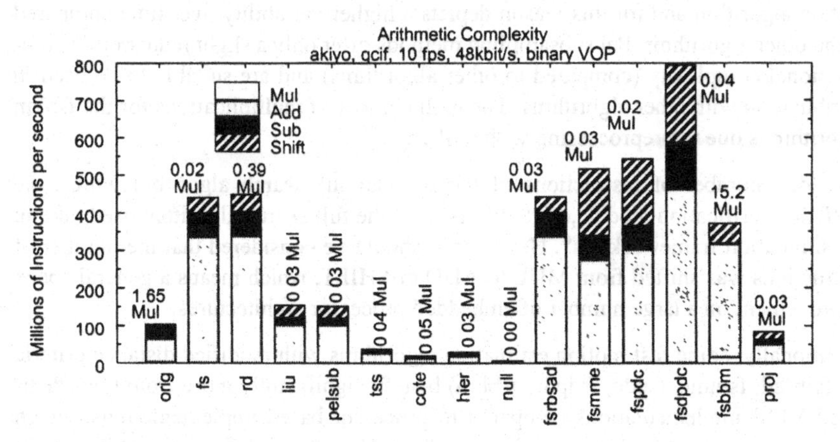


*Figure 5.6: Average arithmetic complexity*

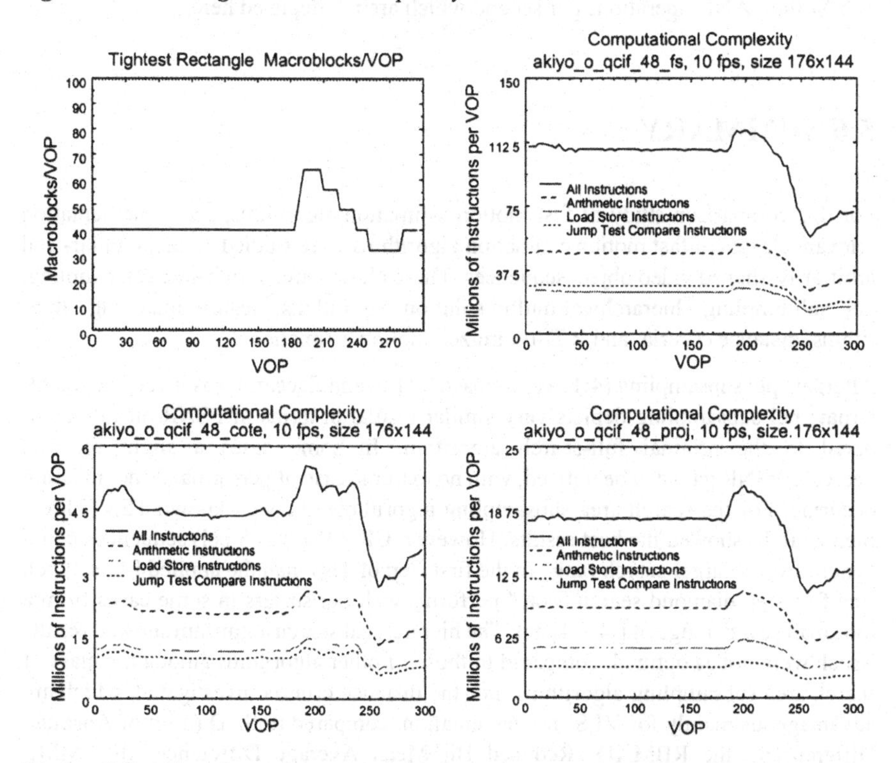


*Figure 5.7: Complexity over time for fast motion estimation algorithms of MPEG-4 test sequence "akiyo", 10fps, 48kbit/s*

From Fig. 5.5, Fig. 5.6, and the annex it can be seen that most of the fast motion estimation algorithms require, beside the arithmetic operations, a significant amount of memory bandwidth and control instructions. TSS, cote and the hierarchical algorithm result in a very low computational complexity. Cote is a content-adaptive motion estimation algorithm and for this reason depicts a higher variability over time compared to the other algorithms. Pel subsampling methods offer only a slight reduction in computational complexity (compared to other algorithms) and are suitable to be used in combination with other algorithms. The high number of multiplications for the fsbbm algorithm is due to preprocessing with a filter.

The total number of instructions of the original full-search algorithm within the MPEG-4 verification model „orig“ differs from the full-search algorithm used here in the simulation framework „fs“. However, it should be considered that the number of instructions was shifted from MUL to ADD or SHIFT, which means a general speed improvement on a large number of embedded processor architectures.

Additionally, some fast motion estimation algorithms with modified distance criteria (as fsrbsad, fsmme, fspdc, fsdpdc, fsbbm) benefit significantly more from their dedicated VLSI implementations compared to processor based implementations, which have been only analyzed for completeness here. Note that fsdpdc also requires about 100 Million AND operations per second which are not depicted here.

## 5.6 SUMMARY

For this comparison study of fast motion estimation algorithms, six implementation relevant classes of fast motion estimation algorithms were studied on rectangular- and arbitrarily-shaped video object sequences. These classes are: search area subsampling, pel subsampling, hierarchical/multiresolution algorithms, feature-matching algorithms, distance criteria, and R/D-optimized motion estimation.

Alternate pel subsampling (4:1) as proposed by Liu and Zaccarin was investigated. Alternate pel subsampling depicts very similar results in terms of PSNR/bit rate compared to the original full-search algorithm. In some cases a slight bit rate decrease/PSNR gain can be noticed, with no extreme case of performance degradation occuring. For the search area subsampling algorithms, the well-known TSS (Three-Step Search) showed the best results. However, CIF (352 x 288 pel) sequences could benefit from a larger search range in the first step of TSS than the applied [-8, +7] pel. The five step diamond search "cote" performs well but suffers in some cases from a too small search range of [-4, +4] pel. The hierarchical search algorithm showed results which were not as optimal, compared to those of other algorithms among the class of search area subsampling algorithms. For the distance criteria investigated, which are advantageous mainly for VLSI implementation, compared to SAD (Sum of Absolute Differences), the RBMAD (Reduced Bit Mean Average Difference), the MME (Min/Max-Error) and the DPC (Different Pixel Count) criteria produced good results when combined with the full-search algorithm. The results of the distance criteria PDC (Pixel Difference Criteria) and BBM (Binary Block-Matching) are considered to be neglectable compared to the other algorithms within this class. Generally, the best results were obtained using the feature-matching method: integral pojection-matching. The PSNR/bit rate is very similar to the full-search method at reduced computational complexity. A general observation is that the sequence "coastguard" benefits from the distance criteria RBMAD, MME, DPC, as well as from integral projection-matching, which is believed to be due to the noise robustness and the smoothing effect of these algorithms.

It is shown that the computational complexity per arbitrarily-shaped P-VOP (video object plane) depicts a significant variation over time for the different motion estimation algorithms, especially when the motion estimation algorithm is scene-adaptive. These results indicate that theoretical estimations, using the number of MBs per VOP, are of limited applicability when being as an approximation for computational complexity over time. This approximation is required e.g. for average system load specification (in contrast to worst case specification), for real-time task scheduling, and for Quality of Service guarantees of several VOs.

# 6

# DESIGN SPACE MOTION ESTIMATION ARCHITECTURES

## 6.1 INTRODUCTION

In this chapter motion estimation architectures are evaluated for the requirements of the visual (video) part of the MPEG-4 standard. Due to the very complex nature of the design space for motion estimation VLSI architectures, there are numerous VLSI architectures and design trade-offs. Proper consideration of these trade-offs can lead to an optimal VLSI architecture design for a selected motion estimation (ME) algorithm or a number of selected motion estimation algorithms under particular application constraints. The aim of this chapter is to evaluate block-matching motion estimation algorithms from a hardware point of view for MPEG-4. This is in contrast to the previous chapter where the block-matching algorithms were evaluated in terms of number of operations and memory bandwidth for software implementation on a programmable processor. It will be shown that the commonly used complexity metric of the number of operations for a processor implementation is not suitable for VLSI implementations.

Therefore, a new design metric is defined here, which not only takes into account silicon area and processing speed (as performed in previous work), but also I/O-bandwidth (which mainly determines the power consumption) and the quality of the algorithm in terms of PSNR (which enables fair comparison of VLSI architectures for different fast ME algorithms). Based on this metric, various VLSI architectures for full-search motion estimation and fast motion estimation architectures are evaluated within the MPEG-4 framework. Upon these results new algorithm-specific architectures were developed for 1.) variable block size motion estimation with luminance correction and 2.) fast motion estimation algorithms for MPEG-4.

### 6.1.1 System reference architecture

Fig. 6.1 depicts an example system reference architecture for MPEG-4 video encoding. This system architecture consists of a system CPU, program memory, an I/O-unit, buffer RAM for the previous and the current VOP, buffer RAM for the alpha-plane (video object shape description for MPEG-4), a function-specific motion estimation accelerator module, and other function-specific accelerator modules (e.g. for DCT/IDCT) depending on the target application. Note that MPEG-4 allows to exploit parallelism on VOP level. Therefore, an MPEG-4 encoder can consist of more than a single motion estimation accelerator, with each working on a different video object (VO) simultaneously.

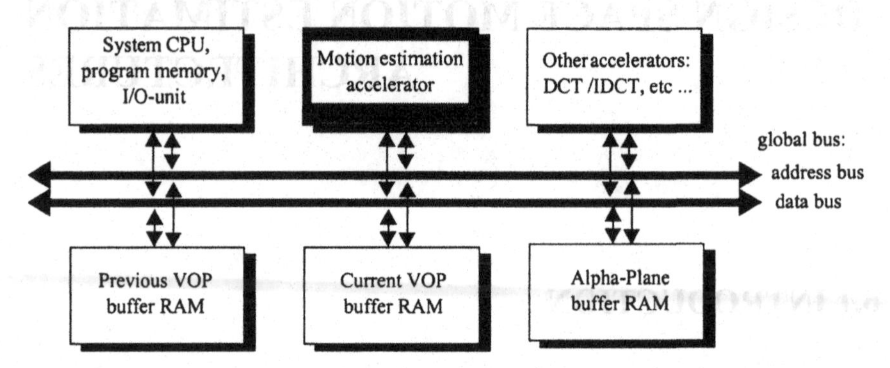


*Figure 6.1: Video encoder system architecture with a motion estimation unit*

## 6.1.2 Motion estimation reference architecture

Fig. 6.2 depicts a typical motion estimation accelerator module in detail.

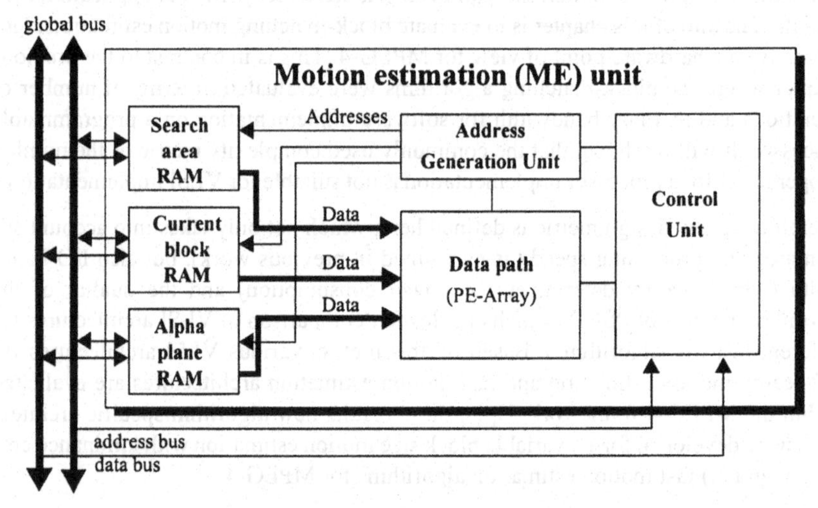


*Figure 6.2: Model of an motion estimation accelerator unit*

The motion estimation accelerator consists of RAM for search area, current block and alpha-plane data, a data path consisting of a PE array, an address generation unit (AGU), and a control unit.

The results presented in the previous chapter showed the high memory bandwidth requirements of a motion estimation processor. To reduce the utilization of the global bus to the VOP memories, the motion estimation module usually provides local memory for the current block, the search area and the alpha-plane data of a particular video object. Within this document the buffer memory of the motion estimation module is referred to as local memory, whereas the VOP buffer is referred to as external memory. The address generator unit (AGU) calculates the address for the next data words in memory, the data path provides the computational core consisting of an one- or two-dimensional array of processing elements (PEs). The control unit initiates the data fetching from the VOP memory, the start of the motion estimation process and the forwarding of the results of the motion estimation unit to the system CPU.

# 6.2 GENERAL DESIGN SPACE EVALUATION

This section on design methodology for motion estimation architectures covers the subjects: 1.) requirements analysis, 2.) (single) design metrics, and 3.) combined design metrics. An abstract overview on the methodology for design space exploration of motion estimation algorithms and architectures can be found in Fig. 1.8.

## 6.2.1 Requirements analysis: real-time conditions

Requirements analysis is the first step of a design space exploration. In this step the real-time conditions are set upon the relevant parameters for a particular target application.

The parameters determining the throughput and memory bandwidth requirements of a motion estimation architecture are defined here for the case of rectangular VOPs:

- Horizontal and vertical rectangular VOP (or frame) size in pel: $N_h$ x $N_v$
- VOP (frame) rate in VOP/s (frames/s): *fps*
- Size of the macro-block (MB) in pel: *N*x*N*
- The number of pel of a MB (= 256 in case no pel subsampling is applied): $MB_{pel}$
- Horizontal and vertical search range in pel: *[-p, +p-1]*
- Local memory bus bandwidth to the RAMs of the ME module: $MemLocal_{bandwidth}$
- External memory bus bandwidth to the VOP buffers: $MemExternal_{bandwidth}$

To meet real-time conditions, the time for processing a macro-block is defined as:

$$MB_{time} = \left(\left\lceil \frac{N_h}{MBpel} \right\rceil \cdot \left\lceil \frac{N_v}{MBpel} \right\rceil \cdot fps\right)^{-1} \quad (6.1)$$

The maximum number of MBs per VOP and sec, $MB_{max}$, may be limited by a maximum time for $MB_{time}$:

$$MB_{time} < (MB_{max})^{-1} \quad (6.2)$$

The macro-block cycle-time is defined as the time of a clock cycle required to process a macroblock and is calculated as:

$$MB_{cycletime} = \frac{MB_{time}}{N_{cycles}} \quad (6.3)$$

with $N_{cycles}$ referring to the number of cycles for the processing of one MB. $N_{cycles}$ is an important criterion to evaluate different motion estimation architectures.

For real-time operation the following condition must be met:

$$\text{Real-time condition: } PE_{cycletime} \leq MB_{cycletime} \tag{6.4}$$

The required parallelism for a motion estimation architecture is determined by the number of operations $Op_{min}$ which have to be calculated in parallel. $Op_{min}$ depends on $N_{cycles}$, the pel rate $f_{pel}$, and the clock frequency $f_{clock}$:

$$Op_{min} = \frac{f_{pel}}{f_{clock} \cdot MB_{pel}} \cdot N_{cycles} \tag{6.5}$$

## 6.2.2 Design metrics

Design metrics are metrics upon which several motion estimation algorithms and architectures can be evaluated. Here VLSI-related metrics are defined and in a forthcoming section these metrics are combined into *combined metrics*, which allow direct comparison of different VLSI architectures, taking several parameters into account.

### 6.2.2.1 Visual quality

The visual quality is determined by the PSNR (peak signal to noise ratio) metric as defined in eq. (6.6):

$$quality = PSNR = 10\log\left[\frac{255^2}{\frac{1}{N_v \cdot N_h}\sum_{m=1}^{N_v}\sum_{n=1}^{N_h}(c_{m,n} - \tilde{c}_{m,n})^2}\right] \; dB \tag{6.6}$$

where $c_{m,n}$ are pels of the original VOP at location *(m, n)* and $\tilde{c}_{m,n}$ are pels of the reconstructed VOP.

The visual quality is an algorithm-specific criterion allowing the comparison of different motion estimation algorithms under the same conditions (bit rate, video test sequence, coding parameters). It is shown in chapter 4 that the subjective visual quality (PSNR) of some fast motion estimation algorithms seems to be acceptable for a broad number of applications. The visual quality metric Q is defined for a single video sequence, as the difference of the visual quality of the original full-search motion estimation algorithm and the motion estimation algorithm with reduced complexity is given as:

$$\text{Quality Metric [dB]:} \quad Q = quality_{org} - quality_{fast} \tag{6.7}$$

The metric for visual quality evaluation for multiple video sequences as used within this work is given in the annex.

### 6.2.2.2 Processing speed

The processing speed is a VLSI-related design criterion and defines the number of clock cycles, $N_{cycles}$, required to perform a specific motion estimation algorithm on a MB using a dedicated VLSI architecture or a programmable processor. The number of clock cycles is often crucial for real-time implementation of motion estimation algorithms. In a more general view the number of clock cycles, $N_{cycles}$ is defined as the design metric time, $T$:

$$\text{Time Metric:} \quad T = N_{cycles} \tag{6.8}$$

### 6.2.2.3 Regularity of address sequence

The address generation unit (AGU) requires considerable area and design time for motion estimation algorithms with irregular data flow [Mira 97], section 6.3.8. It was shown there that the VLSI-area of an address generator depends on the regularity of an address sequence. Therefore, a regularity metric for address sequences is defined by eq. (6.9), with *#adrbits* referring to the number of bits of every address word generated by additional logic, and *#regbits* referring to the number of differing bits of the counter without logic.

$$\text{Regularity Metric: } Regul = \frac{\#regbits}{\#adrbits} \tag{6.9}$$

### 6.2.2.4 Regularity of data flow

The exhaustive motion estimation algorithm has a regular data flow, which allows exploitation of parallelism and the application of systolic array techniques. However, the fast motion estimation algorithms are usually less regular, resulting in various disadvantages for VLSI-design. This will be discussed in more detail in this chapter. As the regularity of the data flow is dependent on the regularity of the address generators, the data flow regularity can be approximated by the address sequence regularity metric.

### 6.2.2.5 Efficiency

To evaluate the efficiency of a motion estimation unit the efficiency of the motion estimation data path is used for comparison, assuming that the overhead for search area buffer filling, etc. is equal for all motion estimation architectures. The efficiency metric can be calculated as:

$$\text{Efficiency Metric: } Efficiency = \eta = \frac{ActiveOperating_{cycles}}{TotalOperating_{cycles}} \tag{6.10}$$

with $ActiveOperation_{cycles}$ referring to the time during which all execution units are busy during the $TotalOperation_{cycles}$.

### 6.2.2.6 VLSI area

The chip size determines directly the cost of a VLSI and depends on the VLSI-technology, as well as on the motion estimation algorithm. Chipsize area cannot be determined precisely until a chip is actually designed, using a specific VLSI technology or design rule. On the architectural level, the size of the silicon area may be estimated by the total number of gates. In this book gate equivalents are used for the calculation of the silicon area size. Wiring costs are not taken into account.

For motion estimation architectures, the VLSI area consists basically of: on-chip RAM, data path (computational core), control unit, decision unit (explained later), pre-processing unit / data flow mapper (also explained later), buses, latches and registers.

The total silicon area which is of unit [gates] or [mm$^2$] can be calculated as:

*Area Metric:*

$$A = A_{RRAM} + A_{CRAM} + A_{Pre} + A_{dpath} + N_{Reg} \cdot A_{Reg} + A_{CU} + A_{Bus} + A_{DU} \quad (6.11)$$

in which $A_{RRAM}$ stands for the silicon area of search area (reference) RAM, $A_{CRAM}$ for the silicon area of current block RAM, $A_{CRAM}$ for the silicon area of the preprocessing and data flow mapping, $A_{dpath}$ for the silicon area for data path and computation core, $A_{Reg}$ for the silicon area for registers/latches, $N_{Reg}$ for the number of registers/latches, $A_{CU}$ for the silicon area for the control and address generation unit, $A_{Bus}$ for the silicon area for data buses, and $A_{DU}$ for the silicon area for the decision unit.

#### 6.2.2.7 I/O-bandwidth

The input/output-bandwidth of a motion estimation chip is determined mainly by the pel access rate of the search window, which may be located off-chip or on-chip. Certain pels need to be accessed repeatedly for some designs, thus increase the I/O-bandwidth. By allowing local (on-chip) memory for search area or for delay buffers, the I/O-cost may be reduced at the cost of increased silicon size.

Note that the I/O-bandwidth can be significantly reduced for certain search algorithms with specific data flow and architectural design, especially when an appropriate number of PEs and on-chip memory is used for a specific fast motion estimation algorithm. These particular architectures and data flow schemes are discussed in the following sections in more detail. Depending on the architecture, throughput and I/O-bandwidth is not proportional to the number of the PEs.

The I/O-bandwidth is mainly responsible for power consumption and often limits the processing speed of the data path. The I/O-bandwidth $B$ is calculated as the number of memory read $Mem_{read}$ and memory write $Mem_{write}$ accesses with *wordsize* bytes each.

$$\textit{I/O-Bandwidth Metric [Bytes]: } B = (Mem_{read} + Mem_{write}) \cdot wordsize \quad (6.12)$$

#### 6.2.2.8 Power consumption

The power consumption with unit [mW] is a design metric depending on various other parameters like: I/O-bandwidth, silicon area, processing speed, algorithm, etc (cf. chapter 1).

*Power Consumption Metric:*

$$P = P_{RRAM} + P_{CRAM} + P_{Pre} + P_{dpath} + N_{Reg} \cdot P_{Reg} + P_{CU} + P_{Bus} + P_{DU} \quad (6.13)$$

in which $P_{RRAM}$ is the power consumption of search area (reference) RAM, $P_{CRAM}$ is the power consumption of current block RAM, $P_{Pre}$ is the power consumption of the preprocessing and data-flow mapping, $P_{dpath}$ is the power consumption for data-path, $P_{Reg}$ is the power consumption of the latches, $N_{Reg}$ is the number of latches, $P_{CU}$ is the power consumption of the control and address generation unit, $P_{Bus}$ is the power consumption of the data buses, and $P_{Bus}$ is the power consumption of the decision unit.

## 6.2.3 Combined design trade-off metrics

Based on the single design metrics presented before combined design metrics are developed here. Related work was conducted by [Schmeck 95] and [He 96].

### 6.2.3.1 Area Time trade-off

An area time trade-off metric to compare VLSI architectures or algorithms is given:

$$\textit{AT Metric:} \quad Cost_{AT} = A^{\alpha} \cdot T^{\tau} \tag{6.14}$$

in which $A$ is the chip size area, e.q. (6.11) and $T$ is the number of clock cycles to perform a specific motion vector search algorithm, e.q. (6.8). The values of $\alpha > 0$ and $\tau > 0$ are parameters to emphasize one of the factors, e.g. for a specific target application. The design metric in eq. (6.14) can be applied with $\alpha = 1$ and $\tau = 1$ resulting in the $AT$-metric for an equally weighting of the chipsize area and the number of cycles. Similarly - e.g. for a real-time target application - the number of cycles spent for the motion estimation can be emphasized by selecting the weights as $\alpha = 1$ and $\tau = 2$, resulting in the $AT^2$-metric. The prerequisites for this design metric to compare ME architectures are: a) equal I/O-bandwidth and b) the same motion estimation algorithm.

### 6.2.3.2 Area Time Bandwidth trade-off

As beyond the area and number of clock cycles, the I/O-bandwidth of a VLSI architecture is an important criterion in terms of cost and power consumption, eq. (6.14) is extended to eq. (6.15) to cover also the different I/O-bandwidth of various VLSI architectures.

$$\textit{ATB Metric:} \; Cost_{ATB} = A^{\alpha} \cdot T^{\tau} \cdot B^{\beta} \tag{6.15}$$

with B being the memory bandwidth, and $\beta > 0$ representing a parameter to constrain the bandwidth factor upon a specific target application. Typical design metrics of this type are: $ATB$, $ATB^2$, $AT^2B$, and $A^2TB$.

# 6.3 DESIGN SPACE MOTION ESTIMATION ARCHITECTURES

## 6.3.1 Overview

Fig. 6.3 gives an overview on the design space for motion estimation architectures.

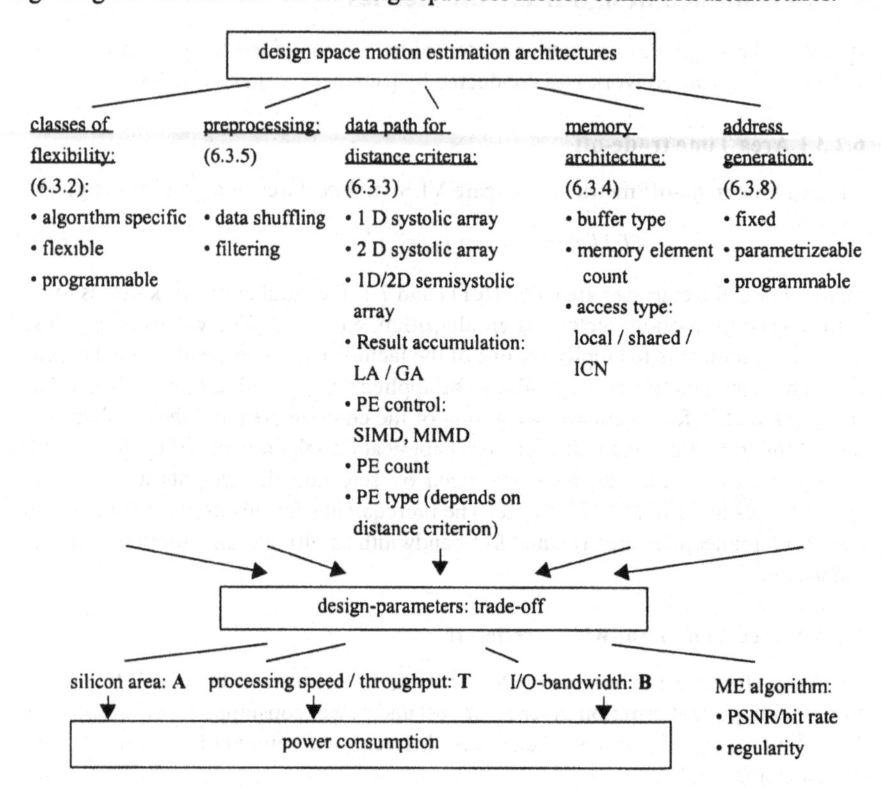


*Figure 6.3: Overview on the design space for motion estimation architectures*

The design space consists of several classes of architectural flexibility, preprocessing options, data path options for the distance criterion, memory architecture options, and different address generation unit solutions. Among these design space options the best trade off has to be found within application and system constraints. The design trade-off is constrained by the design-parameters: silicon area (A), processing time (T), I/O-bandwidth (B) and the quality of the ME algorithm (PSNR/bit rate). The next sections will analyze the first three parameters for MPEG-4 motion estimation architectures.

## 6.3.2 Classes of flexibility

Motion estimation architectures can be classified as different classes of flexibility: 1.) Algorithm-specific architectures, 2.) flexible architectures, and 3.) programmable architectures. An algorithm-specific architecture implements just a single algorithm with usually high efficiency in terms of throughput, area, I/O-bandwidth and/or power consumption. These architectures are also named "dedicated architectures" or "special purpose" architectures. Flexible architectures support a limited set of algorithms or algorithmic variants with some trade-off in terms of throughput, area, I/O-bandwidth and/or power consumption. A virtually unlimited set of algorithms is supported by programmable architectures. However, programmability is often to be paid for by higher area demands and power consumption, as well as lower throughput.

## 6.3.3 Data path for distance criteria

### 6.3.3.1 Introduction

For the description of motion estimation architectures some basic elements are defined, cf. Fig. 6.4. The |a-b| unit calculates the absolute difference of two values, and **ADD** is a simple addition unit.

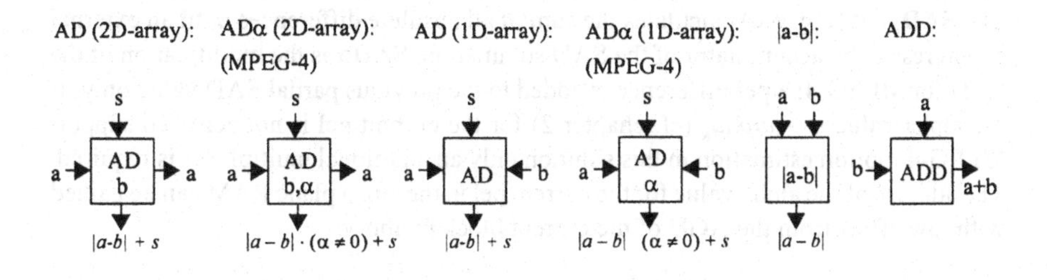


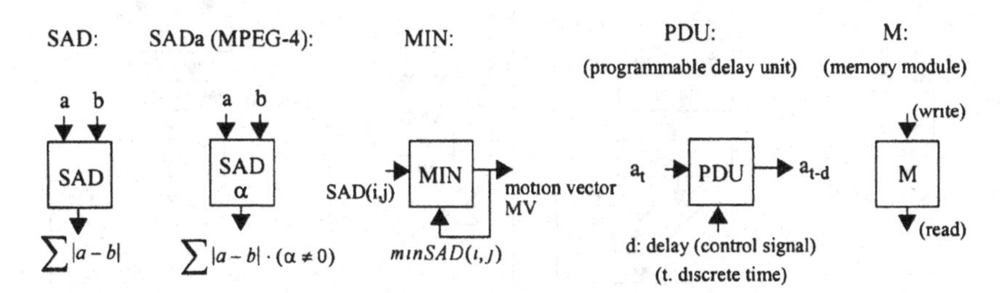


*Figure 6.4: Notation for processing elements (PEs) and other functional units*

***MPEG-4 AD calculation***

The **AD** unit (cf. Fig. 6.5) calculates the absolute difference of the two pels $a$ and $b$ and adds the result to a previously calculated partial sum of absolute differences $s$. The ADα, however, adds a pel difference to the previously calculated partial SAD value $s$ only, if the alpha value $\alpha = alpha_k$ (cf. chapter 2) for the current pel is not zero. The partial sum of absolute differences $s$ is given from one processor element to the next processor element and finally the complete SAD is calculated.

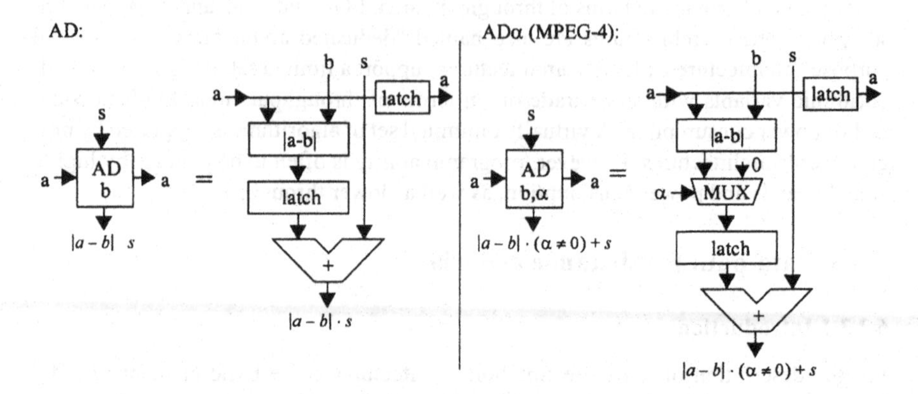


*Figure 6.5: AD as "traditionally" calculated and ADα for MPEG-4*

***MPEG-4 SAD calculation***

The **SAD** unit (Fig. 6.6) calculates the sum of all absolute differences until an external signal resets the accumulator of the SAD summation. **SADα** is the modification of the SAD for MPEG-4: a pel difference is added to the previous partial SAD value only, if the alpha value $\alpha = alpha_k$ (cf. chapter 2) for the current pel is not zero. To support MPEG-4 motion estimation in this solution only an additional multiplexer is required. The address of the alpha value for the current pel in the alpha plane RAM can be gained with low effort from the AGU of the current block memory.

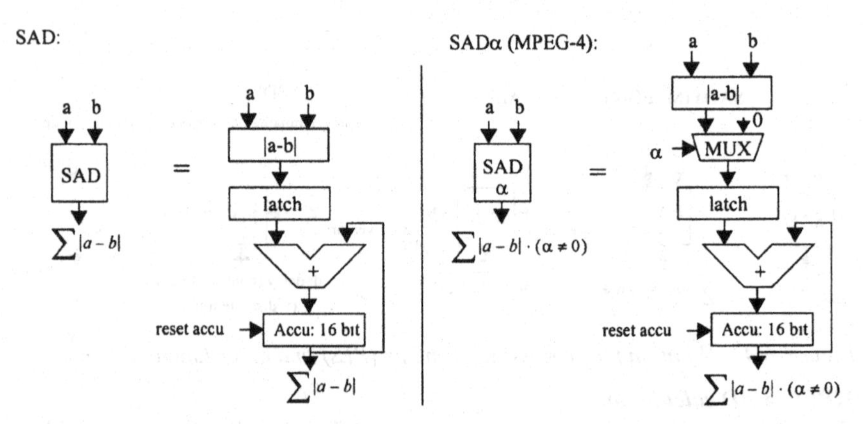


*Figure 6.6: SAD as "traditionally" calculated and SADα for MPEG-4*

***SAD array architecture***

Motion estimation can be parallelized (e.g.) at SAD level with each SAD unit calculating the SAD of a single MV candidate. This architecture is referred to as "SAD array", and can be of the one-dimensional (1D) or two-dimensional (2D) type. However, the SAD units have to be supplied with every clock cycle with the appropriate data from memory.

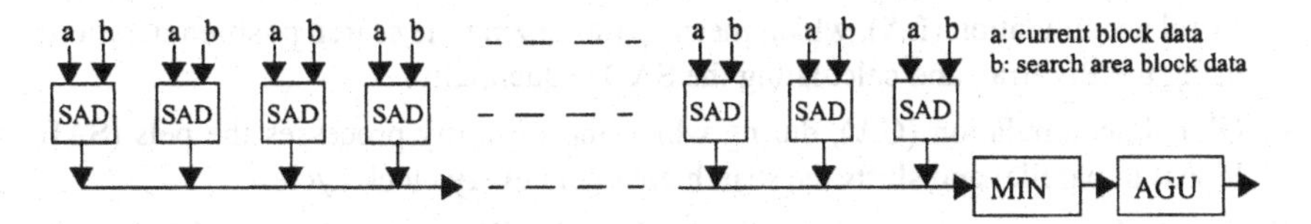


*Figure 6.7: 1D SAD array architecture*

***Time sharing of a single adder unit***

Fig. 6.8 gives a solution to reduce the number of adder elements for a PE array architecture based on the SAD calculation. For MPEG-4 motion estimation both architectures are equally applicable. In case of a motion estimation algorithm with high regularity (e.g. full-search) the partial sums in the result queue can be cyclically accessed and no control overhead for the result buffer queue is required. However, in case of motion estimation algorithms with low regularity or multiple algorithm support, the shared adder architecture reveals the disadvantage of requiring control overhead for the result queue.

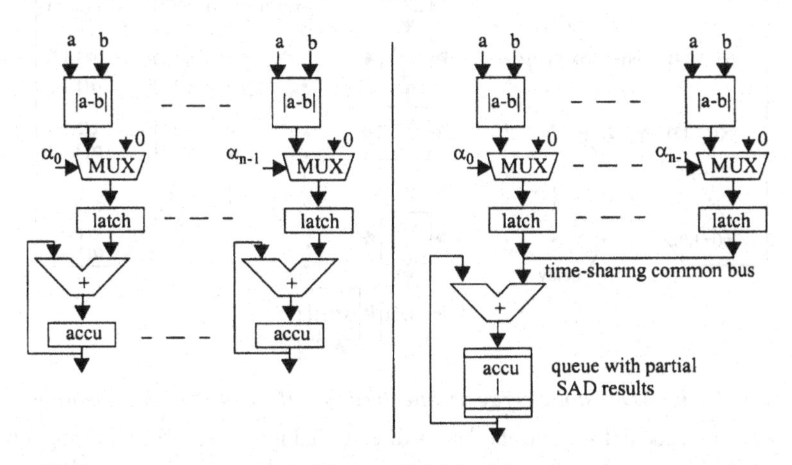


*Figure 6.8: 1D SAD array: sharing of a single result accumulator*

### 6.3.3.2 Systolic and semi-systolic array architectures

For real-time motion estimation numerous systolic / semi-systolic array architectures have been developed so far. In this section, four basic array architectures for motion estimation are analyzed for MPEG-4 motion estimation. An overview on VLSI array architectures was given in [Kung 88], [Kung 93]. Array architectures for motion estimation were discussed e.g. in [Vos 89], [Kom 93], [Vos 94], and [Vos 95] (cf. also the references in section 6.4.2.1). These architectures were basically developed for the full-search motion estimation algorithm, but are also employed for a selected number of fast motion estimation algorithms. A general advantage of systolic array architectures is that there exists only local communication between the PEs, with no global memory access being required for each PE.

Two basic block-systolic matching architectures are compared here:

- Local accumulation (LA), which means parallelizing the search position iterations using the PE array and calculating the SAD sequentially.
- Global accumulation (GA), during which the PE array processes the pels (SAD loops) in parallel and shifts the search area with every clock cycle.

The GA architecture only supports blocks of fixed *NxN* size, summarizes the absolute differences outside of the PE array in an adder tree, and offers the advantage of simple control and flexible search range. The GA-2D-array size is equal to the blocksize.

For the LA architecture the block size is variable (in $2^n$ steps, $n \in \aleph$). However, the LA-2D array of size $(2p+1)^2$ is constrained by the search range *[-p, +p-1]*.

***LA-1D systolic array architecture for MPEG-4 motion estimation***
Fig. 6.9 depicts the LA-1D architecture for MPEG-4 motion estimation.

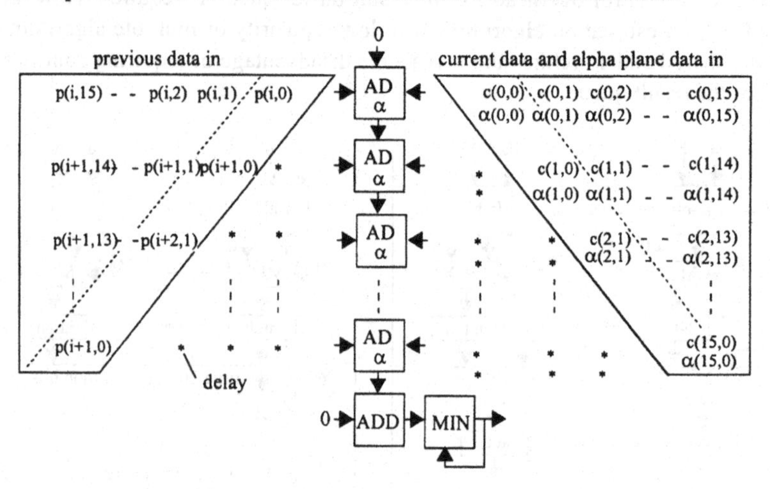


*Figure 6.9: LA-1D systolic array architecture for MPEG-4 motion estimation*

The previous data and the current data is directly fed into every PE, requiring a high I/O-bandwidth. However, for motion estimation algorithms with high regularity (e.g. full-search), data already fetched from memory can be saved in delay lines for later re-use. This limits the local high memory bandwidth requirements of this architecture. The memory bandwidth (without taking delay lines into consideration) and the throughput in arithmetic operations (*Absdiff*= ADD, ABS, SUB) per clock cycle of the LA-1D architecture can be calculated under full efficiency conditions as:

$$\text{Total memory bandwidth (prev) LA-1D: } Mem_{bw} = PE_{count} \cdot 1byte \quad (6.16)$$

$$\text{Total memory bandwidth (cur) LA-1D: } Mem_{bw} = PE_{count} \cdot 1byte \quad (6.17)$$

$$\text{Operations per clock cycle LA-1D: } Op = Absdiff \cdot PE_{count} + ADD \quad (6.18)$$

Each ADα PE calculates the absolute difference of two pels, adds the result to the already calculated partial sum for the same search position given by the neighbor PE, and passes the result to the next PE. At the end of the chain of PEs the SAD calculation is finished. The result is compared with the previous minimum SAD result within the MIN element. This architecture is scalable for search range and blocksize. The number of PEs is equal to the block size *N*. In case of $PE_{count} < N$ a partial sum queue (cf. Fig. 6.8) has to be applied. For MPEG-4 motion estimation, the alpha-plane information has to be led (similar to the current block data) at PE clock rate directly to every single PE.

***LA-2D systolic array architecture for MPEG-4 motion estimation***

The LA-2D architecture, Fig. 6.10, is the two-dimensional extension of the LA-1D architecture, where *N*x*N* PEs are used.

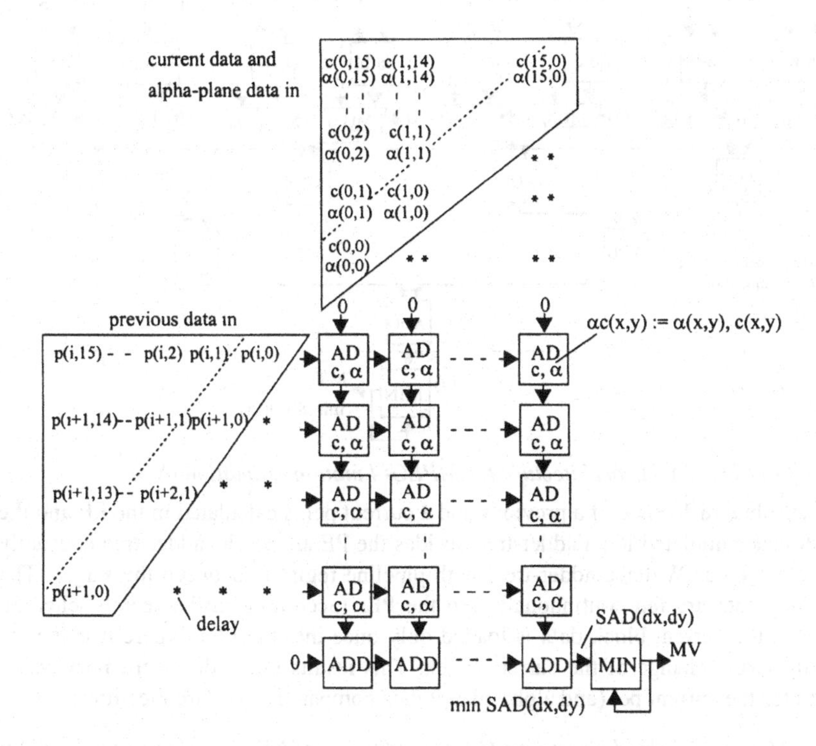


*Figure 6.10: LA-2D systolic array architecture for MPEG-4 motion estimation*

The previous data is passed (in this example) horizontally from one PE to the next PE. The current data and the alpha-plane data is passed vertically from one PE to the next PE. Data reuse is possible for algorithms with high regularity by making use of delay lines and by moving data from one PE to the next PE. This offers the advantage of reducing memory bandwidth compared to LA-1D. The local memory bandwidth and throughput of the LA-2D architecture can be calculated as:

$$\textit{Memory bandwidth previous LA-2D: } Mem_{BWprev} = \sqrt{PE_{count}} \cdot 1byte \tag{6.19}$$

$$\textit{Memory bandwidth current LA-2D: } Mem_{BWcur} = \sqrt{PE_{count}} \cdot 1byte \tag{6.20}$$

*Operations per clock cycle LA-2D:*

$$Ops = (Absdiff + ADD) \cdot PE_{count} + ADD \cdot \sqrt{PE_{count}} \tag{6.21}$$

The number of PEs is equal to the block size $N$x$N$. In case of $PE_{count} < N \cdot N$ a partial sum queue (cf. Fig. 6.8) can be applied. For MPEG-4 motion estimation the alpha-plane data have to be moved vertically from one PE to the next PE (like the current block data) at PE cycle speed.

***GA-1D architecture for MPEG-4 motion estimation***

The global accumulation architecture, GA, Fig. 6.11, is also referred to as "tree architecture", [Jehng 93].

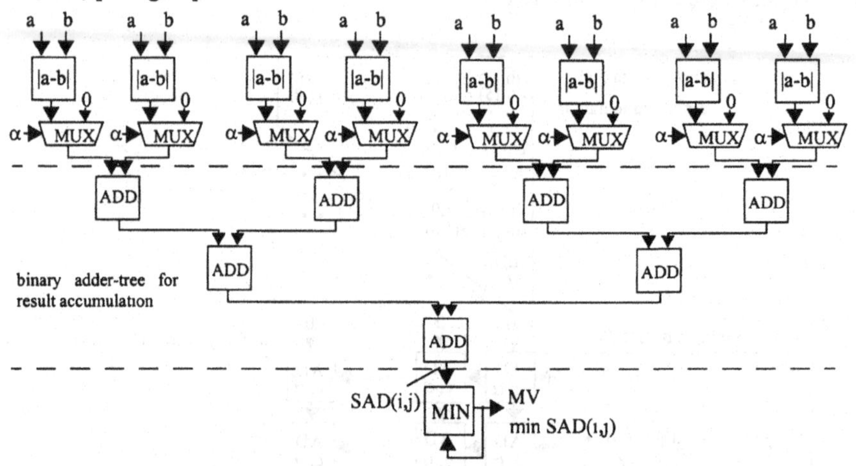


*Figure 6.11: GA-1D architecture for MPEG-4 motion estimation (N=4)*

The absolute difference of a previous and a current pel is calculated in the PE and the result is accumulated in an adder-tree besides the PE array. The adder-tree is usually implemented as Wallace adder-tree, with pipeline registers between the stages. The previous data are fed continuously into the PEs (according to the search scheme), whereas the current block data is loaded only once into the PEs (where it is locally saved) during change of the current block. This results into reduced memory bandwidth for the current pel (and alpha-plane) data compared to LA architectures:

$$\textit{Memory bandwidth previous GA-1D: } Mem_{bw} = PE_{count} \cdot 1 byte \tag{6.22}$$

$$\textit{Memory bandwidth current GA-1D: } Mem_{bwcur} = N / \sqrt{N_{cycles}} \tag{6.23}$$

*Operations per clock cycle GA-1D:* (6.24)

$$Op = Absdiff \cdot PE_{count} + ADD \cdot PE_{count} - 1 \tag{6.25}$$

MPEG-4 motion estimation can be supported by multiplexers between the PEs and adder-tree. The alpha-plane data has to be changed only when the current block data is changed. This results into lower switching activity (and therefore lower power consumption) compared to the LA architecture, where the alpha-plane information has to be led to a new PE with every single clock cycle.

***GA-2D architecture for MPEG-4 motion estimation***

The LA-2D architecture, Fig. 6.12, is the two-dimensional extension of the GA-1D architecture, where *NxN* PEs are used. The first stages of the adder-tree can be also located within the PE array to reduce wiring costs.

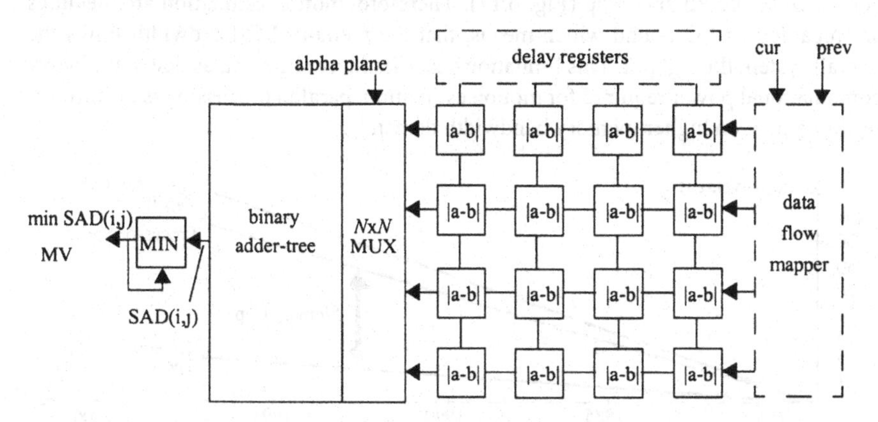


*Figure 6.12: GA-2D systolic array architecture for MPEG-4 motion estimation (N=4)*

The previous data are fed continuously into the PEs (according to the search scheme), whereas the current block data is loaded only once into the PEs (where it is saved locally) during change of the current block. This results into reduced memory bandwidth for the current pel and the alpha-plane data compared to the LA architectures:

$$\text{Memory bandwidth previous GA-2D: } Mem_{bw} = \sqrt{PE_{count}} \cdot 1byte \quad (6.26)$$

$$\text{Memory bandwidth current GA-2D: } Mem_{bwcur} = \frac{N}{\sqrt{N_{cycles}}} \quad (6.27)$$

$$\text{Operations per clock cycle GA-2D:} \quad (6.28)$$

$$Op = Absdiff \cdot PE_{count} + ADD \cdot PE_{count} - 1 \quad (6.29)$$

***Comparison on the LA/GA-array types for MPEG-4***

The GA architecture is well suited for the restriction of motion estimation to pels inside arbitrarily-shaped objects. As the inputs of the adder-tree are directly mapped to single pel positions of a current block, pel values outside the object's shape region can be disabled by multiplexers (MUX, Fig. 6.11 and Fig. 6.12) at the pel difference inputs of the adder-tree.

LA allows to exclude pels outside of a visual object by suppressing the AD results of the pel locations outside of an object. However, the AD units have to be controlled at full PE speed with the alpha plane values. These can be shifted through the PE array in the same way as the current block data, extending the current block data bus by a single bit for the alpha value.

### 6.3.4 Memory architecture

The processing elements (i.e. the arithmetic units) speed up almost linearly with technology improvement, whereas the memory access time shows a minor improvement compared to the PE speed-up (Fig. 6.13). Therefore, motion estimation architectures are regarded as I/O-bound which means, that the memory I/O-bandwidth limits the overall system throughput. This limitation becomes more stringent, as due to the heavy computational power required for motion estimation, parallel processing is performed, resulting in even higher memory bandwidth demands.

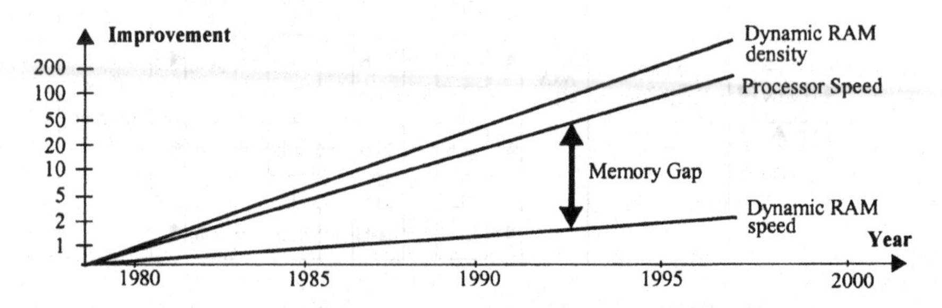


*Figure 6.13: Speed-up of PEs and memory with VLSI technology evolution (cf. [Boxer 95])*

To solve these memory bandwidth limitations several techniques were developed, often in different context, and applied to motion estimation, e.g.: 1.) caching, 2.) memory interleaving, 3.) parallel memory banks, 4.) pipelining, and 5.) hierarchical memory. Caching is basically a local memory as introduced with Fig. 6.2, but with particular cache fill and access strategy. Memory interleaving is used to lower the memory access time by alternatively accessing one of two or more memories, in order to resolve memory access delays and conflicts. Parallel memory banks are used to resolve memory access conflicts and to increase the memory bandwidth. The effective memory bandwidth can be also increased by the pipelining of functional units using the same data. Hierarchical memory is the concept of having a global and a local memory.

***Number of memory modules***

The area cost of a memory module can be approximated by the following equation [NEC 98]:

$$\textit{Memory Area: } Area_{memory} \approx A_{base} + A_{bit} \cdot bits \qquad (6.30)$$

$A_{base}$ is the approximate area required for address decoding and the row and column drivers, which is considered to be constant for small memory sizes (e.g. 32 - 2,048 bit) for this model. $A_{bit}$ is the area size for one bit of memory. From this model it can be derived that a small number of large memories has the advantage of being more area efficient than a higher number of small memories with the same amount of memory capacity in bits, as the VLSI overhead of multiple row- and column-drivers and address decoding is omitted. Besides this advantage, the control and the address generation of a reduced amount of memory modules is less costly. However, larger memories have generally a higher access time than smaller memories, and, with their limited number, they are not able to deliver the same memory bandwidth as multiple memory banks. Larger memories also have a higher power consumption for the access of the same amount of data than a single smaller memory unit. Therefore, the number of memory modules has to be optimized for the employed VLSI technology. A general methodology to evaluate memory architecture trade-offs was presented in [Dutta 98], and buffer size optimization especially for full-size motion estimation was discussed in [Yeh 97].

Using multiport memories instead of several singleport memory modules reduces memory access conflicts, which occur using multiple memory banks, but requires higher area compared to single-port memories at the same size. Motion estimation architectures with multiple memories were analyzed within [Greef 95], [Dutta 95], [Cmar 97] and [Dutta 98] in detail. Pipelining and memory interleaving was discussed e.g. in [Jehng 93].

Other design parameters of the memory system are the size of the memory, depending on the search range and the cost of memory access to the global VOP memory. For extremely costly (in terms of power, access time or throughput) VOP memory access, it is more beneficial to store several rows of MBs in the local search area RAM.

## 6.3.5 Preprocessing and data flow mapping

Typical preprocessing operations are pel subsampling and filtering (cf. chapter 2). Data flow mapping, Fig. 6.14, is often required to align the video data stream to the memory or PE architecture.

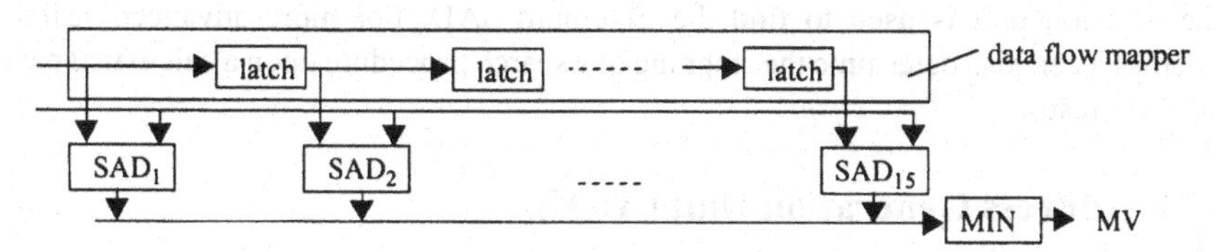


*Figure 6.14: Data flow mapper*

For multiple memory modules, data has to be distributed to these memory modules for conflict-free access. As already mentioned, shift registers can be used to reduce the number of memory modules and the memory bandwidth by storing data for later reuse. Shift registers of variable length with programmable functionality were proposed by [BugW 97]: 1.) length of line buffers, 2.) min. detect on/off, 3.) size of the VOP in rows/cols, 4.) interpolation / decimation on/off, and 4.) number of "idle states" and "idle pixels".

## 6.3.6 Interconnection network (ICN)

An interconnection network is a flexible means to combine a linear or a 2-dimensional processor array with a number of memory modules. This flexibility is sometimes required when more than one motion estimation algorithm has to be supported.

[Lin 96], [Lin 96a], [Dut 96] and [Zheng 97] e.g. used an interconnection network to implement the three-step search algorithm and the full-search algorithm on a linear PE-array. [Pirson 95:1], and [Charlot 95:1] used a crossbar for the full-search algorithm. Note that basically the search area data or the current block data can be broadcast to the SAD-PE elements. For the interconnection network one of several types (Crossbar, Omega, Benes, ...) is possible. A discussion on different interconnection network types for motion estimation architectures can be found in [Dut 96]. Basically, instead of the interconnection network, a shared bus can also be used, which means running the memory bus with a higher clock rate than the processing elements. However, memory access is usually the slowest part of the system. Therefore, a shared-bus concept is only feasible for low-throughput systems.

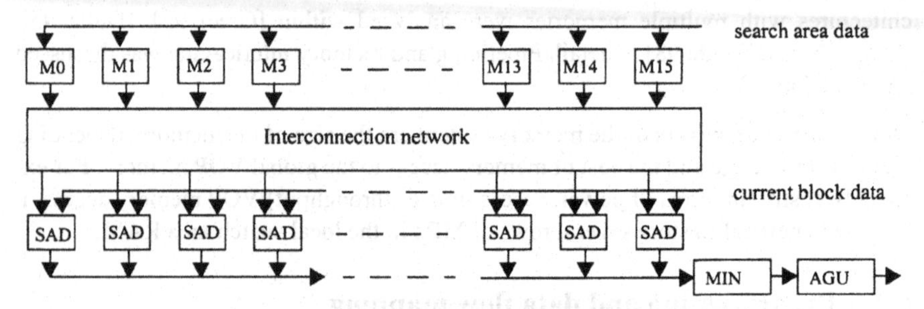


*Figure 6.15: 1-D SAD PE array with interconnection network*

## 6.3.7 Decision unit

The decision unit is used to find the minimum SAD. For more advanced search schemes it can also determine the stopping of a search procedure upon a calculated partial SAD result.

## 6.3.8 Address Generation Unit (AGU)

### 6.3.8.1 Introduction

Access to data memories becomes the limiting factor for high throughput motion estimation, signal processing and other number crunching applications. The basic problem is (besides the memory access time and the parallel access constraints, which were discussed in different sections) to calculate the next address of data in the memory, while fetching the currently required data and feeding it into the data path to keep the data path busy all the time. Data addressing can be accomplished with low effort for regular algorithms (e.g. the full-search motion estimation algorithm), but requires significant amount of computational resources and VLSI area in case of less regular fast motion estimation algorithms. This is even worse if several algorithms have to be supported by a single address generation unit (AGU) with some kind of flexibility or programmability. [Mira 97] observed that programmable AGUs are a dominant factor in terms of area and power consumption for MPEG-type designs. Therefore, it seems to be reasonable to investigate AGUs in detail.

***AE: Address Equation***
AE is defined as the address equation, eq. (6.31), describing a function which maps the parameters $p_1, p_2 \ldots, p_n$ to an address of a single linear addressed memory.

$$AE = f(p_1, p_2 \ldots, p_n) \quad (6.31)$$

The parameters are often indices of nested loops, with AE describing the mapping of these parameters to a memory address. For a system with several distributed memories, several AE may exist. The type of AE can be distinguished in: 1.) affine, 2.) piece-wise affine, and 3.) non-linear.

***Affine AE***
For affine mapping, the AE can be written as a linear expression of the parameters $p_n$ and the constants $C_n$, as depicted in eq. (6.32):

$$AE = C_0 + C_1 \cdot p_1 + C_2 \cdot p_2 + \ldots + C_n \cdot p_n \quad (6.32)$$

Typically, address sequences are generated by a number of nested loops, where the parameters are the loop indices:

```
for (p1=0; p1<p1max; p1++)
    for (p2=0; p2<p2max; p2++)
        B[p1] = f(A[1+p1*2+p2*3])
```

The above example shows an address equation, which calculates the address index for the 2-dimensional array `A[] = A[p1][p2]` as a function of the loop indices `p1` and `p2` and $C_0$=1, $C_1$=2, and $C_2$=3.

***Piece-wise affine AE***
Piece-wise affine mapping is employed, in case only parts of the address sequence can be described by eq. (6.32). In the context of a program, a piece-wise affine AE is gained by altering the AE, using conditional statements.

***Non-linear AE***
The general case is the non-linear address mapping, where no linear relation exists between the loop parameters and address indices. This is for example the case for gradient based fast motion estimation algorithms, where the addresses of the next motion estimation step are calculated, based on the results of the current step. An example for a non-linear address sequence is given in eq. (6.33).

$$AE = C_0 + C_1 \cdot p_1 + C_2 \cdot p^2_2 + \ldots + C_n \cdot p^n_n \quad (6.33)$$

***Predefined vs. dynamic address sequences***
For the scope of motion estimation architectures, basically four situations can be identified for address generation, depending on the range of flexibility required: 1.) predefined address sequences, 2.) customer defined address sequences, 3.) parametrizable address sequences, and 4.) dynamic address sequences, dependent on AGU external events and results. For fast motion estimation algorithms, it can be necessary to support e.g. more than one of the search area subsampling schemes by a single AGU.

***Properties of address sequences for motion estimation algorithms***

For a huge number motion estimation algorithms the following property of address sequences has been observed: most algorithms show a similar scanning of the search area and the current block by a dynamic or predefined address sequence, but with a different starting point (handle).

***Requirements AGU:***

Fig. 6.16 summarizes the 3-dimensional requirement space for AGUs. Note that for every point in this requirement space, there is an optimum AGU-architecture selection chosen from several possibilities, with the additional constraints: design time, power requirements, throughput, VLSI area, and others, depending on the application.

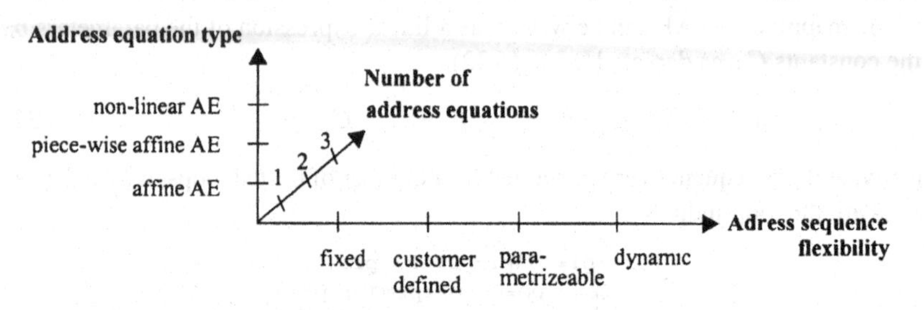


*Figure 6.16: Requirements address generator unit (AGU)*

To meet these requirements of efficient address generation, DSP and microprocessor designers have conceived the Address Generation Unit (AGU), which is sometimes referred to in literature as: Address Calculation Unit (ACU), Address Arithmetic Unit (AAU), Data Address Generator (DAG), or Memory Management Unit (MMU), cf. [Liem 97] p83ff. One or more AGUs usually work in parallel to the data path and solely on address calculation, in order to ensure efficient retrieval and storage of data which are processed by the data path.

***AGU architectures***

In the following sections, several basic types of AGU architectures are analyzed. Dependent on the AEs and the application constraints, a combination of one or more of these basic principles may perfectly fulfill the requirements.

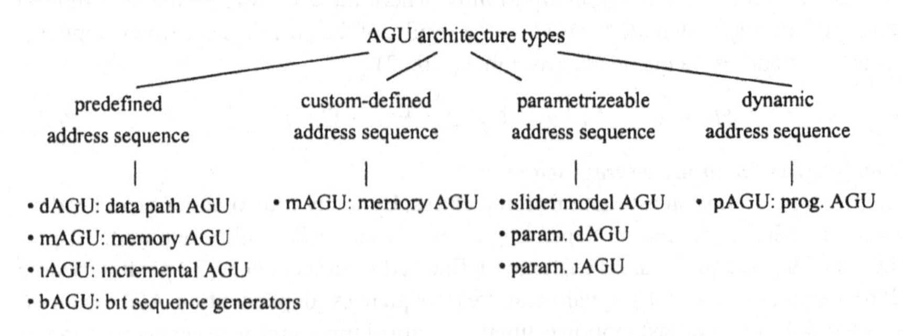


*Figure 6.17: Overview of address sequence types and preferred AGU architecture types*

### 6.3.8.2 Predefined address sequence

For fixed address sequences the addressing pattern is known beforehand during VLSI implementation and can be exploited beneficially. This is especially the case for algorithm-specific architectures, which only support one or a very small range of algorithms. There exist several implementation alternatives which are here discussed in detail.

***dAGU: Data path AGU***

An AGU can be implemented as an application specific unit (ASU) which maps the address equation onto dedicated hardware using a data path approach for address calculation. This approach is advantageous in case an AE is a direct affine function of the loop iterator $i$, especially for highly complex sequences. An advantage of the dAGU is that for long address sequences the arithmetic part only grows logarithmically with the sequence length. However, suboptimal realizations may result in large area overhead (e.g. in case a multiplication of the AE results in a hardware multiplier). Support of multiple AE often leads to direct mapping of each individual AE onto dedicated hardware, resulting in a large area and power consumption overhead. To reduce this overhead, a) transformations of the AE, and/or b) combination of the dAGU with counter/table-based methods, as well as c) bit level optimization of the AGU are performed. An example for an recursive dAGU for the affine address equation, eq. (6.34), is depicted in Fig. 6.18. A dAGU can be further distinguished in 1.) dAGU, consisting of an adder only and 2.) dAGU, consisting of an adder and a multiplier.

$$AE = 2 \cdot i + 1 \tag{6.34}$$

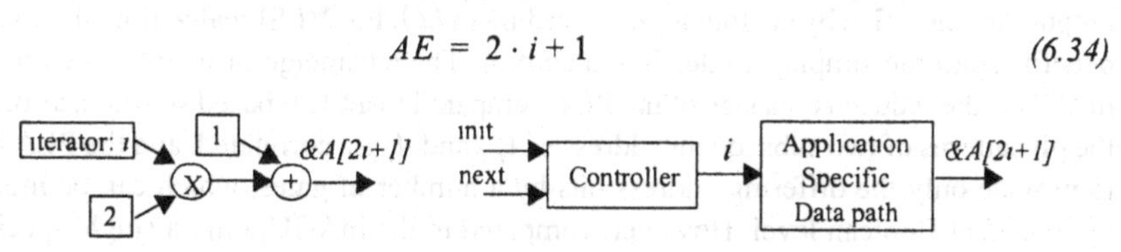


*Figure 6.18: dAGU: data path based Address Generation Unit*

***mAGU: Memory based AGU***

Short address sequences can be directly saved on the chip in a LUT (lookup-table), cf. Fig. 6.19a, which is for example implemented as a ROM (Read Only Memory) either a) uncompressed, or b) in a compressed format. This approach is sometimes reffered to as "counter/table"-approach. Note that the controller of the mAGU may have more features than a simple increment counter, as there are no restrictions on the address sequence fed to the LUT. Compared to other AGUs, the mAGU offers the advantage that address sequences which cannot be represented easily by an address equation, can be generated.

The direct saving of address sequences without compression is, due to it's space, only viable for very short address sequences, or address sequences showing repeated patterns, which is often the case for motion estimation algorithms. The width of the stored address table is a function of the address space, and the length of the table depends on the number of cycles for which the memory addresses have to be stored. For example, a full-search motion estimation implementation by the [Yang 89] architecture requires a total of 4,367 cycles to complete. However, a table of 271 address entries is sufficient, as the same pattern is repeated 17 times.

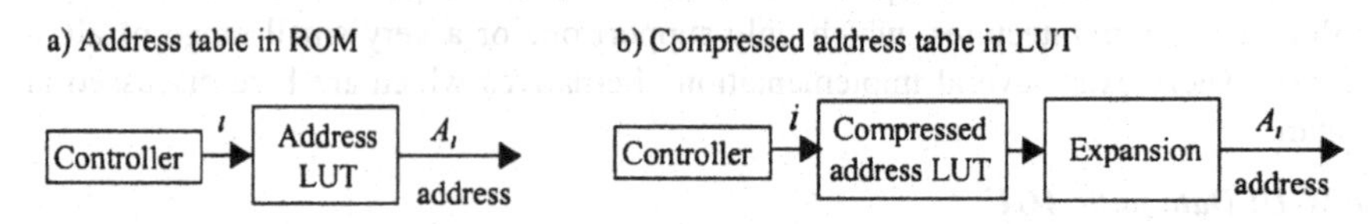


*Figure 6.19: mAGU: Memory based AGU (LUT: look-up table)*

For longer address sequences a compression of the single address entries seems to be sensible, cf. Fig. 6.19b. There exist different methods for entropy-based compression (e.g. [Nel 93]), and the application constraints determine their usefulness here. For the address decompression method, there occurs no timing penalty, as this method can be performed for predefined address sequences in a pipelined fashion, one step before the actual address is required. However, additional VLSI area is required for the decoding of the compressed addresses, which is regarded to be beneficial only for a larger set of address sequences. The address expansion can be, for example, simply the merging of an address offset to the value of the LUT.

***iAGU: incremental AGU***

The iAGU (Fig. 6.20) is a counter-based approach, with the output of simple modulo/binary incremental counter being fed to a PLA (Programmable Logic Array), whose outputs are modified by custom logic according to *h()*. For VLSI realization, the iAGU benefits from the simple counter-based control. The advantage of the iAGU over the mAGU is the reduced area size of the PLA compared to a LUT-based solution, as often the bit patterns of two subsequent addresses $A_{i-1}$ and $A_i$ are correlated, and the PLA has to provide only the differing address bits by a number of gates, which can be highly optimized at Boolean level. However, compared to the mAGU, only a single specific address sequence can be generated, as the counter is allowed only to increment the previous $C_i$ (in contrast to a controller which can provide the LUT with different addresses).

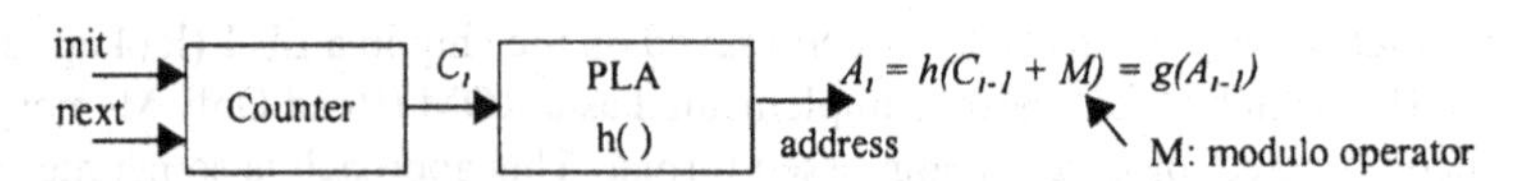


*Figure 6.20: iAGU: incremental AGU*

Compared to the mAGU, the iAGU is advantageous in terms of VLSI area for longer, linear address sequences with an upper limit for loop ranges of about 10E5 iterations [Mira 97], as the sequence addresses of the loop have to be expanded.

Compared to the dAGU, the iAGU can be basically smaller, as no arithmetic units are required. This is especially true for highly regular or short sequences. However, for more complex address sequences, the PLA can require a significant amount of VLSI area and can become much larger than the dAGU. For highly regular address sequences, the amount of logic remains roughly constant with increased sequence length, whereas the counter size increases slightly. An additional advantage of the iAGU is that there are several distributed iAGUs possible in a system, which are controlled each by only two lines (`init, next`), whereas with dAGUs the complete set of loop parameters have to be provided by a central controller.

***bAGU: Bit sequence generators***
The previous described address generation techniques were based on word level AGUs. However, AGUs can be also beneficially implemented on bit level. As the manual design of bit level AGUs is a error-prone task, [Grant 94] proposed a tool "ZIP-PO" to examine a set of predefined address sequences on word- and bit-level, to create a pool of possible hardware solutions, from which a global, optimal bit level implementation is found by an iterative technique to cover all of the predefined address sequences. This work concentrates on the joint optimization of a dedicated implementation to generate *several* address sequences. Fig. 6.21a depicts the situation, where several address sequence generators (e.g. for several fast motion estimation algorithms) are attached (for example) to the two dual-port memory banks, and Fig. 6.21b depicts the approach of using bit sequence generators. With this technique, area savings for the AGU in the range of 10%-60% were reported [Grant 94].



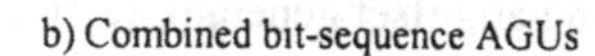


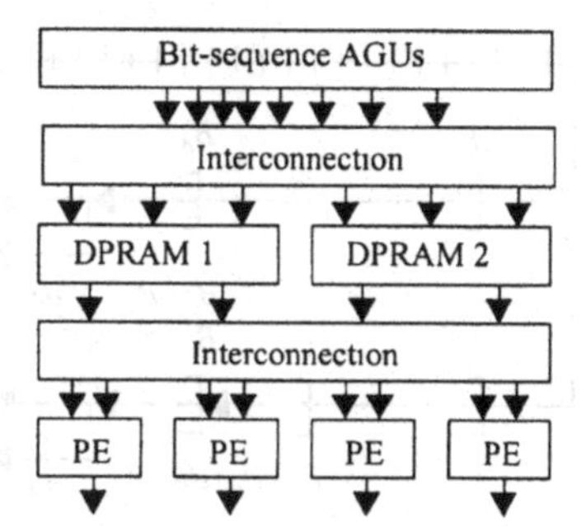


*Figure 6.21: Joint optimization of an AGU for several address sequences [Grant 94]*

### 6.3.8.3 Customer-defined address sequence

The address sequence is decided after the VLSI design, but before the first use of the system by the system integrator. For this situation, basically a mAGU can be used, which can be implemented, e.g. by an (EE)PROM (electrically erasable programmable read only memory) or by a RAM (Random Access Memory).

### 6.3.8.4 Parameterizable address sequence

Parameterizable address sequence generators consist of one or several dAGUs, which are controlled by some input parameters. The advantage of this approach is that the limited number of algorithms which are contained and covered by the dAGUs can be parametrized to the application. However, this approach is limited to the AEs and the number of parameters implemented by the dAGUs. In the following, some implementations are given as an example.

***Slider Model AGU***
A parametrizable approach for address sequence generation was presented by [Hart 97], with a slider model being employed for address generation. For this example a two-dimensional address space is assumed. Fig. 6.22 depicts the slider model for the one-dimensional memory access, with an initial address sequence being initiated and generated in the first step, whereas in the second step the address range is extended and the corresponding address sequence is generated.

The parameters for an one-dimensional slider-model based AGU are: $B_o$: initial base, $L_o$: initial limit, $dB$: modification of base, $dL$: modification of limit, $F$: floor, $C$: ceil, and $dA$: step width for addresses. $B_o$ and $L_o$ describe the initial address range of the first step, with $dB$ and $dL$ extending this address range for the next step, until the current base, $B$, reaches/passes $F$ or the current limit, $L$, reaches/passes $C$. $dA$ sets the step width for the address generation.

The VLSI architecture (Fig. 6.24) of the address generator implementation in [Hart 97] consists basically of a handle position (Fig. 6.23) generator, which generates the start address of the address sequence, and a scan window generator, which consists of an adder unit and an offset generator (a 16 entries lookup-table and a FSM).

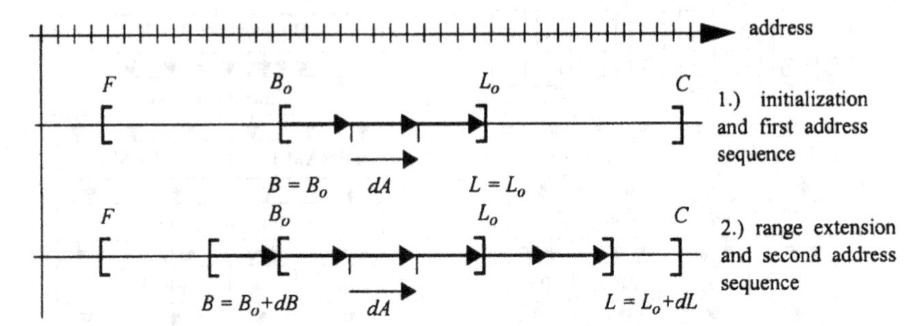


*Figure 6.22: Slider model for parameterizable address sequence generation [Hart 97]*

The handle position generator and the offset generator circuits consist of one or more than one slider-model based AGU. With two-dimensional addresses for image and video processing, two address generators (for x and y) have to be synchronized (e.g. for diagonal scans), and for complex address sequences, several address generators have to be cascaded, which can be beneficially achieved by reconfigurable VLSI technology (e.g. FPGA, Field Programmable Gate Array).

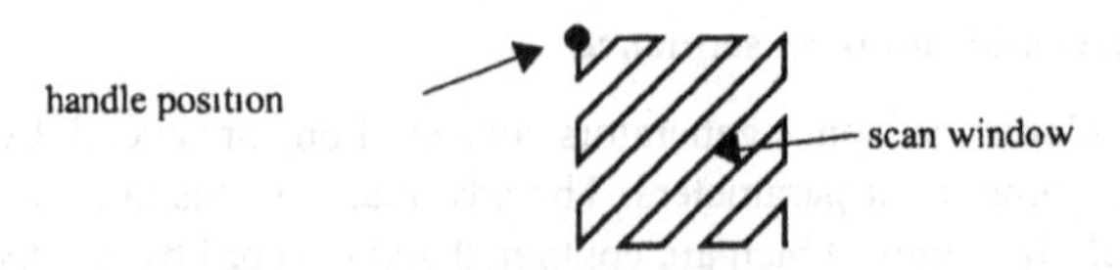


*Figure 6.23: Handle and scan window (example)*

This architecture type may be also employed with non-reconfigurable VLSI technology. If this is the case, the type and complexity of address sequences is limited by the implemented FSM and the number of entries of the lookup-tables. For complex address sequences, this approach requires significant VLSI area. For example, a typical MPEG address sequence (DCT zig-zag scan), requires 9 slider-based AGUs with a total number of 144 parameters.

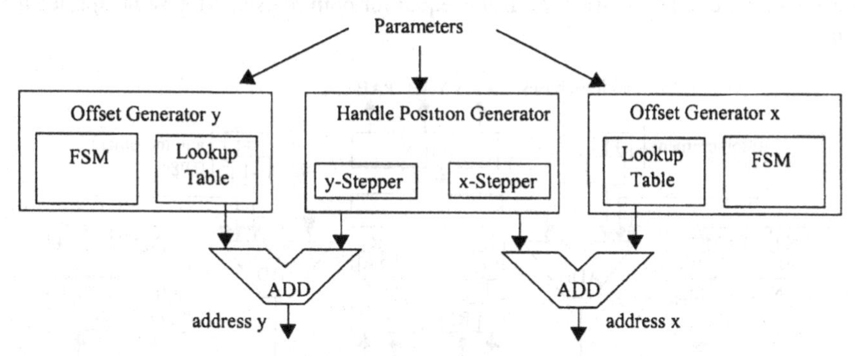


*Figure 6.24: Configurable two-dimensional address generator*

***Parameterizable AGU for motion estimation***

An application specific address generator for motion estimation is contained in most of the existing dedicated motion estimation architectures. For example, [Reven 93] described a parameterizeable address generator for full-search and three step search, which is based on adders, multiplexors and hard-wired control. Different values were added or subtracted for each address calculation, depending on the parameters (search window size and origin, as well as search range) for the modes subsampling or sub-pel accuracy. Generally, address generators can also be used in this context for rearrangement of the sequential video signal into the block-wise order of the algorithm. For parameterizeable AGUs, the types iAGU and dAGU are used in most cases.

### 6.3.8.5 Dynamic address sequence

In this case, the address sequence is dynamically self-modified during run-time, depending on the results caused by external data (e.g. for gradient-based fast motion estimation algorithms). This is the general case, which is similar to adder-based AGUs in programmable digital signal processors (DSP) or microprocessors.

***pAGU: programmable AGU***

Programmable AGUs were developed mainly for DSP applications and are targeted for optimal implementation of the calculation of array indices. Some of the pAGUs support hardware-loop counters with zero timing overhead, having the advantage that no additional clock cycle is required for loop iteration index increment. However, the programmability of pAGUs has often to be paid for by a high VLSI area demand as a result of: a) the number of pAGUs, b) the size of arithmetic blocks (adder), c) address register files, d) custom programmability, e) program ROMs, and f) program instruction decoders.

Fig. 6.25 depicts an adder-based programmable AGU of a commercial DSP of the Motorola 56k series, which has two memories x and y that are addressed by the address busses XAB and YAB. *Nn* are index registers, *Mn* are modulo registers, and *Rn* are general address registers. Increment/decrement operations within the range [-1,+1] can be applied on the *Nn* index registers. The *Mn* register determines the type of address arithmetic of register *Rn*: linear, modulo, or reverse-carry. A register *Rn* can only be indexed by the index register *Nn*, if *n* is equal for both registers (the same applies for *Mn*).

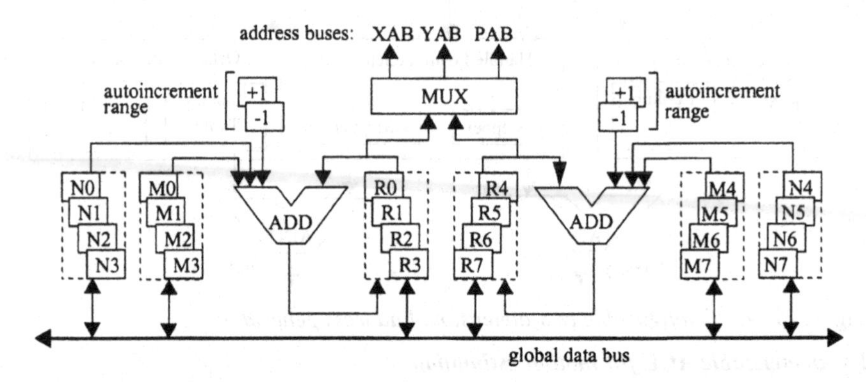


*Figure 6.25: pAGU: programmable AGU (example Motorola DSP 56k series, [Liem 97])*

For these address generators there is an auto-increment/decrement range (e.g. `i++`, `i--` in the C-language), for which the address calculation can be performed within one clock cycle by the AGU, parallel to other data processing. Other address calculations, which are not directly based on the content of the address registers, usually require more than one cycle or cycle costly interaction by the data path processor to modify the address registers. Compared to premodify modes (address increment/decrement, e.g. `--i`, `++i`) postmodify modes (e.g. `i++`, `i--`) can be more beneficially supported by pAGUs without additional cycle penalty.

As there are different pAGU types, the problem of optimal compiler mapping occurs especially for special purpose registers and special purpose operations. An overview of current problems on pAGU and solutions for programmable DSP was given in [Liem 97] p83ff.

***pAGU for image processing***

A specialized, but still programmable AGU, was presented for image processing applications in [Kita 91]. Three of these AGUs are included in a long instruction word (153 bit) DSP. A single AGU supports a number of special addressing modes for image processing applications, which can be configured by parameters such as: 1.) start address, 2.) end address, 3.) increment value, 4.) offset value, 5.) kernel size (for block-scan modes), and 6.) coefficient of matrix (for affine transform). The addressing modes are: one-dimensional raster scan, block-scan, 1-dim. direct mode, FFT mode, 2-dim. raster scan, neighborhood search mode, 2-dim. indirect mode, and affine transform mode.

### 6.3.8.6 Optimization of AGUs

***Introduction***

Optimization and merging of AGUs is usually difficult and error-prone if performed manually. Therefore, several tools for AGU optimization and AGU design space exploration were proposed. Basically, there exist different levels on which optimization can be applied (Fig. 6.26).

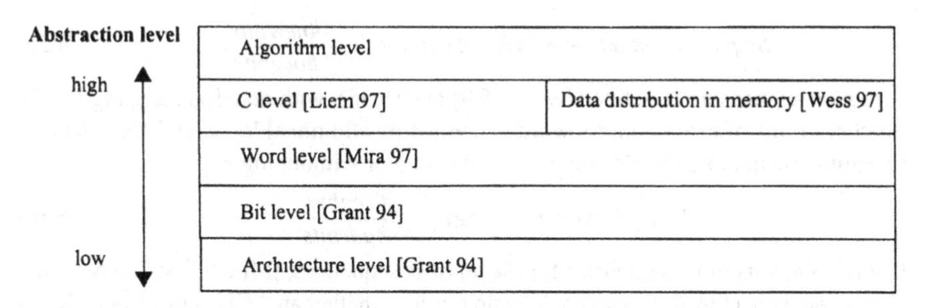


*Figure 6.26: Optimization levels for AGUs*

***Algorithm level optimization***

Optimization on algorithmic level comprises the optimum selection of an algorithm with a convenient AGU representation for a specific task. For example there are several algorithms known for fast motion estimation or DCT/IDCT.

***C level optimization***

[Liem 97] p88ff noticed that state-of-the-art compilers for DSP architectures could be improved in terms of address calculation for data arrays, automatically taking into account the benefits of hardware loop counters employed in DSPs. Based on this observation he presented an address optimizer for data arrays: "ArrSyn". The main observation was that a transformation of the array indices to pointer arithmetic on C-level considerably improves the efficiency of the subsequent compiler in terms of address generation. This technique offers the advantage that *any* commercial (retargetable) compiler can be optimized for array accesses, where the source code is generally not available.

***Data distribution optimization in memory***

An AGU can be optimized for address sequences known in advance by an optimized data distribution in memory. The basic idea is to distribute the data in memory in such a fashion that enables data access within (e.g.) the auto-increment/decrement range of a pAGU, thus minimizing additional clock cycles imposed by the AGU. The optimum data distribution in memory can be found manually or automatically off-line for larger address sequences, e.g. by the "simulated annealing" technique, [Wess 97].

***Word level optimization***

Word level optimization comprises the detection of periodic properties of the address sequence, to reduce the sequence to periodically repeated primitives. The advantages (e.g. for the iAGU) are: smaller counters and higher regularity of the reduced sequence, resulting in less mapping logic for the PLA. However, there is a higher amount of logic required to control the repetition of the periodic primitives.

A sequence reduction measure is defined by eq. (6.35), with *#orgbits* being the number of original bits of every address word before periodic reduction and *#newbits* representing the number of bits after periodic reduction.

$$\textit{Sequence reduction metric} \quad Reduc = 1 - \frac{\#newbits}{\#orgbits} \tag{6.35}$$

A regularity measure is defined by eq. (6.9) and (6.36), with *#adrbits* standing for the number of bits of every address word generated by additional logic, and *#regbits* representing the number of differing bits of the counter without logic.

$$\textit{Regularity metric} \quad Regul = \frac{\#regbits}{\#adrbits} \tag{6.36}$$

Upon the results of these metrics for the address sequences, [Mira 97] stated heuristic rules (based on a few test cases), which determine whether an AGU is more beneficially implemented as dAGU or as iAGU. Basically, for a low *Regul* value, a dAGU is more preferable than an iAGU. This methodology which is automated by the tool "ADOPT", offers the advantage of optimization on system level, without re-iteration of VHDL modeling and synthesis of different AGU alternatives. After having decided on the implementation of iAGU or dAGU, there is still the possibility to merge hardware modules, where different AE sharing the same hardware ressources.

***Bit level optimization***

A bit level address sequence optimization technique was proposed by [Grant 94] and has been already described in the section on bit level AGUs.

***Architecture level optimization: iAGUs***

For counter-based AGUs, the selection of a suitable modulo value for the counter determines the mapping of the logic of the PLA. Therefore, the modulo value can be determined upon the stream of individual address bits.

***Architecture level optimization: pAGU***

With pAGUs, the parameters open for optimization are basically the number of AGUs, the number of address registers ($k$) of a single AGU, and the auto-increment/decrement range $[-l,+l]$. The *Texas Instruments* DSP TMS320C80 for example, employed AGUs with $l$=7 and $k$=8. [Sud 97] analyzes the impact of the auto-increment/decrement range $[-l,+l]$ and the number of registers ($k$) on the performance of typical DSP systems for several application benchmarks, and on the resulting hardware cost (in terms of additional area for program code, as not only the code size but also the instruction width increases with additional address registers). The authors concluded, that $l$=3, $k$=2 and $l$=1, $k$=4 were the best options for a wide range of examples (including JPEG and the image manipulation program *xv*) in terms of code area.

# 6.4 MOTION ESTIMATION ARCHITECTURES FOR MPEG-4

## 6.4.1 Overview

Fig. 6.27 depicts an overview of a selection of VLSI implementations for motion estimation.

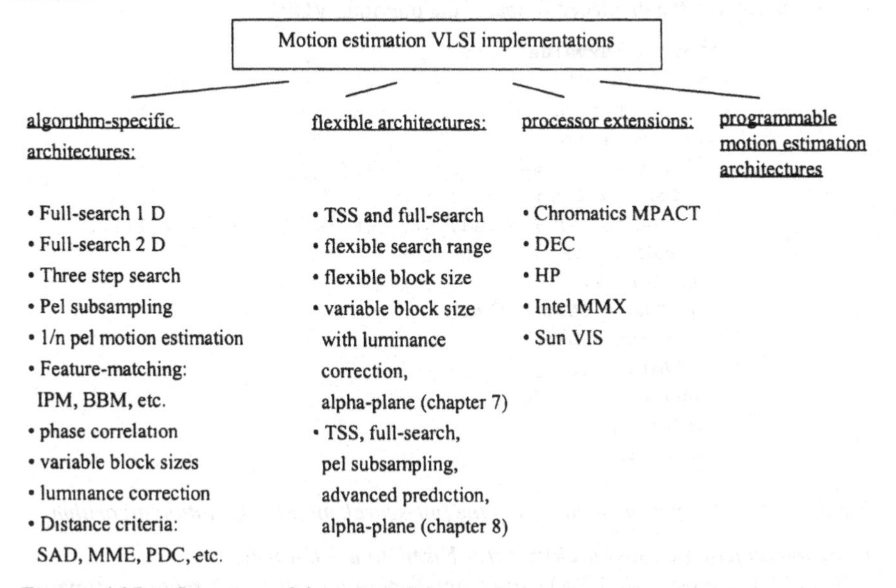


*Figure 6.27: Overview of the state-of-the-art for motion estimation architectures*

## 6.4.2 Algorithm-specific architectures

In this section algorithm-specific VLSI architectures for various full-search and fast motion estimation algorithms are described. Algorithm-specific architectures are optimized for one single algorithm and usually provide high throughput at low VLSI costs. However, this kind of architecture is not flexible enough to adapt this architecture to changing application demands.

### 6.4.2.1 Full-Search

***Introduction***

The advantages of the full-search block-matching motion estimation (FSBME) algorithm over other motion estimation algorithms are: 1) regular structure suitable for VLSI implementation, 2) simple control overhead, and 3.) it yields superior visual quality results in terms of PSNR compared to fast motion estimation algorithms. As the various full-search motion estimation architectures were covered extensively by literature ([Yang 89], [Vos 89], [Kom 89], [Meng 91], [Hsieh 92], [Kom 93], [Sun 93], [Bhas 95], [Mad 95], [Pir 95:1], [Yeo 95], [Kawa 97], [Yeh 97], [Cheng 96], [Cheng

97a]), only a brief overview and some references are given here. Array architectures which are often used for full-search motion estimation, have already been presented in section 6.3.

For a horizontal and vertical image size of $N_h$ x$N_v$, a maximum horizontal and vertical displacement of *[-p, p-1]* the full-search motion estimation algorithm is described in Fig. 6.28 by means of nested loops. `SADmin` is the minimum SAD calculated so far, and `MVmin` is the motion vector for `SADmin`. `x(k,l)` and `y(k+i,l+j)` are pels of the current and the displaced block of the previous VOP.

```
SADmin = maxvalue
MVmin = (0, 0)
for i = -p to p
 for j = -p to p
  for k = 1 to Nh
   for l = 1 to Nv
    SAD(i, j) = SAD(i, j) + |x (k, l) - y(k+i, l+j)|
   endfor (l)
  endfor (k)
  if SAD < SADmin then
   SADmin = D(i, j)
   MVmin = (i, j)
  endif
 endfor (j)
endfor (i)
```

*Figure 6.28: Loop representation of the full-search motion estimation algorithm*

***VLSI implementation and architectures based on a 1-D-array***

[Yang 89] described a 16x1 SAD array architecture for full-search motion estimation which uses a string of latches (i.e. a shift register chain) to reuse already loaded current block data from memory and to broadcast search window data to all PEs. With this architecture 100% efficiency is gained for full-search, with only one memory port required for the current block data and two memory ports for the search area data.

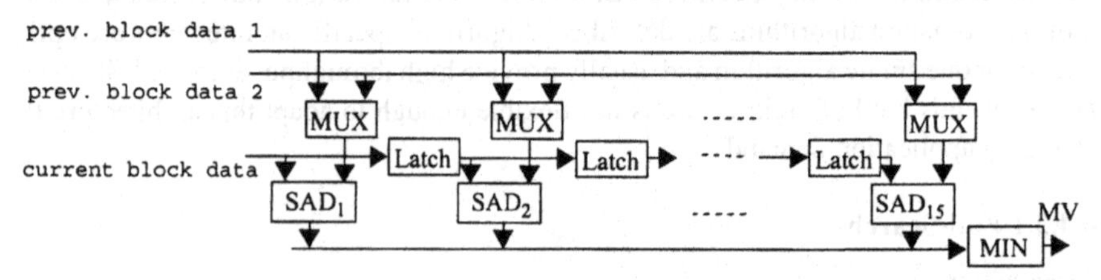


*Figure 6.29: 1D-array for full-search motion estimation using a shift register chain [Yang 89]*

[Dut 96] and [Dut 98] described a flexible block-matching architecture using an interconnection network. [Kwak 95] described a field- and frame-based motion estimation unit. Pipelining and memory interleaving was discussed by [Jehng 93]. Other 1D PE-array architectures for full-search motion estimation were described by [Nam 94], [Nam 96], [Saha 95], and [Ura 94].

***VLSI implementation and architectures based on a 2-D-array***
An implementation with an interconnection network and an external DRAM was given by [Charlot 95:1], [Pirson 95:1]. [Hsieh 97] described an implementation using shift registers. A programmable architecture was described by [Lin 96] and [Lin 96a]. A bit-serial implementation was described in [Bra 94]. A semi-systolic array was proposed by [Lu 95:1] and [LeeLu 97]. VLSI-architectures for full-search variable block size motion estimation were described in [Kuhn 97a], [Kuhn 97b], [Kuhn 97c], [Kuhn 97d], [Kuhn 97e], [Ber 97], [Ber 96], and [Ber 96b]. Other architectures were described e.g. in [BugW 97], [Chan 93], [Chang 95], [Cmar 97], [Col 91], [Greef 95], [He 97a], [Ishi 95], [Ishi 95b], [Lai 97], [Lai 97a], [LeeDL 94], [Meg 91], [Mosh 97], [Mosh 97a], [OOI 97], [Ross 91], [Vos 89], [Vos 94], [Vos 95:1], [Wang 95:1], [Wu 93], [Yeh 97], and [Yeo 95].

### 6.4.2.2 Three step search

For the popular three step search algorithm (TSS, cf. chapter 2) several VLSI architectures were proposed selecting different trade-offs in terms of gate-count, I/O-bandwidth and throughput. The TSS algorithm consists of three steps with nine search positions each, being numbered as depicted in Fig. 6.30. The center of the second (third) step is chosen at the location of the best motion vector position of the first (second) step. The search positions are denoted as `pos0` ... `pos8` and the search steps $i$ are numbered I, II, III. The maximum displacement of a $N$x$N$ sized MB in the search area is $p$. The reduction ratio of the number of matching positions compared to the full-search algorithm is for the TSS: $(2p+1)^2/27$. $D_{h,i}$ is the horizontal and $D_{v,i}$ the vertical distance (and $D_i$ is the distance in general) between the search positions for search step $i$. They are calculated for TSS in general as:

$$D_i = D_{h,i} = D_{v,i} = \left\lfloor \frac{p+1}{2^i} \right\rfloor \tag{6.37}$$

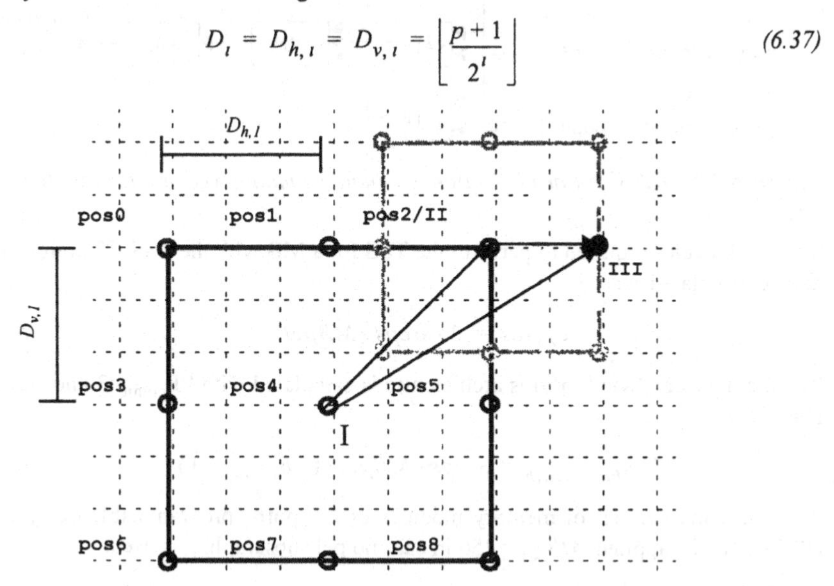


*Figure 6.30: Search positions for the TSS algorithm*

The TSS algorithm consists basically of 3 nested loops, cf. Fig. 6.31:

```
for each of the three steps (loop 1)
    for each of the nine candidate MV positions (loop 2)
        for each of the 256 pel pairs (loop 3)
            calculate the absolute pel difference
```

*Figure 6.31: Nested loops of the Three Step Search (TSS) algorithm*

To parallelize these nested loops, the following architectures are defined:

- TSS-C: calculate loop 2 (candidate MV) in parallel and calculate the other loops sequentially. TSS-C3 uses three PEs and TSS-C9 uses 9 PEs.
- TSS-S: calculate loop 3 (SAD) in parallel and the other loops sequentially.
- TSS-P: Three step search architecture with partial accumulation.

As the search area position of the second step depends on the result of the first step, loop 1 can not be parallelized. For high throughput loop 1 can be basically pipelined, resulting into architecture TSS-P.

***TSS-C3***

Loop 2 can be parallelized using 3 PEs with a single PE calculating the SAD of one of the 3 search positions in the same a) row or b) column. This is a solution with low VLSI area demands but, however, increased number of clock cycles.

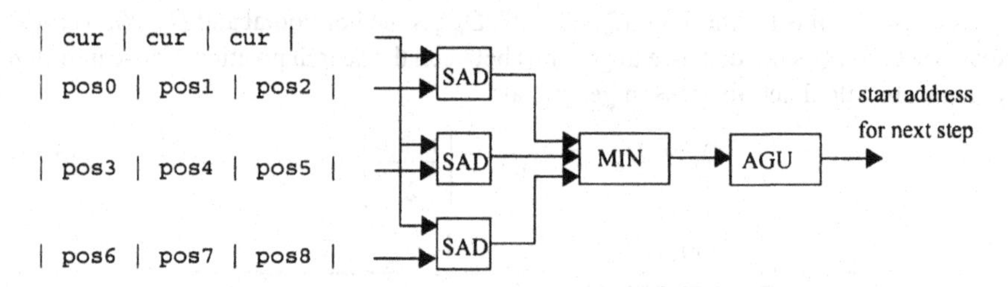


*Figure 6.32: TSS-C3: Parallelization of candidate motion vector calculation with 3 PEs*

The clock cycles required to perform the TSS for a MB with the TSS-C3 architecture can be calculated as:

$$cycles = 3 \cdot steps \cdot MBpel \tag{6.38}$$

The memory bandwidth of this architecture is calculated with $PE_{count}$=3 and steps=3 (for TSS):

$$Mem_{bandwidth} = steps \cdot MBpel \cdot (PE_{count} + 1) \tag{6.39}$$

To reduce the number of memory modules or I/O-ports, programmable delay units (PDUs) can be applied. $MB_{pel} = 256$ in case no pel subsampling is used.

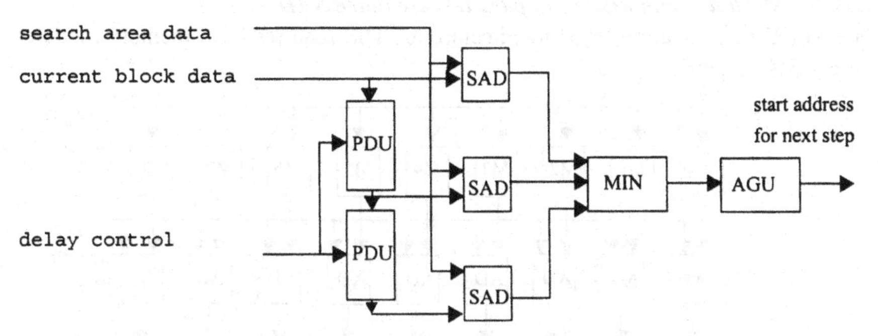


*Figure 6.33: TSS-C3 architecture with programmable delay units (PDUs)*

The delay (and therefore the size) of the PDU can be calculated as:

$$delay_{PDU} = D_l - 1 \tag{6.40}$$

This results in a PDU size of 4 bytes for the TSS. The number of clock cycles required to perform the complete TSS motion estimation with this architecture is:

$$cycles = 3 \cdot \sum_{l=1}^{3} (N + 2D_l) \cdot N \tag{6.41}$$

***TSS-C9: Parallelization of the candidate MV calculation***

Loop 2 can be parallelized more efficiently using 9 PEs with a single PE calculating the SAD of one of the 9 search positions. The MIN unit determines the minimum SAD and the subsequent AGU determines the start address of the next search step.

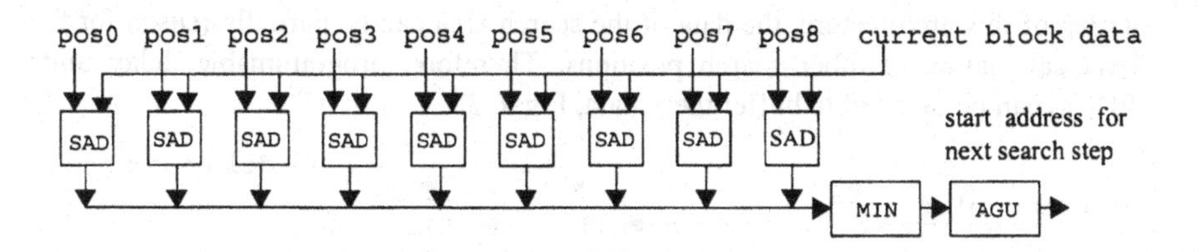


*Figure 6.34: TSS-C9: Parallelization of candidate motion vector calculation with 9 PEs*

The number of clock cycles to perform a three step search for a single MB can be determined as:

$$cycles = steps \cdot MBpel \tag{6.42}$$

with *steps* being the number of search steps (*steps* = 3 for TSS), and *MBpel* being the number of pels of the luminance component of a MB used for the SAD (*MBpel* = 256 in case no pel subsampling is used). The I/O-bandwidth can be calculated as:

$$Mem_{bandwidth} = steps \cdot MBpel \cdot (PE_{count} + 1) \tag{6.43}$$

***TSS-C9: Partial accumulation of parallel calculated SAD***

The partial accumulation VLSI architecture for TSS requires local memory modules for every single PE.

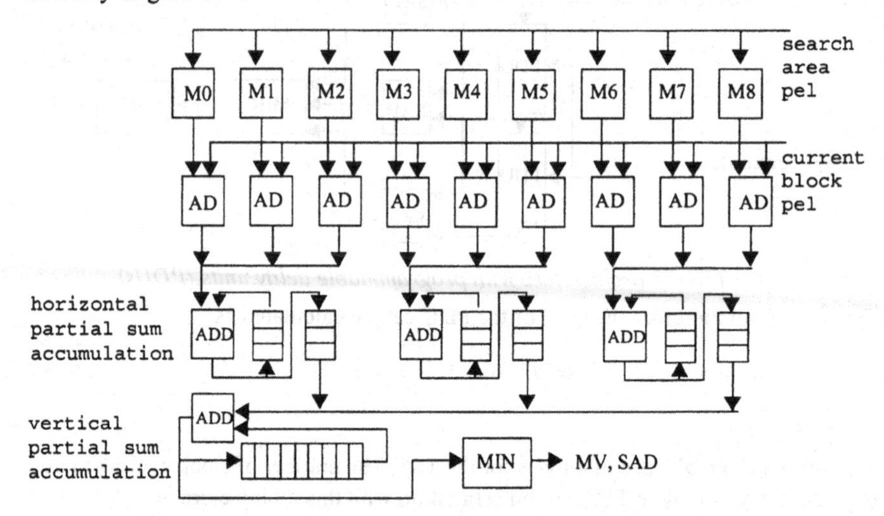


*Figure 6.35: TSS-C9 architecture with partial accumulation*

To avoid memory access conflicts, the search area data are distributed among these memories, using a horizontal and vertical distribution function. The AD processing elements calculate pel differences which are summed up by a partial accumulation scheme. However, the disadvantage of this architecture, compared to other TSS architectures, is the high number of memory modules that result in increased area demands, compared to other architectures (cf. section 6.3.4). To reduce the high external I/O-burden of this architecture, the data of the search area can be partially reused for the SAD calculation at other search positions. Therefore, programmable delay units (PDUs) can be inserted to buffer these data, Fig. 6.36.

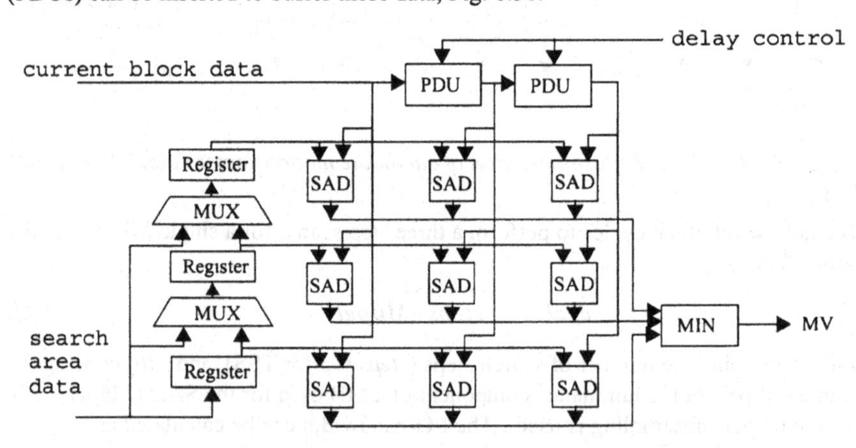


*Figure 6.36: TSS-C9 architecture with programmable delay units (PDUs)*

The required delay of the programmable delay units (PDUs) can be calculated for search step *i* as:

$$delay_{PDU} = D_i - 1 \qquad (6.44)$$

in which $PE_{count}$ is the number of processor elements and i is the search step. However, the PDU units require additional clock cycles for this architecture. The number of cycles to complete step1, step2, and step3 of the TSS algorithms with this architecture can be calculated as (without initial delay etc.):

$$cycles = \sum_{i=1}^{3} (N + 2D_i)^2 \qquad (6.45)$$

***TSS-S: Parallelization of the SAD calculation***

Parallelization of the SAD calculation for the TSS algorithm requires a special data flow which can be gained by an interconnection network (ICN, cf. section 6.3.6). The most common solutions use 16 PEs for SAD calculation.

***TSS: Pipelining of the search steps***

To achieve high throughput it was proposed by [Jong 94] (for MPEG-1) to pipeline the three search steps, cf. Fig. 6.37. For ITU-T H.263 and MPEG-4 this proposed independent search of motion vectors of successive MBs may be not beneficial in all cases, as the predicted motion vector (which can be used as search center for the next motion estimation) of the processed MBs depends on the MV results of the neighborhood MBs. This is referred to as spatial motion vector prediction (cf. chapter 2). However, as the motion estimation algorithm at the encoder is not standardized, this architecture remains still valid.

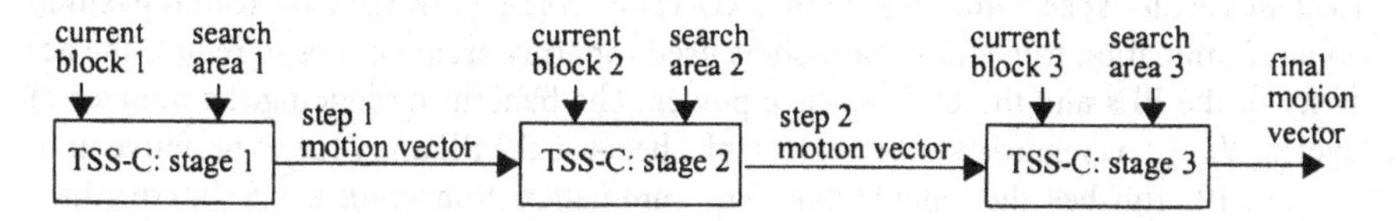


*Figure 6.37: TSS-P: Pipelined three step search architecture with 27 PEs*

***VLSI architectures and implementations:***

[Skim 97] proposed a TSS-C9 VLSI architecture for a computational power constrained TSS similar to Fig. 6.36 with PDUs. The scene-adaptive TSS algorithm allows to stop 1.) after not being performed, 2.) after the first or 3.) after the second search step, in case real-time conditions require to reduce the computational power of the motion estimation. A 9x1 PE array, acting as TSS-C9-architecture with 9 memory modules and an interconnection network similar to Fig. 6.15 was proposed for combined full-search and three step search motion estimation by [Jehng 93] and [Gupta 95]. [Jehng 93] proposed a tree architecture for motion estimation, employing an interleaving of memory banks, which is described in more detail in section 6.3.3.

[Lak 97] described a VLSI implementation for the TSS-C3 architecture, using three PEs and alternatively one or three on-chip memory modules. The on-chip memory modules buffer the data which is reused for the calculation of the SAD for subsequent search positions. This implementation also supports a modified TSS scheme (ETSS:

enhanced TSS) with only four search positions with three pel distance and additional search positions at the center for the first step. [Costa 95] proposed an architecture similar to TSS-C3 with programmable delay units for CIF resolution video.

The VLSI implementation of the TSS-S with partial accumulation as proposed by [Jong 94], was reported to require 794 clock cycles for the 3 PE solution, 2,330 clock cycles for the 9 PE solution and requires an equivalent gate count (including memories) of 36.6 Kgates.

A 16x1 PE array with 16 memory modules and an interconnection network similar to Fig. 6.15, was proposed for combined full-search and three step search motion estimation by [Dut 96]. This approach offers the advantage of potential support for several motion estimation algorithms. However, the size of the interconnection network and the area demands of the 16 memory modules used in this approach are disadvantageous. A similar architecture with 3 16x1 PE arrays for full-search and three step search was proposed by [Zhang 97], requiring 24 kgates (without the memory modules), 204 clock cycles for the TSS (using the pipeline technique) and 1,591 clock cycles for the full-search algorithm with p=8.

[YeoHu] proposed a VLSI architecture for TSS for high throughput and large search range demands, as required for digital high definition TV by a pipelining interleaving technique.

#### 6.4.2.3 SAD summation truncation

Summation truncation (cf. 2.4.2) is used to stop the SAD calculation in case the intermediate result exceeds the minimum SAD value, $SAD_{min}$, of the best search position so far. Summation truncation can also be used for array architectures to reduce the activity of the PEs and therefore to save power. The benefit in reducing the number of cycles, $N_{cycles}$, is regarded to be neglectable, because the PE array has to be active until the last PE finishes the calculations. For summation truncation every intermediate SAD value has to be compared with a threshold (the best SAD value already calculated). There are no direct restrictions for applying SAD summation truncation to MPEG-4 motion estimation.

For the LA architecture class, the $SAD_{min}$ can be broadcast to all PEs (which requires some overhead for wiring), with a local comparator comparing the $SAD_{min}$ with the currently calculated partial SAD, $SAD_{part}$. In case $SAD_{part} > SAD_{min}$, the $SAD_{part}$ is not forwarded to the next PE, thus reducing the overall circuit activity and saving power. Alternatively, a predetermined maximum SAD value could be given to the PE array to save the updating of $SAD_{min}$ whenever a new minimum SAD is calculated.

For the GA-1D architecture class, $SAD_{part}$ can be directly compared with $SAD_{min}$ at the adder-tree, resulting into an efficient implementation. The GA-1D architecture also allows to stop the memory access, in case $SAD_{part} > SAD_{min}$ results in a high amount of memory bandwidth reduction and thus a high power reduction. For the GA-2D architecture, no direct benefit can be gained by summation truncation, as the SAD is calculated within one single cycle of the PE-array.

| Array architecture | CMP | benefit |
|---|---|---|
| LA-1D | N | low |
| LA-2D | *NxN* | low |
| GA-1D | 1 | high |
| GA-2D | 1 | no |

*Figure 6.38: VLSI overhead and benefit to support summation truncation*

***VLSI implementations and architectures***

Implementations of VLSI array architectures using summation truncation are not known so far.

### 6.4.2.4 Distance criteria

There are different approaches to reduce the computational requirements of the distance criteria (cf. 2.4.2). For MPEG-4 motion estimation there are no specific restrictions concerning the choice of the distance criterion. In the next sections of this book, a selection of other distance criteria than the widely used SAD, is discussed in terms of VLSI implementation.

***MME***

Compared to the SAD criterion the MME (Minimized Maximum Error) offers the advantage of a lower VLSI implementation cost, as only an 8-bit comparator is required for MME in comparison to a 16-bit accumulator for SAD. For PE arrays with higher SAD-PE counts, this technique can result in significant area and power savings. [Chen 95:1] implemented the MME criterion using the [Yang 89] architecture and reported a VLSI area saving of about 15% for 0.8 μm CMOS technology. A parallel bit level maximum/minimum selector with low VLSI area was presented in [Lee 94a].

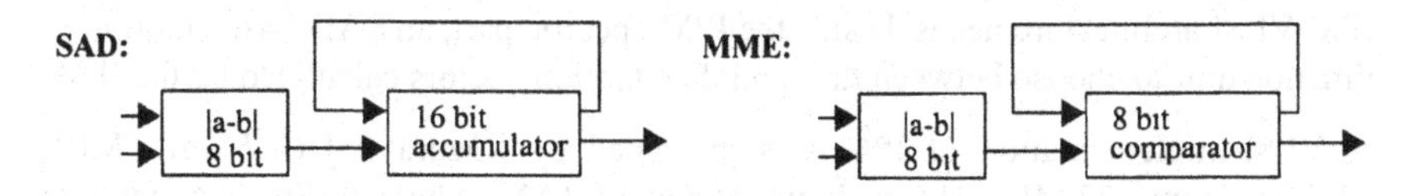


*Figure 6.39: SAD versus MME (Minimized Maximum Error) criterion*

***Other VLSI implementations***

The RBMAD criterion (reducing the number of significant bits used for the SAD calculation) was implemented by [Baek 96], who reported that the area is reduced to 57% for e.g. the 4 bit RBMAD, and a speed increase of 34 % is gained for a 0.8 μm CMOS process. Adaptive bit truncation was implemented by [He 97]. The DPC criterion (motion estimation on a low resolution image) is implemented by [LeeC 96] who reports an *AT* (Area Time) cost reduction by a factor higher than 12.

### 6.4.2.5 Feature Matching

***BBM***

VLSI architectures for binary block-matching (BBM, cf. 2.4.6.2) were presented by [Le 98] (SIMD architecture built on computational RAM), [Nat 96], and [Nat 97]. These techniques use a filter for preprocessing, a simple XOR unit and a lookup-table with 256 entries as the processor element. A hierarchical scheme based on the XOR-unit was presented by [XLee 96].

***Integral projection-matching***

The integral projection-matching (IPM, cf. 2.4.6.1) technique calculates the horizontal sum of the rows and the vertical sum of the columns of the current block, and compares these results with the respective sums of the candidate block of the search area. IPM can be combined with any search area subsampling technique. To reduce the computational complexity, sums already calculated can be reused with the help of a sliding window technique, Fig. 6.40, [WLi 95], [Kom 93].

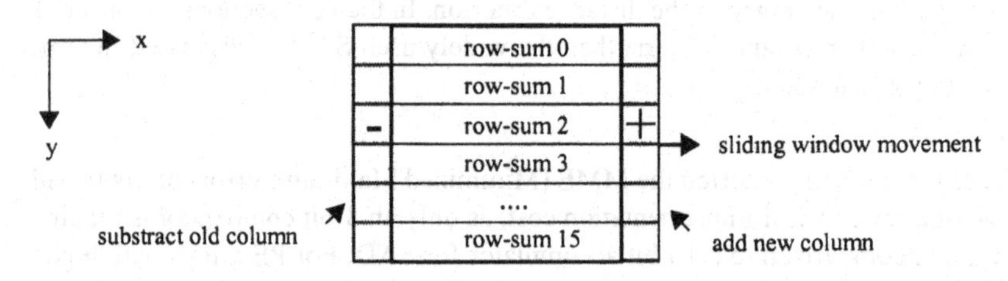


*Figure 6.40: Sliding window for Integral Projection-Matching (IPM)*

However, for a high number of search points a significant number of memory is required to buffer the row (or column) sums. And for a low number of search area subsampling points only a limited amount of sum calculations can be reused. Note that an IPM VLSI architecture needs, beside the IPM-specific part, an SAD-based motion estimation unit to choose between the candidate motion vectors calculated by the IPM.

A VLSI-implementation of IPM was presented by [Ogura 95] (0.7 μm CMOS, 14.7x14.7 $mm^2$, 27 MHz, 185 kG logic, 33 Kbit of RAM) where the Sp-Tp method (cf. 2.4.6.1) was used to reduce the computational complexity. The implementation is straightforward without reusing sums already calculated. In a first step the sums of the rows and columns are calculated and saved, and in the second step the search is performed on these sum norms. [Pan 96] described a two-stage block-matching architecture, where in the first stage the IPM is calculated by a systolic array architecture optimized for IPM, and in the second stage full-search block-matching is performed by a typical systolic array for full-search.

### 6.4.2.6 Pel subsampling

Pel subsampling or pel decimation is a process which takes into account only a limited number of pels of a block for SAD calculation. This technique was described in section 2.4.5.

***2D-GA architecture***

For pel subsampling PEs could be masked [Vos 95], as depicted in Fig. 6.41 (e.g. for 2:1 horizontal subsampling). However, masking of PEs results in reduced efficiency of the architecture and visual quality degradation, while still requiring high external memory bandwidth (and therefore high power consumption). $MB_{pel}$ is the number of pels of a $N$x$N$ block after pel decimation. The efficiency of this approach can be calculated as:

$$\eta = \frac{a}{N^2} \tag{6.46}$$

with $a$ representing the number of the active PEs of the $N$x$N$ processor array while performing pel subsampled motion estimation.

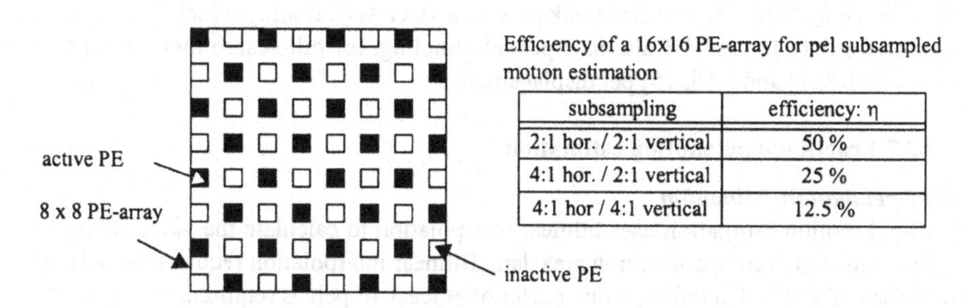


| subsampling | efficiency: η |
|---|---|
| 2:1 hor. / 2:1 vertical | 50 % |
| 4:1 hor. / 2:1 vertical | 25 % |
| 4:1 hor / 4:1 vertical | 12.5 % |

*Figure 6.41: Pel subsampled ME with an 2D-GA array, [Vos 95]*

To compensate efficiency degradation for pel subsampling, [Vos 95] proposed to preprocess the search area and the current block data. Only the pels surviving the pel decimation process are handed over to the 2D PE array, thereby reducing external memory bandwidth and, therefore, power consumption. However, 100 % efficiency for a $N$x$N$ GA-2D array can be gained only for cases, in which the number of subsampled pels matches the size of the PE array. For example a (however, unrealistic) 64x64 block could be subsampled by 4:1 horizontal and 4:1 vertical, resulting in 16x16 pel enabling an $\eta$=100% for a 16x16 PE array.

***2D LA architecture***

For the LA architecture pel subsampling can be achieved by forwarding only the decimated pels to the PE array. However, the same problem as for GA arises if full efficiency is to be achieved.

***VLSI implementations***

[Vos 95] presented a full custom motion estimation VLSI (0.6 µm CMOS, 1 050 kT, 228 mm$^2$ including PADs) based on the 2D-GA architecture. This architecture supports pel subsampling by diagonal PE masking, which is reported to require only a very small number of additional control lines.

[Jung 95] reported an alternate pel subsampling (cf. chapter 2) extension of the [Yang 89] 1D-PE array architecture, requiring 2,561 clock cycles for the alternated pel-subsampled full-search motion estimation with [-8, +7] pel search area. The four subsampling patterns (a ... d) are applied subsequently within the first 2,048 clock cycles, calculating four motion vector candidates. Then, from clock cycle 2,049 to 2,561, the best of these four motion vector candidates is determined, with the help of the full-search block-matching method. This architecture employs two address generators to calculate 1.) row-wise and column-wise reference addresses, and 2.) to calculate the increment addresses for every pel position.

[Saha 95] reports a hierarchical pel decimation scheme implemented on a 64 PE architecture with 4 x 16 1-D PE array clusters. This architecture supports a maximum search range using pel decimation of [-32, +31] pel and can perform full-search around the zero motion vector. [Cheng 96] reported a (theoretical) memory bandwidth and computational complexity analysis of the alternate pel subsampling algorithms implemented on systolic array architectures.

Section 6.6 and chapter 8 of this book present a 16x1 1D-PE array which can be configured for 2:1, 4:1 and 4:1 alternate pel subsampling, for full-search motion estimation with [-8, 7] and [-16, 15] pel displacement.

#### 6.4.2.7 Fractional pel motion estimation

***Half-pel motion estimation***

Half-pel motion estimation uses bilinear interpolation to calculate the values at half-pel positions of current and search area data. Bilinear interpolation requires the pels of two lines of a MB. Therefore, a line buffer of at least 16 pels is required.

For half-pel motion estimation basically these options exist:

- **Pre-interpolation**: As a first step, the current and search area data is bilinear interpolated when loaded into a motion estimation array, and then (integer-pel) motion estimation is performed on these data. This scheme reduces the external bandwidth, as the current block and the search area data is loaded only once. However, the disadvantage is that the number of cycles for motion estimation increases by the factor of four. This method was used e.g. by the VLSI implementation of [Ogura 95] for IPM.
- **Post-interpolation**: As a first step, a full-search (or other fast motion estimation algorithm) is performed, and the half-pel motion vector is estimated in the bilinear interpolated search area around the integer motion vector. This two-step method offers the advantage, that only a small area around the final integer motion vector has to be half-pel interpolated. However, for most 2D array architectures, the search area data around the best integer-pel motion vector is lost after the search is finished. As a result these have to be loaded again from the memory, which in turn results in increased memory bandwidth and cycles cost. This method was used by VLSI implementations of [Ura 94], [Ishi 95], and [Ishi 95b].

***Quarter-Pel and 1/n pel motion estimation***

Quarter-pel motion estimation was (at the time of writing) a topic under discussion for MPEG-4 visual version 2. For quarter-pel motion estimation there already exist some VLSI architectures: [Yang 89], [Sun 93]. A VLSI architecture for 1/n-pel motion estimation is proposed by [Li 96].

#### 6.4.2.8 Variable block size motion estimation (VBSME)

Numerous algorithms and architectures [Pir 95], [Vos 95] have been developed for block-matching motion estimation in different applications and were described so far. However, VLSI architectural evaluations of algorithms, which are under consideration for the next step of standardization, are few, but these are regarded to be essential to give guidelines for implementation feasibility of these emerging algorithms. As shown in section 2.5.2, there are numerous algorithms for variable block size motion estimation. However, VLSI implementations of variable block size motion estimation are rare. This section discusses a new VLSI-architecture for variable block size motion estimation.

***GA-2D array architecture***

Variable block size motion estimation architectures within this section comprise VLSI architectures, which are able to perform motion estimation on several variable block sizes *simultaneously*. This point is essential, as the GA VLSI architecture is basically suited to support motion estimation on variable block sizes, but only *one* block size at a time, i.e. a block size of 16x16 for a 16x16 PE array. The [Vos 95] GA-2D architecture supports variable block sizes (which have to be smaller than the PE array size) by masking of PEs, Fig. 6.43.

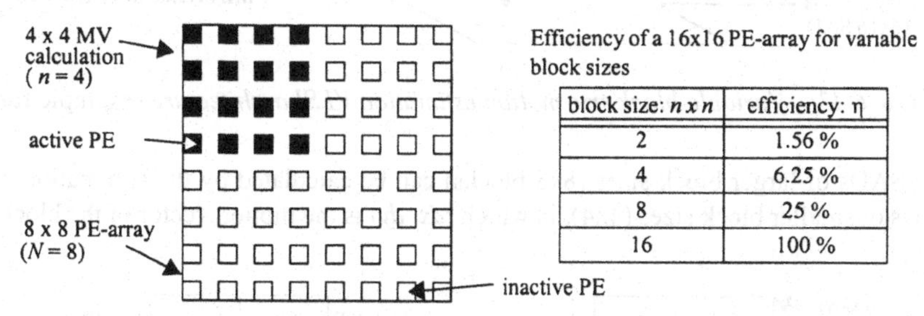


| block size: $n \times n$ | efficiency: $\eta$ |
|---|---|
| 2 | 1.56 % |
| 4 | 6.25 % |
| 8 | 25 % |
| 16 | 100 % |

*Figure 6.42: Support of variable block sizes with the [Vos 95] architecture*

However, PE masking leads to significant efficiency loss. The efficiency of this architecture can be calculated for variable block sizes as:

$$\eta = \frac{n^2}{N^2} \qquad (6.47)$$

with $n$ representing the block size and $N$ standing for the size of the 2D processor array in x- and y-direction. In order to support $i$ different block sizes *subsequently* with this architecture, the efficiency $\eta$ decreases to:

$$\eta = \frac{1}{i} \sum_{n \in [n_1, n_2, \ldots, n_i]} \frac{n^2}{N^2} \qquad (6.48)$$

For example, to support block sizes of $n$ = [2, 4, 8, 16] with a 16x16 PE-array ($N$=16) would lead to a total efficiency of $\eta$ = 33.2 %. To achieve an efficiency of 100% for variable block size motion estimation this section of this book extends the global accumulation (GA) architecture presented by [Vos 89] with an enhanced adder-tree to calculate not only the motion vectors for 16x16 blocks, but also for 8x8 and 4x4 subblocks, cf. Fig. 6.43.

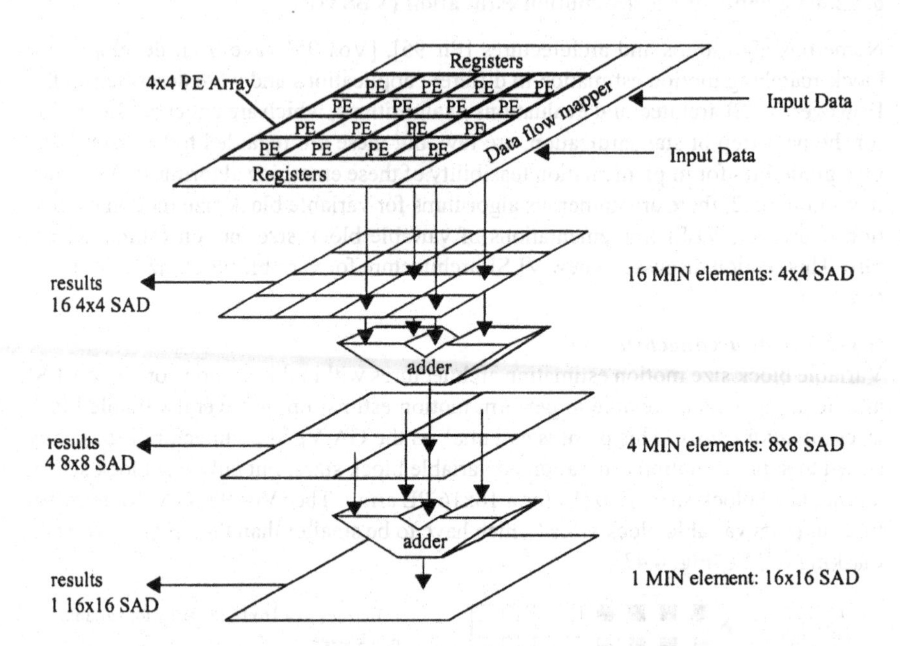


*Figure 6.43: Variable block size motion estimation VLSI architecture* (example for $N$=4)

The SADs of larger block sizes (8x8 blocks) can be calculated by the summation of SADs of smaller block sizes (4x4), in which *(dx, dy)* is the motion vector of the block:

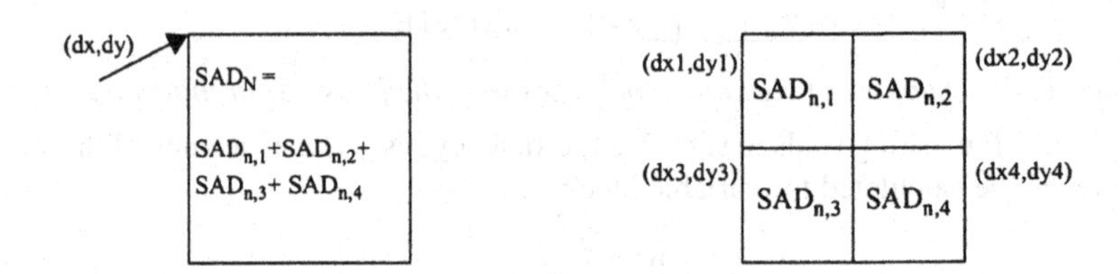


*Figure 6.44: SAD summation from smaller block-sized SADs*

$$SAD_N(dx, dy) = SAD_{n,1}(dx1, dy1) + SAD_{n,2}(dx2, dy2) + SAD_{n,3}(dx3, dy3) + SAD_{n,4}(dx4, dy4) \tag{6.49}$$

The VLSI architectural requirements for this VBSME approach are calculated here. The number of MIN-elements are:

$$N_{MINelements} = \sum_{n \in [n_1, n_2, \ldots, n_i]} \frac{N^2}{n^2} \tag{6.50}$$

The number of 2-input adders in the adder-tree is:

$$N_{adders} = N^2 - 1 \tag{6.51}$$

resulting in 85 MIN elements and 255 2-input adders (within the adder-tree) for $n$ = [4, 8, 16] with a GA 16x16 PE array architecture ($N$=16), compared to a single MIN element and 255 adders for the original [Vos 95] architecture.

***Local Accumulation vs. Global Accumulation Architecture for VBSME***

For VBSME the GA proved to be more effective than the LA architecture, because the SADs for subblocks are already available in the adder-tree and only additional MIN elements are required to gain and save the minimum SAD for these smaller block sizes (subblocks).

In the LA, partial sums of the SAD result are calculated sequentially. Therefore, the pel sequence would have to be reordered and the partial sums of every motion vector candidate would have to be saved in additional registers. By proper pel sequence reordering some of the partial sums could be reused, but difficulties occur with the support of subblock clustering (cf. chapter 7). The summing up of partial sums for VBSME using the LA architecture is data-dependent and requires complex control at PE speed.

***Variable Sized Blocks at the Image Boundary***

To support the full search-range for smaller block sizes than $N$x$N$ at the image boundary, the search area for these blocks has to be extended by a modified control scheme of the search area AGU, [Kuhn 97c]. Without this "virtual" image boundary extension (Fig. 6.45a), e.g. the 8x8 block B45 for the example of $N$=16 of the current frame never matches the 8x8 block B0 of the search area.

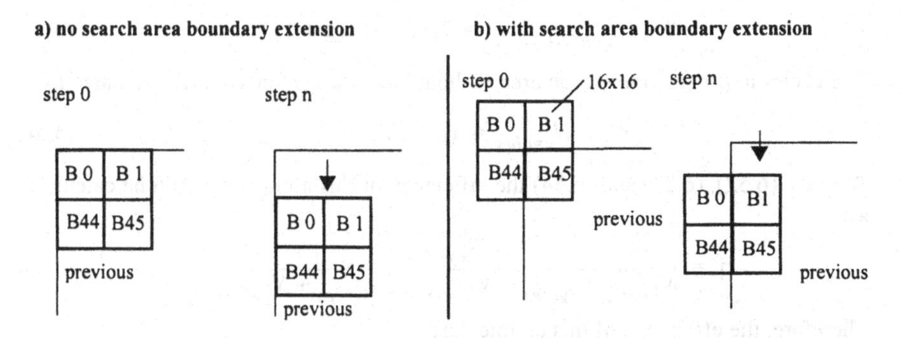


*Figure 6.45: Search area boundary extension for VBSME*

This implies a limitation of the search area which is not acceptable, except in a very few special cases (e.g. with the assumption that there are no long motion vectors at the image boundaries). With image boundary extension (Fig. 6.45b) the original block (e.g. B45) is shifted further to reach the image boundaries. However, this approach limits the efficiency of the PE array (especially at very small image formats) at the image boundaries, as not fully efficient 16x16 block-matching can be performed here. Fig. 6.45c and d depict the general case for VBSME with the maximum blocksize $m$ and the minimum blocksize $n$.

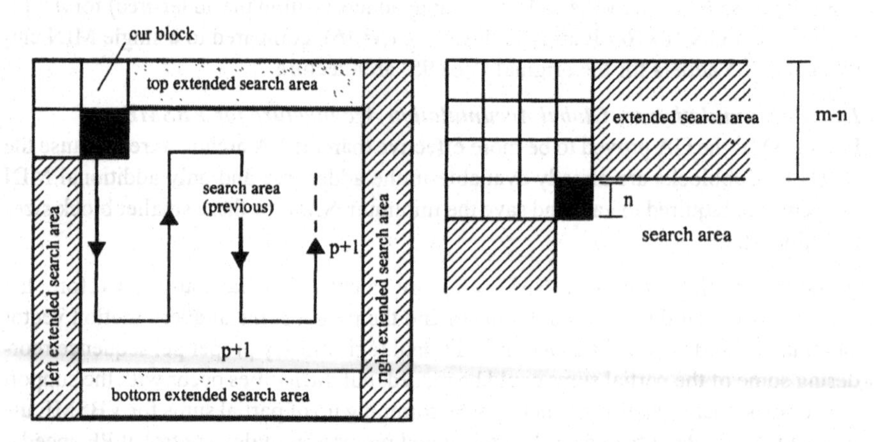


*Figure 6.46: Search area boundary extension for VBSME: efficiency calculation*

The additional search positions at the left and right extended search area can be calculated as:

$$N_{VerticalCycles} = 2(m-n) \cdot (2(m-n) + p + 1) \tag{6.52}$$

Tha additional search positions at the horizontal upper and lower extended search area can be calculated as:

$$N_{HorizontalCycles} = 2(m-n) \cdot (p+1) \tag{6.53}$$

The cycles to process the search area without boundary extension are calculated as:

$$N_{cycles} = (p+1)^2 \tag{6.54}$$

With eq. (6.52), (6.53) and (6.54) the efficiency of this architecture can be calculated as:

$$\eta = \frac{N_{cycles}}{N_{VerticalCycles} + N_{HorizontalCycles} + N_{cycles}} \tag{6.55}$$

Therefore, the efficiency of this architecture is:

$$\eta = \frac{(p+1)^2}{2(m-n) \cdot (2(m-n) + p + 1) + 2(m-n) \cdot (p+1) + (p+1)^2} \tag{6.56}$$

***VLSI implementations***

In [Kuhn 97a], [Kuhn 97b], [Kuhn 97c], [Kuhn 97d], and [Kuhn 97e] the adder-tree as depicted in Fig. 6.43 was used as basis for a 2D-GA VLSI architecture, supporting variable block size full-search motion estimation, luminance correction and other features. Refer to section 6.5 and chapter 7 for details.

A similar architecture, as depicted in Fig. 6.42, targeted for HDTV applications was implemented by [Ber 97], [Ber 96], and [Ber 96b] as a full custom designed 32x32 PE array (200 GOPS, 200 Mhz, 0.5 μm CMOS, 220 mm$^2$, with an estimated power consumption of 15mW at 200 Mhz per PE). A search area of +/-15 pel horizontally and vertically is supported for blocksizes of 8x8, 16x16 and 32x32. This approach is based on similar ideas as presented in 6.5 but was developed independently from this work.

[Hans 96] described a programmable SIMD processor (41 $mm^2$ for a single PE and memory, 50 Mhz, 1 μm) specially designed for real-time CIF 10 fps variable block size motion estimation requiring 12 of these PEs. This VLSI architecture is targeted for an "Adaptive Block-Matching Algorithm" (ABMA), which supports block sizes in the range of 4x4 and 16x16, and includes global and local MC as well as a split/merge procedure for blocks. However, as this SIMD-architecture exploits data parallelism and no special data flow design, a high memory bandwidth of 2 bytes external SRAM memory access per clock cycle and PE is required to supply every PE with new data for every search position.

#### 6.4.2.9 Advanced prediction motion estimation

The support of 16x16 block motion estimation and advanced prediction motion estimation (four 8x8 blocks in the area of [-2, +2] pel around the 16x16 MV vector, cf. chapter 2) is basically a VBSME algorithm with these specific parameters. Therefore, the previous discussion on VBSME architectures is also valid for advanced prediction motion estimation for MPEG-4.

#### 6.4.2.10 Luminance-corrected motion estimation

The block-matching algorithm with luminance correction was described in 2.5.4. The luminance correction coefficient $q$ is added within the absolute difference unit to the difference of the pel values $a$ and $b$, Fig. 6.47. Note that the $q$-coefficient cannot be added after the absolute value calculation. Luminance correction can be applied to the SAD array architecture and the GA architecture beneficially, as the luminance correction coefficient can be broadcast to all PEs. For the LA architecture, the $q$-coefficient has to be moved similarly to the current-block data through the PE-array, causing additional control overhead. However, supporting variable block sizes and luminance-correction with the GA architecture, leads to a dilemma, as different $q$-coefficients have to be applied for different block sizes. A solution for this problem is presented in chapter 7 of this book.

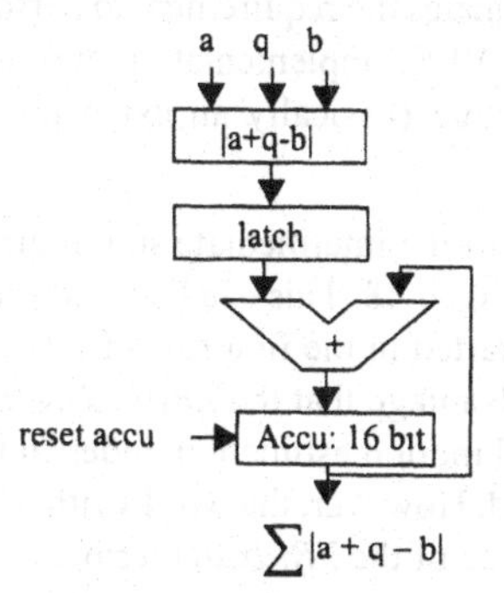


*Figure 6.47: Luminance-corrected SAD calculation*

***VLSI implementations***

A VLSI architecture for block-matching with eliminated DC components was presented by [Yasu 95]. In this work, a luminance pattern for block-matching was obtained, by eliminating the DC components of the current block and the search area pel data *before* the matching process is started. The data used for block-matching only contains AC data and for this reason, this approach differs significantly from the method presented in 2.5.4.

### 6.4.2.11 Preferring the zero motion vector

For MPEG-4 motion estimation, the zero motion vector is preferred by subtracting a value of $N_B/2+1$ from the SAD of MV(0,0), with $N_B$ representing the number of pels inside the VOP within this context (cf. section 2.2.1). For a VLSI implementation two solutions seem to be sensible, which both can be applied basically for the LA-, GA- and the SAD-array architecture:

- Reducing the SAD of the zero MV after conducting the SAD calculation, which has to be done before comparing the SAD in the MIN unit. For some ME architectures this requires time-dependent control of the SAD reduction.
- Preloading of every SAD register, except the zero MV SAD register, with a value of $N_B/2+1$. However, this limits the SAD range for high SAD values, which is considered to be a minor problem though, as these values correspond to poor matching results.

### 6.4.2.12 R/D-optimized motion estimation

For R/D-optimized motion estimation (cf. 2.3) the cost of a matching position can be calculated as following:

$$J(dx, dy) = D(dx, dy) + \lambda \cdot R(dx, dy) \qquad (6.57)$$

The distortion $D(dx,dy)$ equals the SAD for ME architectures within this context. The $\lambda$ parameter is predetermined (e.g. $\lambda = 2\,Q_p$, [TMN 9], with $Q_p$ being the inter-quantization parameter) and the $\lambda R$ product can be precalculated and saved in a RAM for table-lookup. The size of the RAM equals the number of entries of the VLC (Variable Length Code) table. This eliminates the requirement to perform a multiplication within an inner ME search loop. For VLSI implementation two solutions can be applied for the LA- and the GA-architecture (basically an extension of the method of section 6.4.2.11):

- **LA pre-SAD:** Preloading of the intermediate sum registers of the first row of AD elements with $\lambda R(dx,dy)$, Fig. 6.48. This can be performed sequentially, as the systolic SAD calculation is started in the first row of AD elements from left to right. This solution offers the advantage that the $\lambda R(dx,dy)$ values can be applied before the beginning of the actual motion estimation, independently of the PE speed, and that one adder can be saved. However, the wordwidth of the AD elements has to be wider, depending on the size of the $\lambda R(dx,dy)$ term.

- **LA post-SAD:** This method increases the SAD of every MV position after SAD calculation with the $\lambda R(dx,dy)$ value, but before comparing the SAD within the MIN unit. For some ME architectures this requires time-dependent control for addition of the $\lambda R(dx,dy)$ term to the SAD, Fig. 6.48. The advantage of this approach is that the wordwidth of the AD elements has not to be increased, which may be beneficial for large PE arrays or already existing designs. This solution has the disadvantage of requiring a table-lookup (RAM access) for $\lambda R(dx,dy)$ values at PE speed and the usage of an additional adder unit. However, as the sequence of the SAD(dx,dy) results is known beforehand, the table lookup can be initiated in advance.
- **GA:** For GA architectures the only reasonable solution is basically the same as the LA post-SAD method. An adder and the $\lambda R(dx,dy)$ lookup-table is inserted between the adder-tree and the MIN-unit.
- **SAD array**: For SAD array architectures, in which the SAD is calculated locally in a single PE element, $\lambda R(dx,dy)$ can be preloaded in the SAD registers (pre-SAD) before starting the SAD summation or can be applied after SAD calculation, and before the MIN unit (post-SAD) with the advantages and drawbacks as stated above for the method LA post SAD.

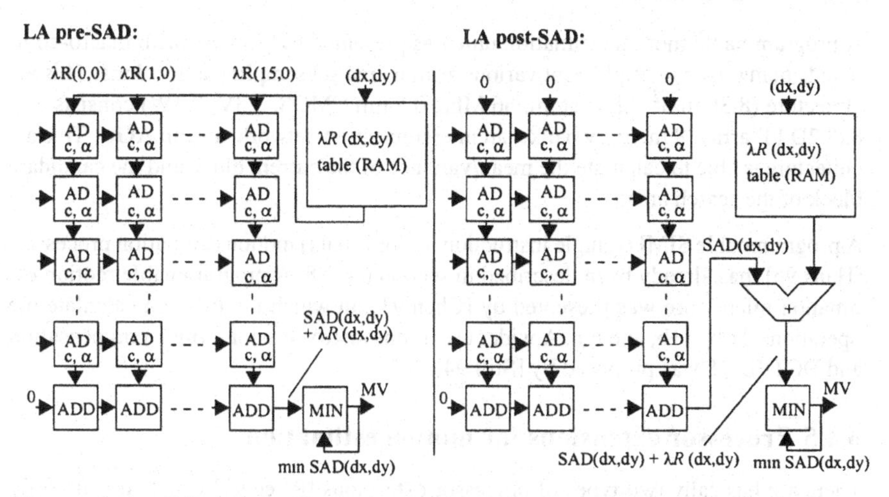


*Figure 6.48: R/D-optimized motion estimation for LA architectures*

## 6.4.3 Flexible architectures

For this work flexible ME architectures are defined as ME architectures which are capable of performing more than one single ME algorithm. Generally flexible architectures require additional logic for memory management, setup of data paths and AGUs. It can be observed that there are only few really flexible motion estimation architectures supporting more than a single algorithm, as flexibility reduces usually efficiency. The [Vos 95:1] 2D-GA architecture is proposed also for pel subsampling (cf. 6.4.2.6), and variable block sizes (cf. 6.4.2.8), which is discussed in these sections.

#### 6.4.3.1 TSS and full-search

VLSI-architectures for TSS and full-search block-matching were proposed by [Jehng 93], [Gupta 95], [Dut 96], [Zhang 97]. These architectures were already discussed in section 6.4.2.2.

A new VLSI-architecture supporting TSS and full-search among other motion estimation algorithms is presented in section 6.6 of this book.

#### 6.4.3.2 Variable block size with luminance correction

A new VLSI-architecture supporting variable block size and luminance-corrected full-search among other motion estimation algorithms is presented in section 6.6 of this book.

#### 6.4.3.3 Other flexible ME architecture approaches

An ME architecture which can be reconfigured for various ME parameters has been presented by [BugW 97] who used programmable delay units (PDUs), cf. 6.3.5.

### 6.4.4 Programmable motion estimation architectures

A programmable motion estimation unit was presented by [Lin 96] with macro-commands being used to implement various search area subsampling algorithms. This architecture (8.36 $mm^2$ full custom, 66 Mhz, 0.5 μm CMOS, 3.3V, 0.3W) consists of a 8x8 2D PE array, 8 memory banks, an interconnection network, and an AGU. This architecture is able to calculate the mean/variance of the current block and the candidate block of the search area.

A programmable SIMD (single instruction multiple data) motion estimation processor, [Hans 96], has already been described in section 6.4.2.8. A programmable motion estimation micro core was presented by [Chen 93], in which the PEs can calculate the operations: |a-b|, a+b, and a-b. A wider use of this micro core for motion compensation and DCT/IDCT was proposed by [Still 94].

### 6.4.5 Processor extensions for motion estimation

There are basically two types of processor extensions [RLee 95] which are currently used to accelerate multimedia applications:

- Subword-parallelism, and
- Arithmetic instruction extensions.

Subword parallelism is also referred to as packed arithmetic and allows to exploit data parallelism by splitting a high precision ALU (e.g. 64 bit) into a number of low precision ALUs (e.g. 8 bit), which are usually controlled together. This is also referred to as SIMD processing and, in this example, results in a maximum speedup of 8.

However, full parallelism can only be achieved in case the data required can be accessed in time, which requires special memory access instruction enhancements. For the above example, the full speed-up of 8 is gained when 16 data bytes can be loaded in parallel. As for fast motion estimation algorithms, these 16 data bytes may not be addressable consecutively in memory for every search position, resulting in a situation a simple load-store unit cannot cope with.

As motion estimation has been identified as a computational demanding task, various processors offer special arithmetic units for motion estimation, resulting in significant performance improvements over the subword-parallelism aproach.

***VLSI-implementations***
The SPARC visual instruction set (VIS) used in the UltraSPARC processors, contains a special instruction (PDIST) for calculating 8 absolute differences in parallel [Kohn 95], [Mou 96], [Ding 98]. This instruction is combined with an efficient memory access scheme which allows to accumulate the SAD for a 16x16 block using 32 of these PDIST instructions.

The Chromatics MPACT VLIW (Very Long Instruction Word), [Kala 96], processor contains an especially efficient arithmetic unit for motion estimation.

[Carl 97] described a motion-estimation processor extension for the DEC Alpha processor, which consists of a single SAD unit, calculating the absolute difference between two pels. However, with this architectural approach only a moderate speed-up is gained by means of pipelining.

The HP PA 7100 processor [Knebel 93], [Bha 95:1] used packed-arithmetic and no special arithmetic unit for motion estimation. The Intel MMX approach is based on packed arithmetic and contains no special arithmetic unit for motion estimation.

# 6.5 DESIGN TRADE-OFFS: SEARCH ENGINE I (2D-ARRAY)

After a more theoretical evaluation of various motion estimation VLSI architectures, in the rest of this chapter the design trade-offs of two VLSI implementations of ME architectures are discussed which are presented in detail in chapter 7 and chapter 8 of this book.

## 6.5.1 Introduction

The VLSI architecture "Search Engine I" is targeted for high throughput full-search MPEG-4 motion estimation. Besides the MPEG-4 compliant full-search motion estimation algorithm, motion estimation algorithms beyond the standard are to be supported with this VLSI-architecture. These algorithms are: 1.) variable block size motion estimation, 2.) segment-matching (cf. 2.5.3) combined with a luminance correction scheme. In this section the design decisions are discussed, and in chapter 7 the algorithm architecture mapping and the VLSI implementation are discussed. The target throughput is 4CIF (704 x 576 pel) at 30 fps per second for the MPEG-4 motion estimation. Support of variable block size motion estimation with luminance correction (cf. 2.5.5) was more of an experimental task, with the throughput set to be at least CIF at 15 fps. The advantages of this algorithm are the high coding efficiency and the high degree of visual quality. Apart from this, decoding with very low computational power is possible, as only motion compensation has to be performed. However, a disadvantage of this algorithm is the high computational power requirements for encoding, which is about 15 minutes encoding time for a single CIF frame on a Ultra-Sparc II/200 Mhz. Therefore, with this VLSI-architecture possible solutions for real-time encoding were investigated.

## 6.5.2 Design Decisions

***VBSME: Variable block size motion estimation***

Variable block size motion estimation, which has already been discussed in detail in section 6.4.2.8 showed as a result, that the GA architecture is better suited for this task than the LA-architecture. The required throughput indicates to favor a two-dimensional GA VLSI architecture.

***Full-search motion estimation with MPEG-4 support***

The support of arbitrarily shapes with the GA architecture is straightforward as discussed before in section 6.3.3.2.

***Luminance correction***

Luminance correction can be directly applied by modifying the PEs as depicted in section 6.4.2.10.

***Subblock clusters***

The support of block combinations (i.e. subblock clusters, cf. 2.5.3) is basically an extension of the adder-tree, developed for VBSME, and fits favorable to the GA-architecture.

***Entropy-constrained iterative VBSME with luminance correction***

However, the support of an entropy-constrained iterative variable block size motion estimation requires the parallelization of an iterative algorithm. This is discussed in detail in chapter 7.

# 6.6 DESIGN TRADE-OFFS: SEARCH ENGINE II (1D-ARRAY)

## 6.6.1 Introduction

In this section a flexible motion estimation accelerator is developed for low-power portable real-time video encoding applications. This VLSI architecture "Search Engine II", or short "SE-II", also supports, besides exhaustive search motion estimation, several motion estimation algorithms with reduced computational complexity, in order to meet the high computational demands and the low-power consumption requirements. In this section the basic PE array architecture for this application scenario is developed, in chapter 8 the algorithm-architecture mapping is performed and the results are presented.

## 6.6.2 Motivation

While for previously standardized video coding schemes usually a single motion estimation algorithm was employed, MPEG-4 now introduces the support of multiple arbitrarily-shaped video objects (VOs), in which for every single VO different quality constraints and, therefore, different motion estimation algorithms can be selected. This is especially useful for handheld applications, for which, as an example, the coding options and the motion estimation algorithm, for example for the background VO (e.g. a landscape), can be optimized in terms of low complexity, and for the foreground VO (e.g. a person talking) a high visual quality can be obtained by spending more computational power on motion estimation. Therefore, for a low-power MPEG-4 motion estimation architecture the requirements arise to support 1.) multiple VOs, 2.) several motion estimation algorithms with different complexity/distortion trade-off, and 3.) the alpha-plane defined within MPEG-4.

As a high memory access bandwidth is mainly responsible for high power consumption in various motion estimation algorithms [Nacht 98], we aimed to minimize memory bandwidth to on-chip buffer memories. An additional aim was to take into account a low number of memory modules (or memory ports) in order to reduce the area overhead for multiple address generation units (AGUs), as well as address decoders and row-and-column-drivers within the RAM modules. Reduction of the area demands of the AGUs is regarded to be as important as the support of several ME algorithms and multiple memory modules, since the AGU size reaches a significant share of the total area.

The aim of the presented VLSI architecture was therefore, to gain high efficiency at low memory bandwidth requirements for the computationally demanding algorithms, as well as the support of several motion estimation algorithmic features with less additional area overhead. Possible architectural solutions were discussed and evaluated with an Area-Time (=clock cycles)-Bandwidth (ATB) metric. Based on these results, a VLSI-architecture is derived, which is targeted for mobile MPEG-4 video compression.

## 6.6.3 Requirements

### 6.6.3.1 Target throughput

Based on the requirements of profiles and levels currently under development within the standardization of MPEG-4 visual [MPEG], the Core Profile at Level 1 and 2 is anticipated to be supported by this architecture with high quality, using the full-search ME algorithm at different search ranges. However, fast motion estimation algorithms which are also supported by this architecture, allow higher throughput or lower computational load and reduced memory bandwidth requirements, resulting in reduced power consumption in exchange for a slight loss of visual quality of the encoded visual sequence.

The target throughput for the MPEG-4 profiles under discussion is:

- Core Profile @ Level 1 (2 QCIF, 30 fps): full-search [-16, +15]
- Core Profile @ Level 2 (2 CIF, 30 fps): full-search [-8, +7], fast search [-16, +15]
- Simple Profile @ Level 2: CIF, 30 fps [-16, 15] full-search
- Simple Profile @ Level 1: QCIF, 30 fps [-16, 15], full-search

#### 6.6.3.2 Design constraints

The basic goal for the motion estimation algorithm selection was a) to support a motion estimation mode with very high quality, b) to support motion estimation with high quality, and c) one or two modes with drastically reduced computational complexity and memory bandwidth demands.

***Very high quality mode***

The very high quality mode resulted in the requirement of supporting full-search with [-16, +15] pel search area, half-pel motion estimation, R/D-optimized motion estimation and advanced prediction mode.

***High quality mode***

For a high quality mode full-search is still required with reduced search area and optional pel subsampling. Half-pel motion estimation, advanced prediction and R/D-optimized motion estimation can be used optionally.

***Low-power and low-complexity mode***

For the low-power and low-complexity mode an algorithm with the quality and computational complexity in the range of the three step search is to be used. The options half-pel, advanced prediction mode, or pel subsampling may be used optionally.

***Architectural design constraints***

Apart from the algorithmic design constraints, there are also some architectural design constraints. The design constraints for this architecture are identified from the target application:

- Low memory bandwidth to reduce the power consumption
- Low VLSI area size to reduce the VLSI costs
- Processing time to meet the real-time constraints

These design parameters have already been discussed in section 6.2 and resulted in the *ATB* metric.

### 6.6.3.3 Discussion on algorithmic features to be supported

***Integral projection-matching***

From the visual quality evaluation in chapter 5 integral projection-matching (IPM) seemed to be a good candidate for fast motion estimation. However, as explained in section 6.4.2.5, IPM requires a memory size too high to save the immediate row (or column) sums. Therefore, IPM was not investigated further for this VLSI architectural concept.

***Supported algorithmic features***

Based on extensive algorithmic and VLSI implementation studies on numerous fast motion estimation algorithms within the previous chapters, the following base features are required to be implemented by this flexible motion estimation architecture:

- alpha-plane (MPEG-4)
- full-search motion estimation: [-8, +7], [-16, +15] pel
- advanced prediction mode: full-search motion estimation [-2, +2] pel around predictor with 16x16 and 8x8 MV result
- pel subsampling: 2:1, 4:1, 4:1 alternate pel subsampling
- half-pel motion estimation
- Three Step Search (TSS)

### 6.6.3.4 Flexibility

To support a wide range of fast motion estimation algorithms, increased flexibility and programmability of the VLSI architecture are required, compared to the algorithm-specific architecture presented above. However, increased flexibility has usually to be paid for with a loss of efficiency. Therefore, for the VLSI architecture developed within this chapter, besides a processor enhancement of a single SAD instruction, a set of hardwired primitives of several commonly used motion estimation techniques is provided to support high efficiency and high throughput at low-power consumption. Comparing an 1D and 2D array architecture, it can be noted that besides the lower VLSI area requirements the 1D architecture offers benefits in terms of flexibility. However, the 2D architecture offers higher throughput.

For reasons of low die size and flexibility, a 16x1 ME architecture was chosen. However, at this step of the design flow it was not clear, if a 1D architecture was able to deliver the required throughput. Therefore, in the subsequent section a performance evaluation of the 1D array architecture is performed.

### 6.6.3.5 Performance Evaluation

In this section the 16x1 ME-architecture is investigated in terms of throughput for real-time applications under worst case conditions. From the above algorithms the full-search algorithm requires the highest memory bandwidth and computational demands and is therefore investigated in detail.

***Cycle time estimation for the full-search algorithm without data flow design***

The theoretically most favorable scenario of a 16x1 SAD-array (Fig. 6.7) without any special data flow design is assumed. I. e. it is assumed, that every processor element (PE) accesses two bytes (1 byte cur. and prev. each) from memory (conflict free) within a single clock cycle. The memory bandwidth per clock-cycle can be calculated as:

$$Mem_{bandwidth} = 2 \cdot PE_{count} \cdot 1byte \qquad (6.58)$$

with $PE_{count}$ being the number of PEs employed. The arithmetic operations per second for exhaustive motion estimation can be calculated as:

$$Op = 3 \cdot 2p \cdot 2p \cdot N_h \cdot N_v \cdot f \qquad (6.59)$$

with *3* representing the three operations per SAD (SUB, ADD, ABS) and *f* being the number of frames per second (fps). The $PE_{cycletime}$ is defined as:

$$PE_{cycletime} = \left(\frac{Op}{stages \cdot PE_{count}}\right)^{-1} \qquad (6.60)$$

with *stages* being the pipeline stages per PE. Depending on the VLSI technology, *stages* is equal to three or two. A value of *stages*=3 is assumed for the subsequent calculations.

***MB processing time with data flow design***

As the reduction of memory bandwidth is crucial in terms of power consumption, this work tries to limit the bandwidth to external memory and to on-chip current and search area RAMs. The other goal was to limit the number of memory modules, as multiple small memory modules (as shown in 6.3.4) require more VLSI area as a single memory module with the same size in bytes. Therefore, this scenario takes data flow design (e.g. Fig. 6.29) into account. 3 bytes per clock-cycle (2 bytes for prev. and 1 byte for cur) are required in case of 16 PEs. The maximum memory bandwidth per clock cycle for 16 PEs can be calculated as:

$$Mem_{bandwidth} = 3 \cdot 1byte \qquad (6.61)$$

The macro-block cycle-time, $MB_{cycletime}$, is calculated here as defined in eq. (6.3). The advantage of the data flow design is a memory bandwidth reduction of a factor of about 9. However, a higher fanout for prev, prev' and cur is required to drive a maximum of 16 MUX (to direct the data from memory to the appropriate PEs), as well as a special addressing scheme for the search area memory and the current memory.

| Full-search motion estimation Search Area | Array Size | $N_{cycles}$ clock cycles | Memory Access search area (2 RAM banks) | Core @ L1 (2 QCIF 30 fps) | *for comparison* Core @ L1: 16 PEs, no data-flow design |
|---|---|---|---|---|---|
| [-8, 7] | 16 x 1 | (16 * 16 *1 6) + 15 = 4,111 | 7,936 Bytes/MB | 48 MByte/s | 389 Mbyte/s |
| [-16, 15] | 16 x 1 | (32 * 32 * 16) + 15 = 16,399 | 31,744 Bytes/MB | 194 Mbyte/s | 1,557 Mbyte/s |

*Figure 6.49: Worst case performance evaluation: Full-search motion estimation algorithm*

***MPEG-4 Profiles and Levels: Performance Requirements***

For a configuration of 16x1 PEs and a number of 3 pipeline stages per PE, the PE cycle time is calculated for several levels and profiles of MPEG-4 (as they are currently under discussion):

| Visual Profile | Level | Visual session size | fps [Hz] | Max. total reference memory (MB units) | Max. number of MB/sec | $MB_{time}$ [usec] | Full-search, [-16, 15] $N_{cycles}$: 16,399 | | | Full-search, [-8, 7] $N_{cycles}$: 4,111 | | |
|---|---|---|---|---|---|---|---|---|---|---|---|---|
| | | | | | | | Arith GOPS | PE cycle-time [ns] | MB cycle-time [ns] | Arith. GOPS | PE cycle-time [ns] | MB cycle-time [ns] |
| Main | L3 | CCIR 601 | 30 | 3,240 (2 CCIR601) | 194,400 progres. | 5.1 | 76.44 | (0.62) | (0 31) | 19 11 | (2 51) | (1 2) |
| | L2 | CIF | 30 | 792 (2 CIF) | 23,760 | 42 | 18.86 | (2.54) | (2.56) | 4.67 | 10.27 | 10.2 |
| Core | L2 | CIF | 30 | 792 (2 CIF) | 23,760 | 42 | 18.86 | (2.54) | (2.56) | 4.67 | 10.27 | 10.2 |
| | L1 | QCIF | 30 | 198 (2 QCIF) | 5,940 | 168 | 4.671 | 10.27 | 10.2 | 1.16 | 41.37 | 40 8 |
| Simple | L2 | CIF | 30 | 396 (CIF) | 11,880 | 84 | 9.342 | 5 138 | 5 1 | 2 33 | 20.60 | 20.4 |
| | L1 | QCIF | 15 | 99 (QCIF) | 1,485 | 673 | 1.167 | 41.13 | 41.0 | 0 291 | 164 | 163 |

*Figure 6.50: Performance requirements of several MPEG-4 Profiles and Levels*

### 6.6.3.6 Design decisions

***1D- or 2D-array architecture***

Based on these results the throughput is considered to be sufficient for a 16 x 1D PE array. Therefore, it was decided to implement a 1D motion estimation array architecture using a special data flow design to reduce the bandwidth to the local memory.

***LA-, GA- or SAD-array architecture***

Finally a SAD array architecture was chosen as architectural basis for futher investigations for the other algorithms, as for this architecture the full-search algorithm which depicts the highest computational complexity, can be implemented very efficiently [Yang 89].

***Advanced prediction mode***

The advanced prediction mode would have been easier to implement using a GA architecture. However, with the advantages of the selected basis architecture as depicted above, it was decided to implement the advanced prediction mode using the SAD array architecture.

***SAD reduction to prefer the zero motion vector***

As discussed earlier in section 6.4.2.11, SAD reduction for the (0,0) MV is considered to be implementable both for the LA- and GA- architecture.

***Low complexity distance criteria***

The SAD distance criterion was required for the high quality motion estimation mode. The support of other (simpler) additional distance criteria was considered to be too costly in terms of VLSI area, as additional multiplexers would have also been necessary to switch between the two distance criteria.

## 6.7 SUMMARY

In this chapter design metrics were defined to evaluate motion estimation architectures. The design space for motion estimation VLSI architectures was defined, based upon several basic architectural principles. In the subsequent section, motion estimation VLSI architecture implementations for a set of commonly used motion estimation algorithms were discussed.

Based on these results, the basic decisions for a new 2D PE array full-search motion estimation architecture, supporting variable blocksize motion estimation and luminance correction (Search Engine I) were made.

These results also led to the basic design decisions of an 1D PE array architecture (Search Engine II) also supporting, apart from full-search motion estimation, several fast motion estimation algorithms, as well as R/D-optimized motion estimation, half-pel motion estimation, and advanced-prediction motion estimation with high efficiency.

# 7

# VLSI IMPLEMENTATION: SEARCH ENGINE I (2D ARRAY)

## 7.1 INTRODUCTION

This chapter describes the VLSI implementation of a flexible 100 GOPS (Giga Operations Per Second) exhaustive search segment-matching architecture to support evolving motion estimation algorithms, as well as block-matching algorithms of established video coding standards.

The architecture is based on a 32x32 GA array (cf. 6.3.3.2), a VBSME adder-tree (cf. 6.4.2.8) and a 10,240 byte on-chip search area RAM. This architecture allows concurrent calculation of motion vectors for 32x32, 16x16, 8x8 and 4x4 blocks and partial quadtrees (called segments) for a +/-32 pel search range with 100% PE utilization. This architecture supports object-based algorithms (MPEG-4) by excluding pixels outside of video objects from the segment-matching process, as well as advanced algorithms like variable block size segment-matching with luminance correction. The VLSI has been designed using VHDL synthesis and a 0,35 μm CMOS-technology and has a clock-rate of 100 Mhz (min.), allowing the processing of 23,668 32x32 blocks per second with a maximum of +/- 32 pel search area.

## 7.2 ALGORITHM ARCHITECTURE MAPPING

### 7.2.1 Algorithms for feasibility evaluation

The support of MPEG-4 ME using the GA-2D architecture was already described within section 6.3.3.2. In this chapter the algorithm architecture mapping of the iterative, entropy constrained variable block size motion estimation algorithm is described.

The original algorithm [Cal 97] described in section 2.5.5, employs the MSE (Mean Square Error) for distortion calculation. However, to enable a fair evaluation of the algorithm architecture mapping the "original" results and the results based on the new architecture were gained using the SAD-criterion. The PSNR difference to the original, resulting from the use of the SAD criterion instead of the MSE criterion was found to be negligible. For the simulation results within this chapter, the first rectangular

VOP was intra coded and all of the following frames were coded by prediction only (IPPPPP...), using the "original" algorithm or this architecture. The results were obtained with the MPEG-4 QCIF sequence "container" and a search area of +/- 16 pel for Y and +/-8 pel for UV. No OBMC was employed and MPEG-4 video VM VLC (variable length code) tables were used.

### 7.2.2 Parallelization of the Iterative Algorithm

The parallelization of the iterative algorithm is described as follows. Fig. 7.1 depicts a simplified overview of the iteratively executed loops of the original algorithm. For every block size level (Fig. 7.1), as well as for each of the 16 segment shapes and for every search area step the luminance correction coefficient q has to be calculated, as well as the prospective bit rate, together with the distortion and the cost for Lagrange optimization. In case the calculated cost of a luminance corrected segment at a specific level is lower than for a previous segment at this level of the search area, the cost, shape, MV and q are saved. Since this algorithm works iteratively, a straightforward VLSI implementation seemed to be impossible, as there would be a very high loss of efficiency.

```
for (level=0; level<4; level++)
  for (shape=0; shape<16; shape++)
    for (dy=0; dy<2y+1; dy++)
      for (dx=0; dx<2x+1; dx++){
          q=calc_q(segment_cur,segment_ref);
          rate=calc_rate(MV,q,shape);
          err=SAD(segment_cur,segment_ref,q);
          cost=err+lambda*rate;
          if(cost < best_cost){
              best_cost = cost;
              best_MV = MV;
              best_q = q;
          }
      }
```

*Figure 7.1: Iterative inner loops of the algorithm (simplified)*

One of the main difficulties of mapping this iterative algorithm to a VLSI architecture was that the luminance coefficient q had to be calculated for a segment ("L"-shape, Fig. 2.26) of a 32x32 block <u>before</u> the SAD of this subblock cluster could be calculated to find the optimum MV for this segment.

Without q correction, the distortions calculated on every block size level could be simply summarized to gain the distortions for higher block size levels. For q corrected blocks, reuse of already calculated distortions is only possible for blocks with equal q coefficients. Therefore, and to achieve efficient parallelization and compatibility to existing block-matching architectures, the calculation of q had to be moved from the inner loop (Fig. 7.1) to a loop outside of the SAD calculation (Fig. 7.2).

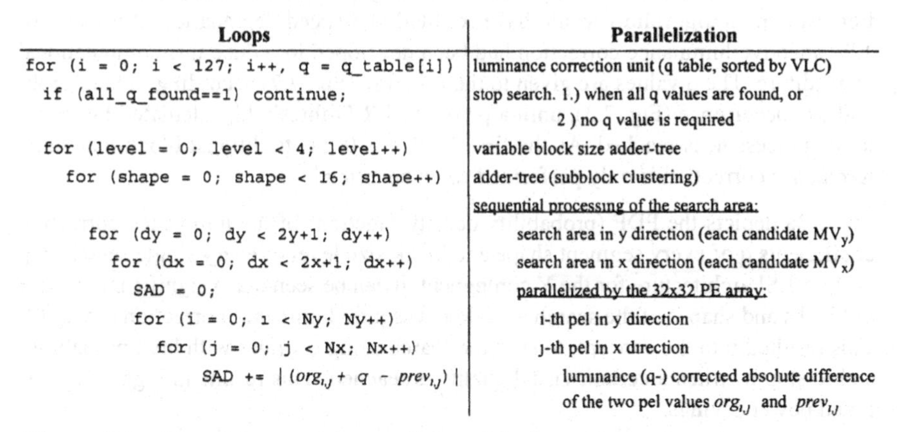

| Loops | Parallelization |
|---|---|
| `for (i = 0; i < 127; i++, q = q_table[i])` | luminance correction unit (q table, sorted by VLC) |
| `if (all_q_found==1) continue;` | stop search. 1.) when all q values are found, or 2 ) no q value is required |
| `for (level = 0; level < 4; level++)` | variable block size adder-tree |
| `for (shape = 0; shape < 16; shape++)` | adder-tree (subblock clustering) |
| | <u>sequential processing of the search area:</u> |
| `for (dy = 0; dy < 2y+1; dy++)` | search area in y direction (each candidate $MV_y$) |
| `for (dx = 0; dx < 2x+1; dx++)` | search area in x direction (each candidate $MV_x$) |
| `SAD = 0;` | <u>parallelized by the 32x32 PE array:</u> |
| `for (i = 0; i < Ny; Ny++)` | i-th pel in y direction |
| `for (j = 0; j < Nx; Nx++)` | j-th pel in x direction |
| `SAD +=` $\mid (org_{i,j} + q - prev_{i,j}) \mid$ | luminance (q-) corrected absolute difference of the two pel values $org_{i,j}$ and $prev_{i,j}$ |

*Figure 7.2: Nested loops of the parallelized block-matching algorithm*

However, this parallelization could only be achieved at the expense of performing iterations on different q values, but this solution finally proved to be effective (Fig. 7.4). The basic idea behind this new architecture (Fig. 7.7) is to give a (preselected) q value to the SAD calculation (PE array), calculate all SADs (of different segment shapes at different levels) with this given q value in parallel during a single clock cycle, and also to calculate the q' values from the mean values of *prev* and *cur*, for every single segment at every block size level during the same clock cycle. Only the SAD and MV results are saved, when the given q value and the calculated q' value are equal.

## 7.2.3 Luminance coefficient q-iteration

The q values are multiples of 4 within the range of [-252, 252] and their probabilistic occurrence is reflected by a VLC table with 128 entries.

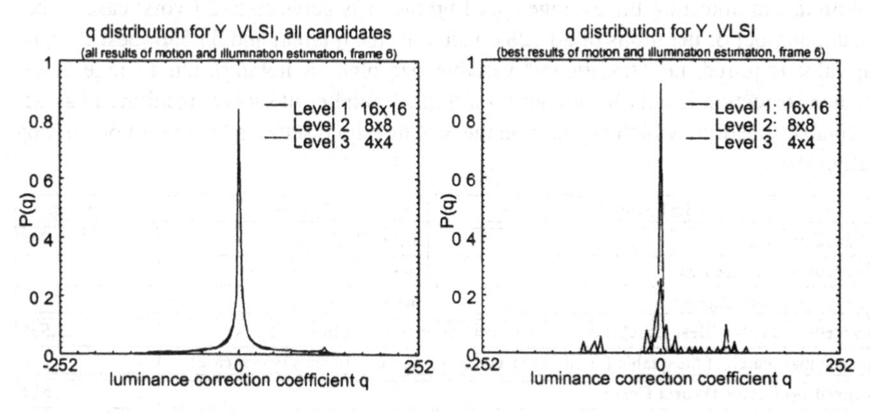


*Figure 7.3: PDF of the luminance correction coefficient q: a) original algorithm, b) VLSI result*

As an exhaustive search on all q values (i.e. adding a third search dimension to the motion estimation algorithm) would have resulted in a speed degradation of a factor of 128, adaptive luminance correction had been developed to reduce this computational expenditure. The q values are given to the PE array (Fig. 7.7) according to their probabilistic occurrence (Fig. 7.3), until a predefined R/D-threshold, calculated by an external processor, is reached. As the first q value is 0, i.e. traditional block-matching, luminance correction is only performed when beneficial.

Fig. 7.3a depicts the PDF (probability density function) of the luminance correction coefficients q of every segment shape and block size level of every single search step of the VLSI architecture for the Y component. It can be seen that a significant number of blocks and shapes of the search area do not require luminance correction (i.e. q=0). This resulted into the concept, to start the search using q values with high probability and to stop it, when a certain visual quality is reached. This results in high computational power savings.

With this 32x32 architecture, the best 295 results of MV, SAD, and q for every segment shape at every block size level (cf. section 7.4.1), are found within the full-search motion vector area, where some results are reusable in case the q value is constant for the whole PE array. If two MV candidates of the search area for the same segment shape at the same level show an equal SAD value, the smaller MV is saved as the result. Lagrange optimization is performed here on the best 295 MVs by a programmable external processor, taking preset R/D constraints into account. However, compared to the "original" algorithm, quadtree partitioning is performed here using a suboptimal (but parallelizable) method. Fig. 7.3b shows the distribution of q values for the best blocks of every block size level as determined by the VLSI architecture before Lagrange optimization.

## 7.2.4 Speed-up of the proposed architecture

With this architecture the average speed-up factor is between 512 (worst case, maximum number of iterations of all 128 q values of the q table) and 65,536 (best case, no q value required, i.e. "traditional" variable size block-matching). On average, luminance correction is only beneficial for a limited number of blocks, resulting in an acceptable speed-up, which depends on the R/D threshold settings of the post processing algorithm.

| VLSI Module | Function | Speed-up |
|---|---|---|
| 32x32 PE-array | SAD calculation | 1,024 |
| variable block size tree | levels | * 4 |
| subblock clustering tree | shapes | * 16 |
| **Speed-up Factor (Best Case)** (no luminance correction required for this block) | | **= 65,536** |
| sequential search of the q table for all luminance correction coefficients (worst case) | | / 128 |
| **Speed-up Factor (Worst Case)** | | **= 512** |

*Figure 7.4: Speed-up for the iterative algorithm using this architecture*

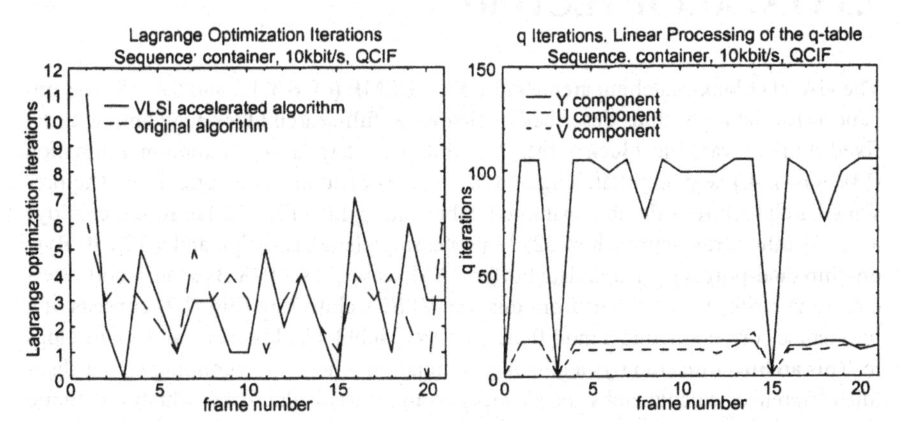


*Figure 7.5: a) Lagrange optimization iterations, b) number of q iterations for VLSI*

Fig. 7.5a depicts the number of Lagrange optimization iterations employing the bisection search *[Press 92]*. Fig. 7.5b depicts the number of q iterations necessary for linear full processing (no R/D threshold setting) of the q table. Note that the probability of larger |q| values is very small (Fig. 7.3). This property can be used to stop the q table iterations with reaching a predefined R/D-threshold. However, the number of Lagrange optimization iterations and q iterations depends on the sequence content and the rate control algorithm.

## 7.2.5 Coding results

Fig. 7.6 depicts the PSNR of the Y component and bit rate of the "original" and the VLSI-accelerated algorithm, meeting the conditions described before. Note that the parallelization (which is, however, suboptimal) of the iterative algorithm leads to an average PSNR performance degradation of only 0.95 dB, at the same bit rate of this sequence.

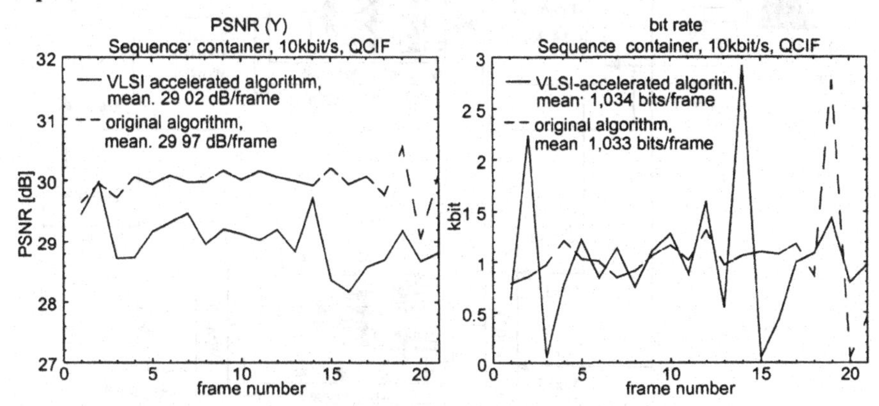


*Figure 7.6: Results: a) PSNR (Y) and b) bit rate of the "original" and the VLSI-accelerated algorithm*

# 7.3 VLSI ARCHITECTURE

The GA-2D block-matching architecture for VBSME (cf. 6.3.3.2 and 6.4.2.8) was enhanced for the algorithms described in chapter 2: full-search block-matching with 1.) fixed and 2.) variable block sizes, 3.) arbitrarily-shaped object motion estimation (MPEG-4), 4.) segment-matching and 5.) support of luminance correction. The presented architecture is flexible, parametrizable and highly efficient, because it employs a 2D systolic array approach of 32x32 processing elements (PEs), and a 10,240 Byte on-chip dual-port search area data buffer RAM, as well as 2,048 Byte dual-port reference data buffer RAM. Apart from this, the VLSI architecture (Fig. 7.7) consists of a luminance correction unit, a modified adder-tree, subblock clustering, and control logic. This architecture employs a global accumulation scheme for summing up the absolute differences calculated by the PE array in an external adder-tree, which was found to be advantageous for these four extensions of the underlying architecture.

## 7.3.1 Luminance Correction

As full processing of all q values would have resulted in a considerable speed degradation, adaptive luminance correction was developed to reduce this expenditure. The q values are stored in a table (q table, Fig. 7.7) and sorted according to the VLC table.

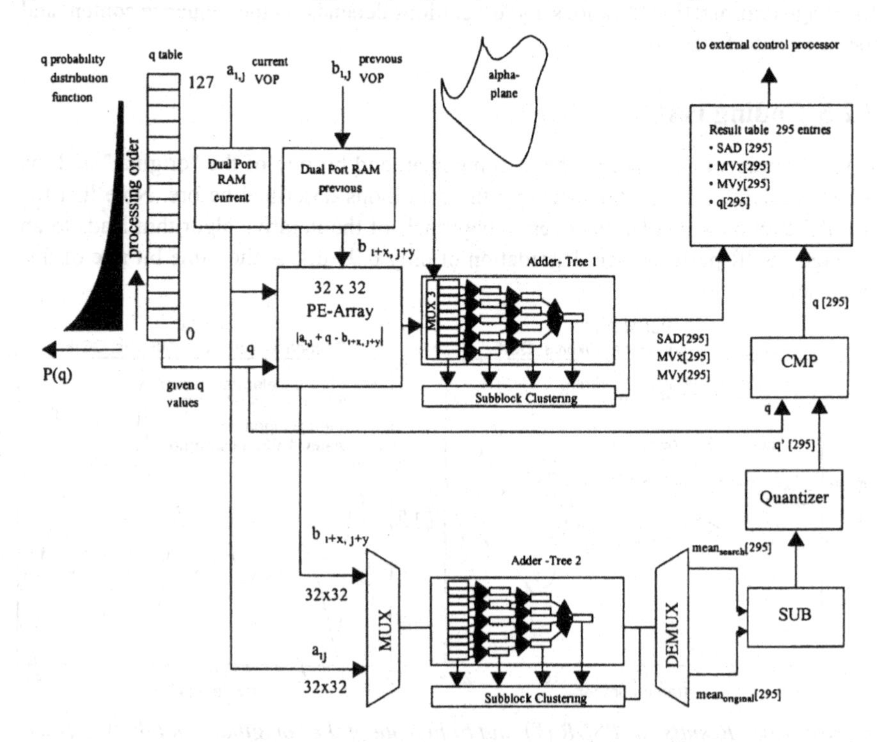


*Figure 7.7: VLSI architecture for luminance-corrected variable block size ME*

After a full-search with q=0 ("traditional" block-matching without luminance correction), the SAD results are compared with a threshold in an external processor and the processor decides, whether a further search with a different luminance correction coefficient is required. In case of a further search with a different luminance correction coefficient, the q coefficient with the highest probability (according to the VLC table) is passed on as the first q value to the PE array, and the full-search procedure is started for this q coefficient. This procedure is iterated until a) the q table is fully processed, or b) the external processor decides to stop the search if the results are within the given limits.

With every shift in x- or y- direction of the search area, adder-tree 2 generates a q' value (q'[295]) for every of the 295 subblock clusters of the currently processed 32x32 search window. These q'[295] values are calculated from the quantized (module "Quantizer") differences (module "SUB") of the mean values of the subblock clusters of the search area of the previous frame ($mean_{search}$[295]) and of the mean values of the subblock clusters of the original frame ($mean_{original}$[295]). Whenever one of these q'[295] values matches (module "CMP") the current q value of the q table, the corresponding SAD, $MV_x$ and $MV_y$ are passed to the result table. This procedure is repeated with several q coefficients until the SAD values are within a predefined threshold range. Besides linear processing of the q values according to the VLC table, other search schemes could be applied for faster convergence.

With this adaptive architecture, luminance correction is only performed where necessary (which is for a few blocks) and the average computation overhead is kept low. As the $mean_{original}$[295] values are calculated only once at the beginning of the full-search process, only one single adder-tree 2 is used, which can be multiplexed between $a_{ij}$ and $b_{i+x, j+y}$. This makes a third full adder-tree, with an additional overhead of only three clock cycles for each macro-block unnecessary.

# 7.4 PROCESSOR ELEMENT ARRAY

The PE array consists of 1,024 PEs. A single PE is depicted in Fig. 7.8a.

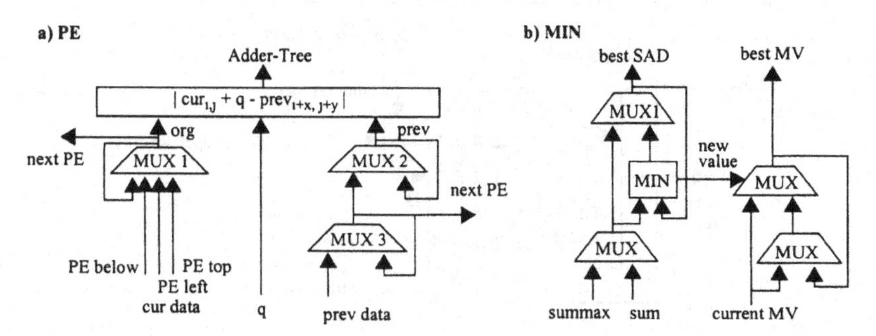


*Figure 7.8: a) Single processor element and b) MIN element (comparator / register)*

The PEs calculate the absolute difference of every single pel of a 32x32 block (and 16x16, 8x8, 4x4 subblocks) of the current and previous VOP during a single clock cycle, taking into account the luminance coefficient q. The PEs are connected to each neighborhood PE and are able to shift data to the PEs on the left or on the right side, as well as to those positioned above or below. However, every PE shifts the data in the same direction. This configuration allows to shift the previous VOP data in a "maeander"-like data flow through the PE array, Fig. 7.9, similarily to [Vos 95].

This local communication scheme significantly reduces the memory bandwidth of this architecture. As each PE sums up the same q value, q can be broadcast to all PE elements and the additional amount of chip size to support luminance correction is very small for the PE array. The absolute pixel differences are summed up and are compared with the previous results in the adder-tree. The data of the current VOP is moved in a "zig-zag" data flow through the PEs.

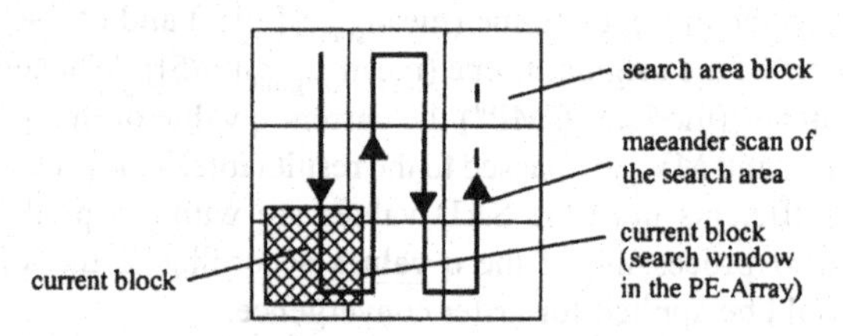


*Figure 7.9: Meander scan of the search area*

Fig. 7.10 depicts the PE array and the memory access scheme. Port A delivers the search area data from the on-chip dual-port RAM, when the search window moves up, and port C delivers the data, when it moves down. To avoid memory access of a second address dimension for the column data, the pel values of the next row (33 th byte) are stored in a shift register and are handed over to the PE array in case of a search window movement to the right. However, 100% PE utilization can be gained only for search ranges above 33 pel in horizontal direction.

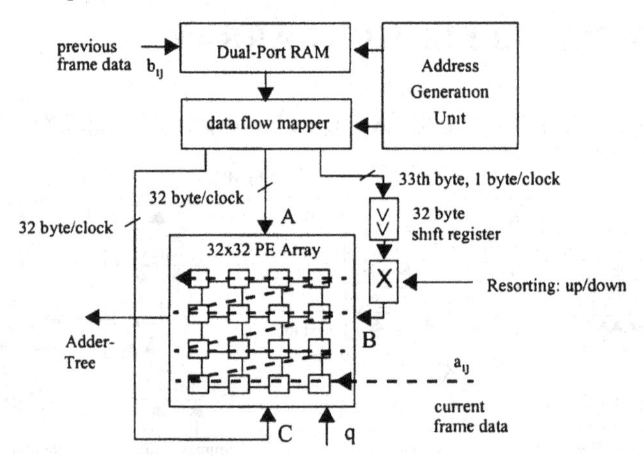


*Figure 7.10: 32x32 PE array and search area memory access*

## 7.4.1 Modified Adder-Tree

In this section the support of the different algorithmic features by the adder-tree are discussed.

***Arbitrary-Shaped Objects (MPEG-4)***

Support of arbitrarily-shaped objects is provided by an extra multiplexer array (MUX 3) input to the adder-tree (cf. Fig. 6.11), in which, with an alpha-plane value of 0, cf. eq. (2.9), the absolute difference of the corresponding pel is excluded from the accumulation process of the adder-tree.

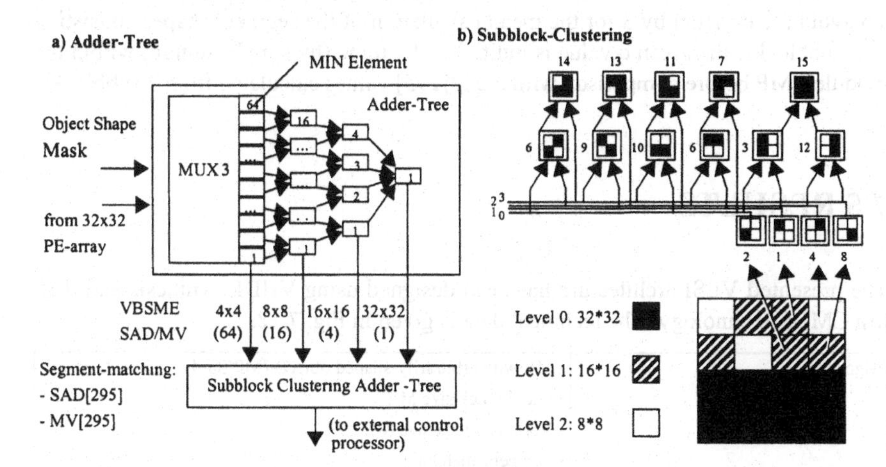


*Figure 7.11: a) Adder-tree for MPEG-4, VBSME and segment-matching, b) segment-matching*

***VBSME***

The adder-tree (depicted in Fig. 7.11) sums up the current SADs of the small (4x4, 8x8, etc.) blocks to obtain the SADs of the larger blocks or subblock clusters. For VBSME e.g., one $SAD_{8x8}$ is calculated from four $SAD_{4x4}$ results. This results in a total of 85 SAD, MV result values for the VBSME. Every single block size level consists of MIN elements (i.e. comparator/register elements) to save the best SADs and MVs of the whole search area, resulting in an additional cost of only 84 comparator/adder elements to support VBSME. By additional control circuitry this architecture allows a separate predefinition of the MV range for every single block size.

***Segment-Matching***

For segment-matching, two or three of the subblocks of a 32x32, 16x16 or 8x8 block for the luminance component (Y), or of a 16x16, 8x8 or 4x4 block for the chrominance components (U,V) are clustered (Fig. 2.27, Fig. 7.11) and described more efficiently by a single MV and q value. For this purpose, all MV, SAD and q values of the best matching blocks of the 15 (14 for level 2) possible clusters on each level are saved in the subblock clustering adder-tree. Fig. 7.11 depicts the subblock clustering of a single level: already calculated SAD results are reused for subsequent SAD calculation. This leads to a total of (16*14 clusters (8x8 blocks) + 4*14 clusters (16x16) + 1*15 clusters (32x32)=) 295 subblock cluster results for Y in the 32x32 block, with only the best subblock cluster results (SAD, MV and q) of the search area requiring to be saved in comparator/register elements for the 32x32 PE array. Support of segment-matching requires 295 extra MIN elements and the accompanying adders.

An external control processor selects from these data the best MV-, q- and subblock cluster-combination of the three levels using Lagrange optimization, by taking into account the distortion (SAD) and the coding costs (external lookup-table) of the MV and q of different block sizes.

To avoid the division by 3 for the mean calculation of the segment shapes consisting of 3 subblocks, the given q value is multiplied by three (by simply adding) within the module CMP before comparison with the q'[295] values calculated from 3 subblocks.

# 7.5 RESULTS

The presented VLSI architecture has been designed using VHDL synthesis and 0,35 μm CMOS-technology. The technical data is given in Fig. 7.12.

| Algorithms | • ME with arbitrarily-shaped objects (MPEG-4)<br>• Fixed block size ME<br>• Variable block size ME<br>• Segment-matching<br>• Illumination correction |
|---|---|
| Block sizes | 4x4, 8x8, 16x16, 32x32 |
| Processing power | 100 GOPS @ 100 Mhz |
| Max. throughput | 23,668 32x32 blocks/second at +/-32 pel search range |
| Clock frequency | 100 Mhz (min.) |
| Library | 0.35 μm Compass Passport |
| Design style | Standard cell synthesis, VHDL |
| On-chip memory | 10,240 byte search area buffer<br>2,048 byte reference block buffer |

*Figure 7.12: Features of the 32x32 PE segment-matching VLSI*

The VLSI was described using VHDL and initially synthesized as a 16x16 PE array with a 0.5 μm CMOS library, for which the results were given in [Kuhn 97a]. The VHDL synthesis approach allowed a migration to a 0.35 μm CMOS library with acceptable effort, enabling also several architectural optimizations which lead to a reduced chip size. Interleaved RAM access (cf. 6.3.4), which was required with the 0.5 μm technology due to speed constraints, could be dropped leading to a more compact design. The 4x4 SAD calculations are now included within the PE array, and the adder-tree is now built as a Wallace adder-tree leading to a more compact design [Kuhn 97c].

| **Module** | **Size without luminance correction** [mm$^2$] | **Size with luminance correction** [mm$^2$] |
|---|---|---|
| 32x32 PE-Array | 25.455 | 37.495 |
| Adder-Tree with Subblock Clustering | 0.222 | 2 x 0.237 |
| Pipelining between Adder-Tree levels (3168 FFs) | 1.341 | 2 x 1.434 |
| Dual-Port RAM Search Area (10 blocks of 32 x 256 Bit = 10240 Byte) | 10.512 | 10.512 |
| Dual-Port RAM Reference Block (16 blocks of 32 x 32 Bit = 2048 Byte) | 2.780 | 2.780 |
| **Sum** (without wiring area) | 40.310 | 54.129 |

*Figure 7.13: Chipsize area: 32x32 PE segment-matching VLSI* ***without*** *and* ***with*** *support of luminance correction (0.35 μm, 100 Mhz min.)*

VLSI implementation results are depicted in Fig. 7.13 for a clock-rate of 100 Mhz (min.) and the 0.35 μm Compass Passport library. With control logic, I/O and wiring area, the full chip size is expected to be slightly above 100 mm$^2$. As a result, a 32x32 PE array seems to be possible on a single chip, using the VHDL synthesis approach.

# 7.6 COMPARISON OF THE RESULTS WITH OTHER IMPLEMENTATIONS

[Berns 96] described a VBSME architecture for HDTV applications. [Hans 96] described a programmable SIMD processor for VBSME requiring 12 of these PEs. These architectures were described in section 6.4.3.2.

The main differences of the presented architecture to the architectures mentioned above, are the extension of the adder-tree (which sums up the absolute differences calculated by the PE array) to support variable block sizes, as well as segment-matching and the inclusion of a luminance correction unit. And, however, the support of MPEG-4 motion estimation. A VHDL synthesis approach was used for this design, contrary to the full custom design methodology used by [Ber 96], which offered the advantage of fast migration to newer VLSI technology libraries.

| | Universite Catholique de Louvain, Belgium 1996 | RWTH Aachen, Germany 1997 | Search Engine I TU Munich, Germany 1997 |
|---|---|---|---|
| Search strategy | programmable | full-search | full-search |
| Number of PEs | 12 | 32 x 32 | 32x32 |
| Clock Frequency | 50 Mhz | 200 Mhz (worst case) | 100 Mhz (min.) |
| Area | 41 mm$^2$ per PE | 220 mm$^2$ | 54.1 mm$^2$ (incl RAM) |
| Design Style | full custom | full custom | VHDL synthesis |
| Process | 1 μm | 0.5 μm, 2LM | 0.35 μm Compass |
| Processing Power | n.a. | 200 GOPS @ 200 Mhz | 100 GOPS @ 100 Mhz |
| Power Consumption | n.a. | 16W @ 200 Mhz est. | n.a. |
| On-Chip Memory | 512 Bytes | no | 12,288 Bytes |
| MV search area | programmable | +/-15 pel | [-16, +15], [-32, +31] |
| MPEG-4: alpha-plane | no | no | yes |
| Variable block size (n) | 16, 8, 4 (prog.) | 32, 16, 8 | 32, 16, 8, 4 |
| Segment-matching | no | no | yes |
| Luminance correction | no | no | yes |
| References | [Hans 96] | [Ber 96], [Ber 96a], [Ber 97] | chapter 7 of this book [Kuhn 97a-e] |

*Figure 7.14: Motion estimation VLSI designs supporting VBSME (n.a.: data not available)*

# 7.7 PROCESSOR FLEXIBLE ASIC COMPARISON USING THE ATB METRIC

The computational complexity of an unoptimized software implementation of the reference algorithm [Cal 97] using the MSE distance criterion, is given for the sequence "container", 10kbit/s in Fig. 7.15. Only the luminance corrected variable block size block-matching process is taken into account and not the Lagrange optimization, bit stream generation, Intra-frame coding, etc.

| **Memory Bandwidth** MByte / s | | | **Mem. Access** | **Arithmetic** MIPS | | | | | | **Control** | **Others** | **Total** |
|---|---|---|---|---|---|---|---|---|---|---|---|---|
| Read | Write | All | MIPS | MUL | ADD | SUB | SHIFT | DIV | Total | MIPS | MIPS | MIPS |
| 55,836 (78.1 %) | 15,643 (21.8 %) | 71,480 (100%) | 23,511 (29.9 %) | 1,900 (2.42 %) | 11,432 (14.5 %) | 665 (0.84 %) | 1,389 (1.77 %) | 113 (0.14 %) | 21,726 (27.6 %) | 25,019 (31.8 %) | 5,857 (7.46 %) | 78,469 (100%) |

*Figure 7.15: Computational complexity of the software implementation of the reference algorithm*

For the comparison of the processor implementation (software) and the flexible VLSI implementation (hardware) of the different motion estimation algorithms, the Area-Time-Bandwidth metric (ATB) as defined in 6.2.3.2 is used. The data for the processor based implementation was gained using the iprof tool (c.f. chapter 3), where a single clock cycle is approximated by one RISC instruction. For the Ultra-SPARC processor a silicon area of 149 $mm^2$ (0.35 μm CMOS technology) is used for this comparison [SPARC]. For these model calculations, the search area data and the previous block data are regarded to be accessed in the processor's on-chip cache.

To process one single MB requires $33 \cdot 33$ clock cycles. QCIF contains 99 MBs per frame, and for 7 frames (corresponding to 1 second here) the motion estimation is performed. The clock cycles to process all MBs of one second of video is calculated for VLSI implementation with a [-16, +16] search-area:

$$clocks = 33 \cdot 33 \cdot 99 \cdot 7 = 754,677 \qquad (6.60)$$

With every clock cycle 33 bytes are loaded from the search area RAM into the PE array, and after every $33 \cdot 33$ clock cycles, the current block data ($32 \cdot 33$ pel) is loaded from the current-block RAM into the PE-array. With every single of the 32 shifts of the search window to the right no data is loaded from the previous-RAM into the PE-array, therefore $32 \cdot 33$ has to be substracted.

$$\text{Memory bandwidth} = ((33 \cdot 33 \cdot 33 - 32 \cdot 33) + 32 \cdot 32) \cdot 99 \cdot 7 = 24,882,165 \quad \text{Byte/s} \qquad (6.61)$$

For the worst-case situation, these results have to be multiplied with the maximum number of q-iterations (128). The Area-Time-Bandwidth metric (ATB) is applied to both the software- and the hardware-implementation:

| | Area<br>[mm²] | Time<br>[million clock cycles] | Bandwidth Memory<br>[Mbyte/s] | ATB-Metric |
|---|---|---|---|---|
| Software implementation | 149 | 78,469.00 | 71,480.00 | $8.35 * 10^{11}$ |
| VLSI implementation (best case) | 100 | 0.75 | 24.88 | 1,866 |
| VLSI implementation (worst case) | 100 | 96.59 | 3,184.64 | $30.76 * 10^{6}$ |

*Figure 7.16: Area-Time-Bandwidth (ATB) comparison of the hardware- and the software- implementation*

The results of Fig. 7.16 depict that the flexible VLSI-implementation is in terms of the ATB-metric by numbers of magnitude favorable to a processor based implementation.

# 7.8 SUMMARY

This chapter presented an efficient VLSI architecture for 1.) fixed and 2.) variable block size full-search motion estimation, 3.) segment-matching, 4.) support of arbitrary shaped video-objects (MPEG-4) and 5.) luminance corrected, variable block size motion estimation. With a new adaptive luminance correction technique real-time CIF luminance corrected variable block size motion estimation video coding is possible, without employing any intra frame encoding/decoding technique. The absence of intra frame coding allows a very fast decoder with low computational complexity, best suited for mobile applications. It is shown that with this architecture, the additional chip size area costs for the support of block-matching with arbitrarily-shaped objects (MPEG-4) is very low, and luminance correction can be implemented with about 15 $\mu m^2$ additional chip size, using the 0.35 μm CMOS technology. The total chip size area is 54.1 $mm^2$ without wiring. The presented architecture allows variable block size motion estimation at 23,668 32x32 blocks per second with +/- 32 pel search area. Parallelization of an iterative partial quadtree motion estimation algorithm was shown with less than 1 dB PSNR (Y) loss. It was shown that the flexible VLSI implementation is more favourable in terms of the ATB (Area-Time-Bandwidth) metric by numbers of magnitude, compared to a processor based implementation.

# 8

# VLSI IMPLEMENTATION: SEARCH ENGINE II (1D ARRAY)

## 8.1 INTRODUCTION

This chapter describes the VLSI implementation of a flexible motion estimation (ME) accelerator for low-power portable video real-time encoding applications. For low-power motion estimation the main design aim was to optimize the VLSI-architecture for low on-chip memory bandwidth, as the memory access bandwidth is directly related to power consumption. The second design goal was to take into account a low number of memory modules (or memory ports), cf. section 6.3.4. The presented VLSI architecture "Search Engine II", [Kuhn 99], supports, besides the exhaustive search motion estimation, several ME algorithms with reduced computational complexity, in order to meet computational demands and power consumption requirements.

## 8.2 ARCHITECTURE

### 8.2.1 System level overview

Fig. 8.1 depicts the top level architecture of the motion estimation accelerator module.

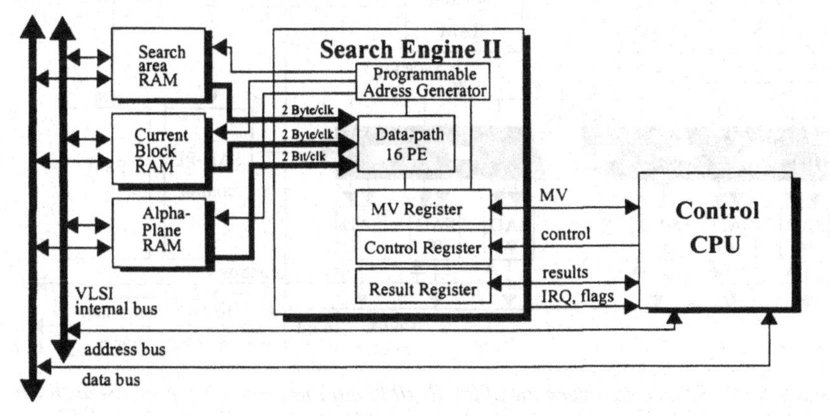


*Figure 8.1: Top level architecture of the motion estimator*

The motion estimation accelerator core is surrounded by a search area RAM, a current block RAM, an alpha-plane RAM and a control CPU. The CPU is required for motion vector predictor calculation, coding decisions and other tasks. However, these tasks can also be performed by a system processor. The motion estimation accelerator consists of a configurable data path of 16 PEs for SAD calculation, a programmable adress generator unit (AGU), and three registers for processor communication. The control register is used by the processor to configure the data path and the address generator. The motion vector register holds a candidate motion vector given by the CPU to the Search Engine II (SE-II), where the SE-II calculates the SAD for this MV candidate and, using the result registers, returns this SAD to the CPU.

The search engine can also perform a search strategy like exhaustive search fully self-controlled, with the best MV forwarded by the MV registers (and the SAD by the result registers) back to the CPU. To indicate the end of an SAD calculation or a search strategy performed by the search engine, an IRQ (interrupt request) is sent to the processor and a flag bit is set accordingly. Note that the SAD registers of the PEs can be preloaded with a predetermined value by the CPU to assist algorithms like R/D-optimized motion estimation.

## 8.2.2 Details

### 8.2.2.1 Overview

The processor array consists of 16 SADα-PEs, cf. Fig. 8.4, which calculate the SAD of the current block data and the previous data (search area). A special data flow design is used which allows to perform up to 16 absolute difference calculations in parallel, while loading only up to 4 bytes from memory per clock cycle. For the search area, current block and alpha-plane data, a local Dual-Port RAM with two ports each is provided, Fig. 8.2, Fig. 8.3.

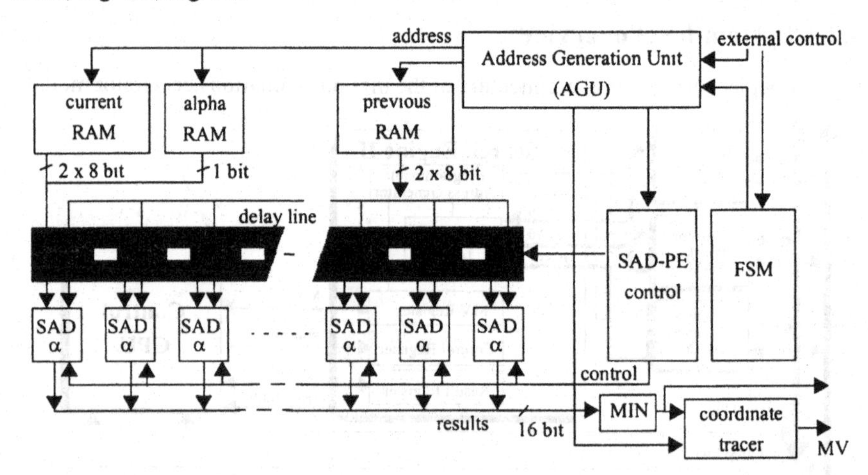


*Figure 8.2: Schematic overview of the flexible motion estimation core (Search Engine II)*

### 8.2.2.2 PE array and delay line

Fig. 8.3 depicts the PE array consisting of 16 SADα-PEs, with P0 and P1 being the ports for the previous search area RAM and C0 and C1 being the ports of the current block RAM respectively. The data from the search area ports, P0 and P1 are broadcast to all PEs. The data from the current block RAM is shifted with every single clock cycle from PE to PE, using a delay line with port C0 as input, for which, optionally, the port C1 of the delay line can be selected as input for the PEs P8 ... P15, Fig. 8.3.

Control is performed by a hierarchical scheme, consisting of an address generation unit (AGU), together with the SAD-PE control on the lower level, and a finite state machine (FSM) controlling several modi on the top level.

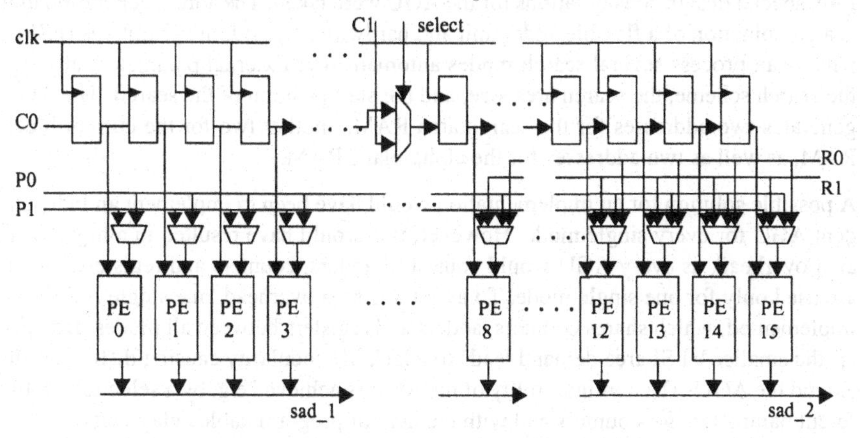


*Figure 8.3: Architecture of the PE array using a delay line*

### 8.2.2.3 Processor element

Fig. 8.4 depicts a modified SADα-PE element, with *a* being the previous data input port and *b* being the current data input port.

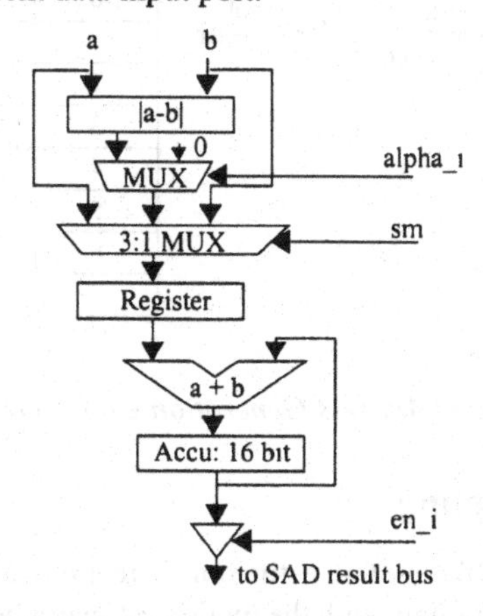


*Figure 8.4: SADα PE element*

The *alpa_i* signal determines if the current absolute difference, calculated by the |*a-b*| element, has to be summarized or not. This feature is required for MPEG-4 alpha-plane and for PE bypassing. With the signal *sm,* the summation of the absolute difference or the summation of the current or previous pels (for mean calculation) can be selected. The output of the result is enabled by signal *en_i.* The delay of the PEs implemented for this VLSI-architecture is 2 clock cycles.

#### 8.2.2.4 Address Generation Unit (AGU)

The AGU, Fig. 8.5, has to support several motion estimation modes and thus became an important part of the design in terms of VLSI area and development time. Therefore, special design considerations for the AGU were taken. The adress generation unit is a combination of a flexible adder unit and hardwired algorithmic primitives (FSM), which can process several search modes automatically. External parameters are, e.g. the search scheme, the search area size, and the start position of the search. The AGU generates two addresses for the search area RAM, another two for the current block RAM, as well as two addresses for the alpha-plane RAM.

A possible solution for an implementation would have been to implement an independent AGU for every single mode. However, this would have resulted in a high VLSI area overhead, as every AGU would consist of adders, counters and registers, which are used only for one single mode. To avoid this area overhead, one single AGU was implemented, which shares counters, adders and registers between all modes. However, the smaller VLSI area demand leads to a lack of modularity causing difficulties to extend the AGU. The configurability of the AGU is achieved e.g. by a selection of different parameters for counters and with the help of programmable delay units.

**a) Flexible address generation unit (AGU) overview**

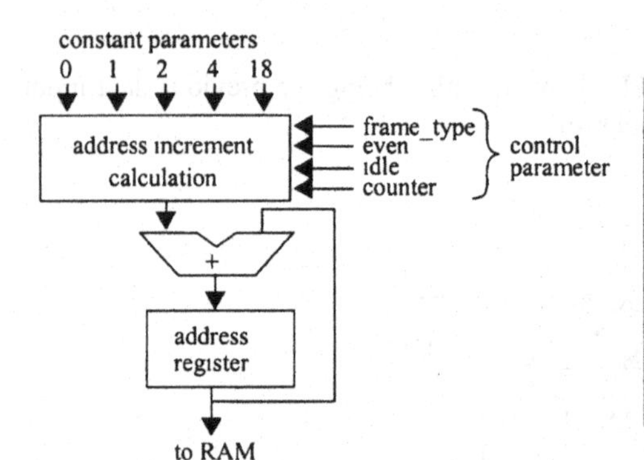

| mode | functionality |
|---|---|
| 0 | no subsampling |
| 1 | 2:1 subsampling |
| 2 | 4:1 subsampling |
| 3 | 4:1 alternate pel subsampling |
| 4 | advanced prediction mode |
| 5 | 16x16 half-pel search |
| 6 | 8x8 half-pel search |
| 7 | Three step search |

*Figure 8.5: a) Adder-based Address Generation Unit (AGU) and b) AGU modes*

### 8.2.3 Processor Coupling

For processor coupling of the motion estimator there are two supported modes which are: 1.) loose processor coupling, with the motion estimator being fully self-controlled and 2.) tight processor coupling, with all actions of the motion estimator being directly controlled by the processor. However, the loose processor coupling mode offers advantages, using the hardwired motion estimation algorithms in terms of throughput, whereas the tight processor coupling mode benefits in terms of flexibility for more advanced motion estimation algorithms.

#### 8.2.3.1 Loose Processor Coupling

For loose processor coupling the Search Engine II can be seen as a dedicated, self-controlled motion estimation accelerator. The steps of operation are explained as follows:

- 1: The CPU calls a multicycle SE-II instruction (e.g. performing a full-search motion estimation for one single MB).
- 2: The CPU proceeds with other tasks or VOs, and SE-II executes multicycle instructions simultaneously.
- 3: The SE-II sends IRQ, MV and SAD to the CPU when the job is finished.
- 4: The CPU continues operation using this SAD and MV.

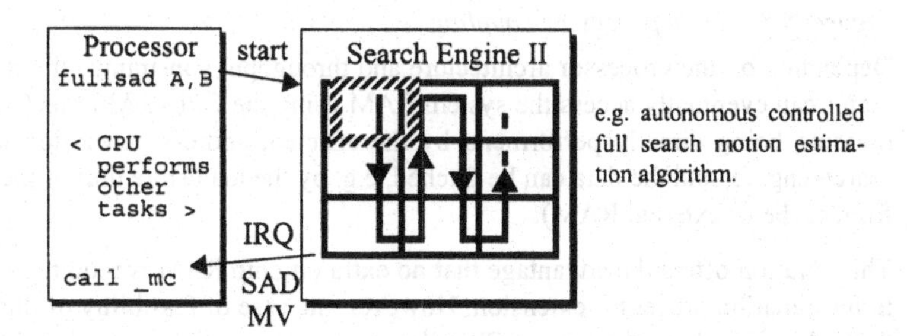


*Figure 8.6: Loose processor coupling*

For the loose processor coupling local RAM for the SE-II is required to obtain high efficiency.

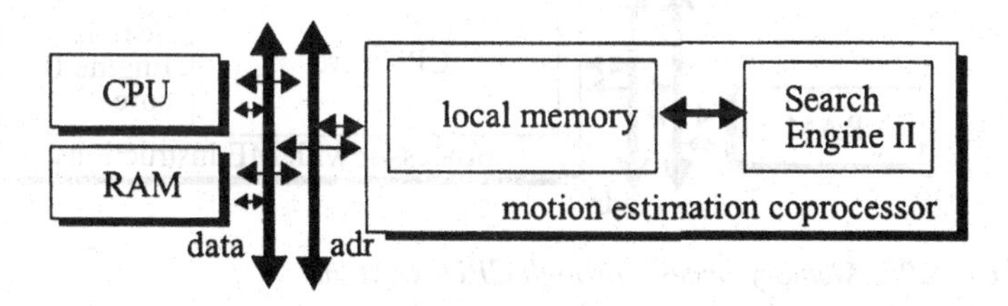


*Figure 8.7: Example of an architecture for loose processor coupling*

#### 8.2.3.2 Tight processor coupling

Using tight processor coupling, the motion estimator can be seen as a multimedia processor extension, taking full advantage of the embedding in the processor environment.

The fundamental steps of communications are depicted as follows:

- 1: The CPU calls a multicycle SE-II instruction and gives a candidate MV (CMV) to SE-II (e.g. calculate the SAD for a given MV).
- 2: The CPU waits for SE-II results (CPU stalls).
- 3: The CPU calculates a new candidate MV (CMV) based upon the SE-II results (MV results).
- 4: The final MV is selected by the processor using e.g. rate/distortion-criteria.

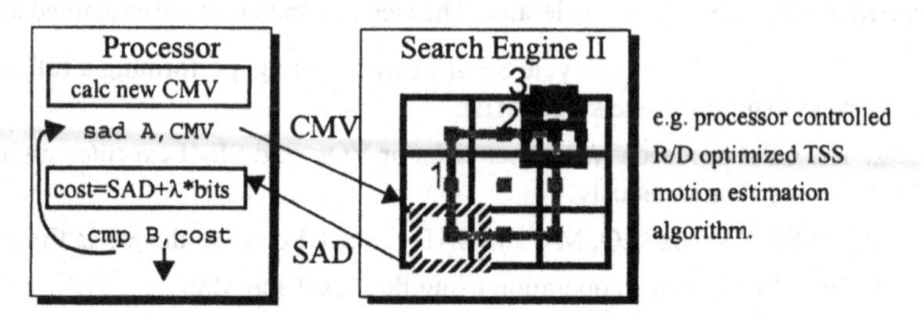


*Figure 8.8: Tight processor coupling*

Depending on the processor architecture and throughput constraints, the motion estimator can eventually access the system RAM using the CPU-RAM bus (with the addressing being directly performend by the efficient address generation unit of the search engine, and the data can be fetched, e.g. by the load/store unit of the processor from cache or external RAM).

This solution offers the advantage that no extra on-chip RAM is required for the motion estimation processor extension. However, the ease of feasibility of this approach depends primarily on the type of CPU choosen, and it may be not feasible in some cases. Therefore, in the most common cases, the motion estimator will require additional local RAM for the search area, as well as for the current block and the alpha-plane.

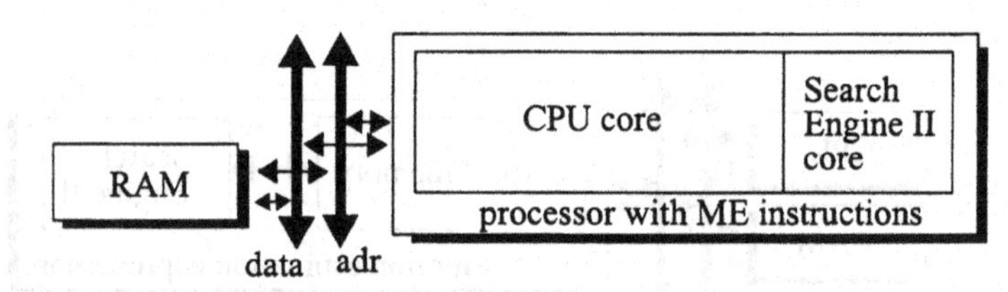


*Figure 8.9: Memory access through CPU-RAM bus*

## 8.3 ALGORITHM ARCHITECTURE MAPPING

The hardwired algorithms are optimized for high throughput and effective data flow, and serve either as fully self-controlled motion estimation scheme or as a low-level VLSI-supported algorithmic primitive, used by the control processor.

The algorithms and algorithmic primitives with optimized data flow are:

- full-search ME [-8, +7], [-16, +15] with search around the predictor
- 2:1 pel subsampling and full-search [-8, +7], [-16, +15]
- 4:1 subsampling and full-search [-8, +7], [-16, +15]
- 4:1 alternate pel subsampling and full-search [-8, +7], [-16, +15]
- half- pel motion estimation
- advanced prediction mode
- Three Step Search (TSS)
- alpha-plane (MPEG-4)
- SAD reduction
- R/D-optimized motion estimation

In the next sections the implementation of these algorithmic primitives is described.

## 8.3.1 Full-search

The full-search mode (cf. 2.2.1) supports a search range of [-8, +7], [-16, +15] pel around a given MV predictor. Fig. 8.10 gives the data flow for the 16 PEs, with the current block data being denoted by *a(x, y)*, and the previous data being denoted by *b(x,y)*.

| clocks | PE0 | PE1 | PE2 | PE3 | ... | PE14 | PE15 |
|---|---|---|---|---|---|---|---|
| 0 | a(0,0)-b(0,0) | | | | | | |
| 1 | a(0,1)-b(0,1) | a(0,0)-b(0,1) | | | | | |
| 2 | a(0,2)-b(0,2) | a(0,1)-b(0,2) | a(0,0)-b(0,2) | | | | |
| 3 | a(0,3)-b(0,3) | a(0,2)-b(0,3) | a(0,1)-b(0,3) | a(0,0)-b(0,0) | | | |
| 4 | a(0,4)-b(0,4) | a(0,3)-b(0,4) | a(0,2)-b(0,4) | a(0,1)-b(0,4) | | | |
| ... | | | | | | | |
| 14 | a(0,14)-b(0,14) | a(0,13)-b(0,14) | a(0,12)-b(0,14) | a(0,11)-b(0,14) | | a(0,0)-b(0,14) | |
| 15 | a(0,15)-b(0,15) | a(0,14)-b(0,15) | a(0,13)-b(0,15) | a(0,12)-b(0,15) | | a(0,1)-b(0,15) | a(0,0)-b(0,15) |
| 16 | a(1,0)-b(1,0) | a(0,15)-b(0,16) | a(0,14)-b(0,16) | a(0,13)-b(0,16) | | a(0,2)-b(0,16) | a(0,1)-b(0,16) |
| 17 | a(1,1)-b(1,1) | a(1,0)-b(1,1) | a(0,15)-b(0,17) | a(0,14)-b(0,17) | | a(0,3)-b(0,17) | a(0,2)-b(0,17) |
| ... | | | | | | | |
| 254 | a(15,14)-b(15,14) | a(15,13)-b(15,14) | a(15,12)-b(15,14) | a(15,11)-b(15,14) | | a(15,1)-b(15,14) | a(14,15)-b(14,30) |
| 255 | a(15,15)-b(15,15) | a(15,14)-b(15,15) | a(15,13)-b(15,15) | a(15,12)-b(15,15) | | a(15,1)-b(15,15) | a(15,0)-b(15,15) |
| 256 | a(0,0)-b(1,0) | a(15,15)-b(15,16) | a(15,14)-b(15,16) | a(15,13)-b(15,16) | | a(15,2)-b(15,16) | a(15,1)-b(15,16) |
| 257 | a(0,1)-b(1,1) | a(0,0)-b(1,1) | a(15,15)-b(15,17) | a(15,14)-b(15,17) | | a(15,3)-b(15,17) | a(15,2)-b(15,17) |
| 258 | a(0,2)-b(1,2) | a(0,1)-b(1,2) | a(0,0)-b(1,2) | a(15,15)-b(15,18) | | a(15,4)-b(15,18) | a(15,3)-b(15,18) |

*Figure 8.10: Data flow for full-search motion estimation*

Every single PE calculates the SAD for a specific MV position, whereas PE0 calculates the SAD for the upper leftmost search position, and PE1 calculates the horizontal neighboring MV position. Therefore, with 16 PEs 16 MV displacements can be calculated concurrently, which corresponds to a [-8, +7] pel search area. For [-16, +15] pel motion estimation, first the left and then the right half of the search area are processed subsequently. For clock instance 16 (Fig. 8.10), PE0 accesses different previous data than the other PEs (the next line has begun). Therefore, two previous memory ports P0 and P1 are required, which give the broadcasted data by multiplexers to the specific PEs. The current block data is input by port C0 and shifted with every clock cycle to the neighboring PE, while the current port C1 is not required for this mode.

At clock cycle 255 the first SAD result is completed. When a SAD calculation is finished, the output-buffers give the result to the result bus, where it is compared with the previously gained best SAD result in the MIN-unit. The first SAD result is compared with the hex-value 0xFFFF. At clock cycle 256, the PE0 calculates the next absolute difference for a horizontal displacement by one in y-direction.

## 8.3.2 Alpha-plane (MPEG-4)

The consideration of the alpha-plane for MPEG-4 within the PE array has already been described in section 8.3.2. It has to be noted that the alpha-plane RAM is addressed with the same address generator as the current data RAM, and that the alpha-plane data are shifted through the delay line in the same way as the current block data.

## 8.3.3 Pel Subsampling

### 8.3.3.1 2:1 pel subsampling

2:1 pel subsampling (cf. 2.4.5) is combined in this mode with full-search motion estimation. Fig. 8.11 depicts the two phases of the current block pels and the search area pels. The pels, which are used to calculate the absolute pel difference are marked black. Note that for the current block two pel subsampling pattern exist, depending on the search position. This is denoted in Fig. 8.11c, where the matching position 0 of the search area (in which the upper left pel is marked as black) requires data of phase 0 of the current block. Matching position 1 of the search area (where the upper left pel is marked as white) requires data of phase 1 of the current block.

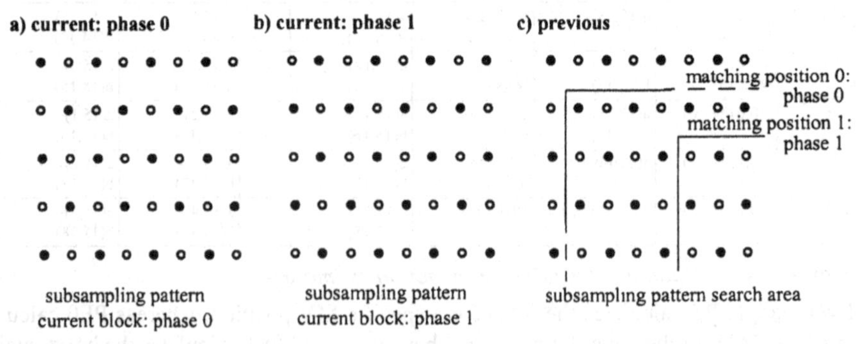


*Figure 8.11: 2:1 pel subsampling patterns depending on the displacement (for N=8)*

For current phase 0 and current phase 1, totally different data is used. For this reason the PE array architecture is required to support two current data ports to support phase 0 and 1. This second data port was implemented as additional port C1 at the delay line after PE7 (Fig. 8.3), which can be selected by a multiplexer. The data port C1 can be used optionally, thus it does not interfere with the efficient data flow for full-search motion estimation without pel subsampling. For example, PE0 ... PE7 calculate the "even" horizontal displacements (0, 2, 4, ... 14 pel), and PE8 ... PE15 calculate the "odd" horizontal displacements (1, 3, 5, ... 15 pel) for line 0 (zero vertical displacement).

The previous data is accessed by ports P0 and P1 in a similar way as described for full-search motion estimation. With the difference, however, that only the pels at every second pel position are used.

Fig. 8.12 depicts the data flow for 2:1 subsampling. Similarily to the full-search algorithm, the current block data (which is input by the ports C0 and C1) is shifted with every clock cycle from PE0 to PE7 and from PE8 to PE15. Note that PE0 to PE7 and PE8 to PE15 use different current block data.

| clock | PE0 | PE1 | | PE7 | PE8 | PE9 | | PE15 |
|---|---|---|---|---|---|---|---|---|
| 0 | a(0,0)-b(0,0) | | | | | | | |
| 1 | a(0,2)-b(0,2) | a(0,0)-b(0,2) | | | a(0,1)-b(0,2) | | | |
| 2 | a(0,4)-b(0,4) | a(0,2)-b(0,4) | | | a(0,3)-b(0,4) | a(0,1)-b(0,4) | | |
| ... | | | | | | | | |
| 7 | a(0,14)-b(0,14) | a(0,12)-b(0,14) | | a(0,0)-b(0,14) | a(0,13)-b(0,14) | a(0,11)-b(0,14) | | |
| 8 | a(1,1)-b(1,1) | a(0,14)-b(0,16) | | a(0,2)-b(0,16) | a(0,15)-b(0,16) | a(0,13)-b(0,16) | | a(0,1)-b(0,16) |
| 9 | a(1,3)-b(1,3) | a(1,1)-b(1,3) | | a(0,4)-b(0,18) | a(1,0)-b(1,1) | a(0,15)-b(0,18) | | a(0,3)-b(0,18) |
| 10 | a(1,5)-b(1,5) | a(1,3)-b(1,5) | | a(0,6)-b(0,20) | a(1,2)-b(1,3) | a(1,0)-b(1,3) | | a(1,5)-b(0,20) |
| ... | | | | | | | | |
| 127 | a(15,14)-b(15,14) | a(15,12)-b(15,14) | | a(15,0)-b(15,14) | a(15,11)-b(15,12) | a(15,9)-b(15,12) | | a(14,12)-b(14,27) |
| 128 | a(0,1)-b(1,1) | a(1,5)-b(1,1) | | a(15,2)-b(15,16) | a(15,13)-b(15,14) | a(15,11)-b(15,14) | | a(14,14)-b(14,29) |
| 129 | a(0,3)-b(1,3) | a(0,1)-b(1,3) | | a(15,4)-b(15,18) | a(0,0)-b(1,1) | a(15,13)-b(15,16) | | a(15,1)-b(15,16) |
| 130 | a(0,5)-b(1,5) | a(0,3)-b(1,5) | | a(15,6)-b(15,20) | a(0,2)-b(1,3) | a(0,0)-b(1,3) | | a(15,2)-b(15,18) |
| 131 | a(0,7)-b(1,7) | a(0,5)-b(1,7) | | a(15,8)-b(15,22) | a(0,4)-b(1,5) | a(0,2)-b(1,3) | | a(15,4)-b(15,20) |

*Figure 8.12: Data flow 2:1 pel subsampling*

The second half PE array, PE8 to PE15, starts computation one clock cycle after the first half PE array. Therefore, the previous data *(b1,1)* at PE8 (cf. clock cycle 9 in Fig. 8.12) had to be delayed by a single clock cycle towards the previous data for PE0 ... PE7. This delay of the previous data was implemented by a delay element of a single clock cycle, using the additional ports R0 and R1 of PE8 ... P15. The advantage of this additional delay element is that the 2:1 pel subsampling mode could be implemented efficiently, using only two ports for previous and current each. As the two half PE arrays deliver their SAD results nearly simultaneously, a second MIN unit had to be implemented.

#### 8.3.3.2 4:1 pel subsampling

The 4:1 pel subsampling mode (cf. 2.4.5) is combined with full-search motion estimation. 4:1 subsampling is implemented similarily to 2:1 subsampling, the four phases for the current block data are depicted in Fig. 8.13. PE0 ... P7 use current block data phase 0 for even horizontal displacements (x= 0, 2, ... 14) and PE0 ... P7 use current block data phase1 for odd horizontal displacements (x= 1, 3, ... 15) in the first line. For a displacement shift in y direction, this scheme is repeated, with the difference being that current block data phase 2 and phase 3 are used. Both half PE-arrays use again different current data.

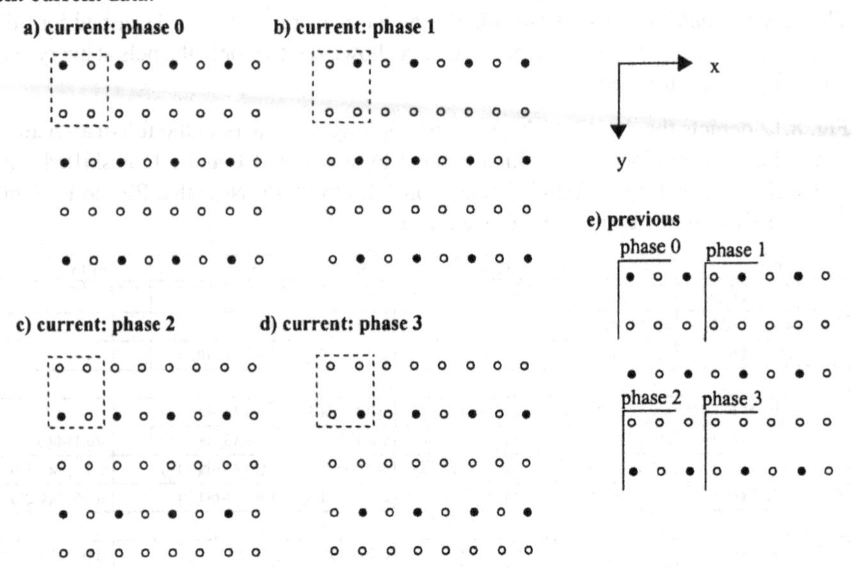


*Figure 8.13: 4:1 pel subsampling patterns (for N=8)*

#### 8.3.3.3 4:1 alternate pel subsampling

The 4:1 alternate pel decimation mode (cf. 2.4.5) is combined with the full-search mode. Fig. 8.14a depicts the current data phases *a* ... *d* for 4:1 alternate pel subsampling. Fig. 8.14b depicts the usage of one of four current block phases *a* ... *d* upon the search position in previous.

For example for a search area position *a* in Fig. 8.14b (which corresponds to Fig. 2.20a), the current block subsampling pattern *a* in Fig. 8.14a is used at this position, which results in the current block data phase *a*. Fig. 8.14c shows (in a simplified view) the usage of the subsampling patterns for the search area of the previous frame dependent on the search position.

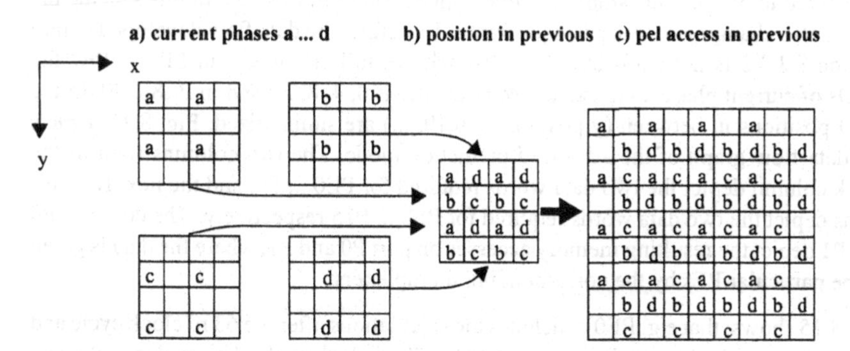


*Figure 8.14: 4:1 alternate pel subsampling: pel access patterns*

| clock | a+(0,0), PE 0-7 1. data word | a+(0,0), PE 0-7, 2. data word | a+(2,0), PE 8-15 1 data word | a+(2,0), PE8-15, 2. data word | P0 | P1 |
|---|---|---|---|---|---|---|
| 0 | (0,0) | | | | (0,0) | |
| 1 | (0,2) | | | | (0,2) | |
| 2 | (0,4) | | | | (0,4) | |
| 3 | (0,6) | | | | (0,6) | |
| 4 | (0,8) | | | | (0,8) | |
| 5 | (0,10) | | | | (0,10) | |
| 6 | (0,12) | | | | (0,12) | |
| 7 | (0,14) | | | | (0,14) | |
| 8 | (2,0) | (0,16) | (2,0) | | (0,16) | (2,0) |
| 9 | (2,2) | (0,18) | (2,2) | | (0,18) | (2,2) |
| 10 | (2,4) | (0,20 | (2,4) | | (0,20) | (2,4) |
| 11 | (2,6) | (0,22) | (2,6) | | (0,22) | (2,6) |
| 12 | (2,8) | (0,24) | (2,8) | | (0,24) | (2,8) |
| 13 | (2,10) | (0,26) | (2,10) | | (0,26) | (2,10) |
| 14 | (2,12) | (0,28) | (2,12) | | (0,28) | (2,12) |
| 15 | (2,14) | | (2,14) | | | (2,14) |
| 16 | (4,0) | (2,16) | (4,0) | (2,16) | (4,0) | (2,16) |
| 17 | (4,2) | (2,18) | (4,2) | (2,18) | (4,2) | (2,18) |
| 18 | (4,4) | (2,20) | (4,4) | (2,20) | (4,4) | (2,20) |
| ... | | | | | | |
| 56 | (14,0) | (12,16) | (14,0) | (12,16) | (12,16) | (14,0) |
| 57 | (14,2) | (12,18) | (14,2) | (12,18) | (12,18) | (14,2) |
| 58 | (14,4) | (12,20) | (14,4) | (12,20) | (12,20) | (14,4) |
| 59 | (14,6) | (12,22) | (14,6) | (12,22) | (12,22) | (14,6) |
| 60 | (14,8) | (12,24) | (14,8) | (12,24) | (12,24) | (14,8) |
| 61 | (14,10) | (12,26) | (14,10) | (12,26) | (12,26) | (14,10) |
| 62 | (14,12) | (12,28) | (14,12) | (12,28) | (12,28) | (14,12) |
| 63 | (14,14) | | (14,14) | | | (14,14) |
| 64 | | (14,16) | (16,0) | (14,16) | (16,0) | (14,16) |
| 65 | | (14,18) | (16,2) | (14,18) | (16,2) | (14,18) |
| 66 | | (14,20) | (16,4) | (14,20) | (16,4) | (14,20) |
| 67 | | (14,22) | (16,6) | (14,22) | (16,6) | (14,22) |
| 68 | | (14,24) | (16,8) | (14,24) | (16,8) | (14,24) |
| 69 | | (14,26) | (16,10) | (14,26) | (16,10) | (14,26) |
| 70 | | (14,28) | (16,12) | (14,28) | (16,12) | (14,28) |
| 71 | | | (16,14) | | (16,14) | |
| 72 | (4,0) | | | (16,16) | (16,16) | (4,0) |
| 73 | (4,2) | | | (16,18) | (16,18) | (4,2) |
| 74 | (4,4) | | | (16,20) | (16,20) | (4,4) |

*Figure 8.15: Data flow 4:1 alternate pel subsampling*

Compared to 4:1 pel subsampling without alternating (cf. 8.3.3.2), in this scheme the data is also changing for the previous frame. Therefore, the data flow developed within section 8.3.3.2 is not applicable here. To achieve full efficiency, in PE0 ... PE7 the SADs of current phase *a* are calculated (y positions 0, 4, 8, ..), and in PE8 ... PE15 the SAD positions in between (y positions 2,6,10, ...) are summarized. Fig. 8.15 depicts the data flow graph of the advanced prediction mode. The two columns right of the clock column depict the two data words required for PE0 ... P7, and the next two columns depict the two data words required for PE8 ... P15 respectively. The columns P0 and P1 depict the resulting memory accesses on port P0 and P1, where the data is given to the particular PEs by the corresponding multiplexers.

Fig. 8.15 shows, that e.g. PE0 concludes the calculations after the 63 rd clock cycle and can start new calculations, beginning with the 72nd clock cycle. This is due to the fact the two previous busses have already been used for the transportation of other data. However, this reduction of efficiency is a tribute to a design with only two memory ports for current and previous each.

### 8.3.4 Advanced prediction mode

The advanced prediction mode performs full-search with 8x8 blocks around the MV position of the best 16x16 MV result with a search area of [-2, +2] pel. The search area of [-2, +2] pel results into totally 25 search positions, which have to be evaluated for each of the four 8x8 blocks. Using the delay line, the current block data is passed from PE to PE with every single clock cycle. This offers the advantage that, with the two ports C0 and C1, the SAD of 10 MV positions can be calculated in parallel, while only 2 bytes of the current block RAM have to be accessed in parallel.

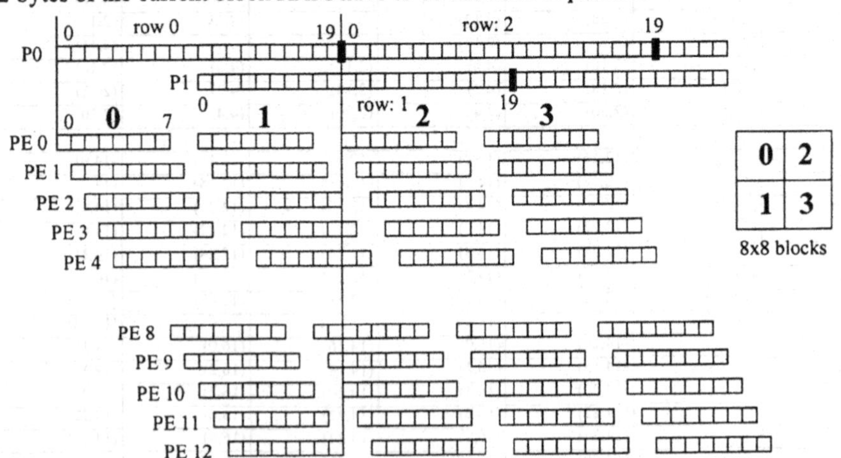


*Figure 8.16: Data flow graph of the advanced prediction mode*

Fig. 8.16 gives an overview on the data flow-scheme, in which two ports for the previous VOP data are used, P0 and P1. A single line of the previous phase has a length of 20 pel (16 pel for the current block, each 2 pel for the [-2, +2] pel shift), resulting in access of the pels 0 ... 19 of row 0 using port P0. Port P1 loads the next line, row 1. To gain high data flow regularity, the data of port 1 is delayed by 10 clock cycles towards the data of port 0 (Fig. 8.16). Using 8x8 blocks for the advanced prediction mode, the phases are only 8 clock cycles long. As the next row of data is available after 10 clock cycles at P1, PE0 is paused for two clock cycles (cf. the blanks in Fig. 8.16).

The PEs are assigned to 8x8 blocks as following: PE0 ... PE4 calculate the SAD for the left two 8x8 blocks and PE8 ... PE12 process the right two 8x8 blocks. PE5 ... PE7 and PE13 ... PE15 are not required for the advanced prediction mode. This particular assignment scheme was selected, as for the calculation of the two blocks on the right different data of the current block are required. This data is accessed by the C1 port, which is connected to PE8 ... PE15 only. Each of the PEs is assigned to a horizontal search position, e.g. PE0 calculates the SAD for $MV_x$=-2, PE1 for $MV_x$=-1, etc.

SAD calculation of the leftmost MV position of the right two 8x8 blocks starts immediately after the first 8-pel row of the left half is processed, i.e. PE8 starts SAD summation after 8 clock cycles. The first two SAD result calculations (for the $MV_x$ = -2 positions) of the two 8x8 blocks of the upper half, are finished after 80 clock cycles (including 16 PE idle clock cycles for prologue and data shifting). The SAD results of the other MV positions are finished according to their delayed start of SAD summation with every clock cycle. For the five SAD results on the left and the five SAD results on the right, the minimum is determined and saved for the left and right 8x8 block each, and the SAD accumulation of all PEs is reset.

After the processing of these two upper 8x8 blocks, the two bottom 8x8 blocks are calculated and their minimum SADs are also saved, requiring a total of four SAD result registers for the advanced prediction mode. After every SAD calculation of a new block position, the result is compared with the SAD value of the respective SAD result register, and the new position and the SAD value is saved, in case the SAD is lower.

An additional accumulator and MIN unit is provided, which calculates from the 8x8 SAD results for a single search position the 16x16 SAD for this particular SAD position. In case this SAD value is better than the SAD value for a previously calculated 16x16 block, the position and the SAD value are saved. With this approach, additional processing power for calculation of the 16x16 MV within the search range of the advanced prediction mode is not required.

The results of PE0 are finished after 800 clock cycles:

$$AdvancedPrediction_{clocks} = N \cdot N_{row} \cdot N_{hor} = 16 \cdot 10 \cdot 5 = 800 \qquad (8.1)$$

in which $N$ is the (vertical) blocksize, $N_{row}$ is the number of clock cycles to process a 8x8 row, and $N_{hor}$ is the number of horizontal search positions. After these 800 clock cycles, 20 clock cycles epilogue for the results of other PEs and 10 clock cycles for minimum evaluation are required, resulting in a total number 830 clock cycles for the advanced prediction mode.

Optionally, the advanced prediction mode implementation can be directly succeeded by the half-pel motion estimation mode, in order to gain half-pel motion vectors for the 8x8 and 16x16 blocks.

## 8.3.5 Half-pel motion estimation

For half-pel motion estimation the current block data and the search area data are filtered to half-pel resolution by bilinear interpolation. Motion estimation is performed on this half-pel data within a [-1, +1] pel search range. As half-pel motion estimation is often used in conjunction with the advanced prediction mode, both modes can be perfomed subsequently with high efficiency.

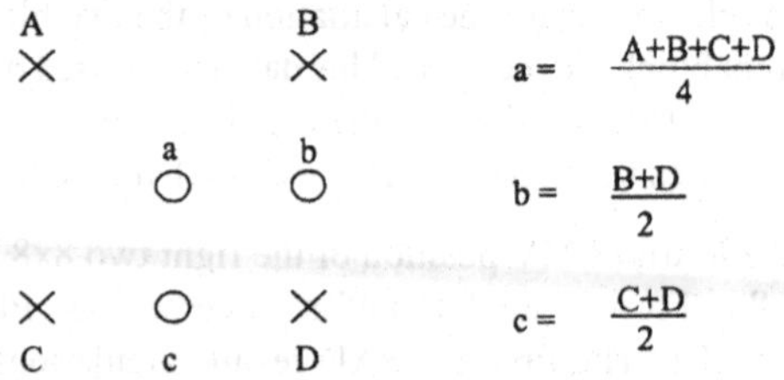

A B a b C c D

$$a = \frac{A+B+C+D}{4}$$

$$b = \frac{B+D}{2}$$

$$c = \frac{C+D}{2}$$

*Figure 8.17: Calculation of pel values at half-pel positions*

The data path unit also consists, beside the delay line, of a half-pel filter, Fig. 8.18. The half-pel filter output is connected with PE0 ... PE7. PE8 ... PE 15 are not required for this mode. Fig. 8.18a depicts the assignment of PEs to half-pel positions, in which "x" marks the integer-pel position at the upper left block boundary. Every single PE calculates the SAD for one of these half-pel positions by subsequently summing up the differences between half-pel interpolated values and current block data.

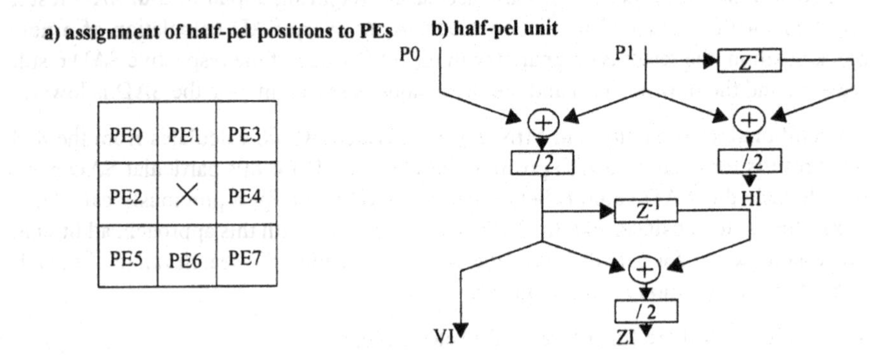


*Figure 8.18: a) Assignment of half-pel positions to PEs and b) details of the half-pel unit*

Fig. 8.18b gives details of the half-pel filter unit. Port ZI outputs the central interpolations, port VI gives the vertical interpolations, and output HI gives the horizontal interpolations according to the positions a, b, c of Fig. 8.17. Division by two is calculated with a simple shift of one bit position to the right. A delay element by a single clock cycle is denoted by $Z^{-1}$.

Fig. 8.19 depicts the data flow for half-pel processing. In this mode, the AGU addresses two vertically neighboring pels during a single clock cycle, using the previous memory ports P0 and P1.

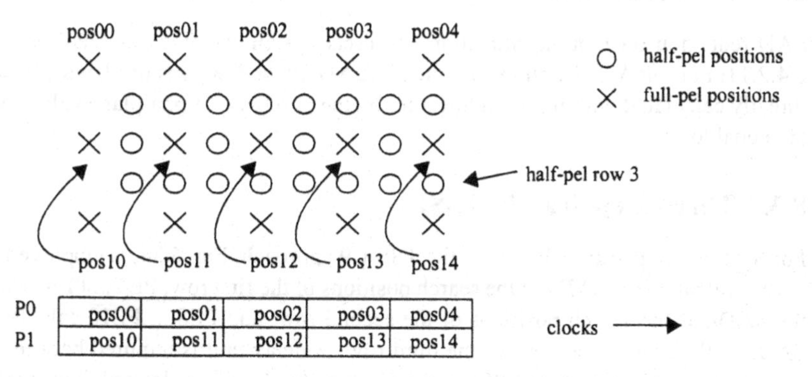


*Figure 8.19: Data flow for half-pel positions*

The data of a third integer-pel row is required for calculation of the half-pel row 3 in Fig. 8.19. Therefore, PE5 and PE6 are delayed by 10 clock cycles, and PE7 is delayed by 11 clock cycles, as 10 clock cycles are required to process the first and second row. PE3 and PE4 calculate the upper right and the middle right position and are delayed by a single clock cycle, as they require additional data from port P0 and P1 simultaneously. The data for PE7 is delayed, relative to PE5 and PE6, by a single clock cycle. The data flow is given in Fig. 8.20.

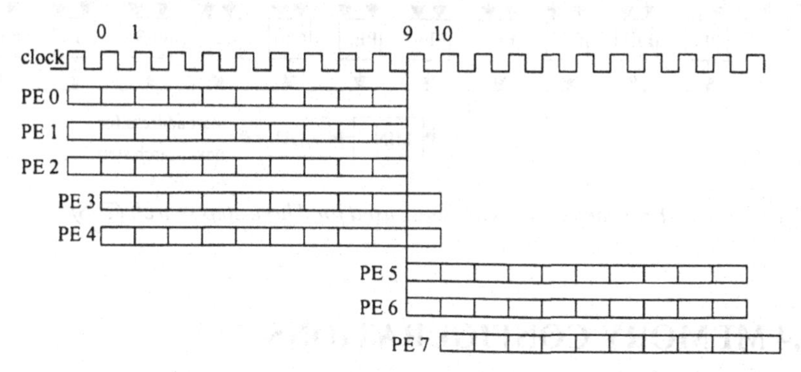


*Figure 8.20: PE utilization for half-pel motion estimation (first 8x8 block row each)*

The delay of the current block data is implemented by additional ports to the shift register line. For half-pel calculation at the block boundary, the pels inside the block are mirrored at the block border to the outside, which is implemented by means of a special addressing mode.

## 8.3.6 R/D-optimized motion estimation

R/D-optimized motion estimation has already been explained in detail in section 6.4.2.12. For this VLSI architecture, the pre-SAD methodology is used because of its advantage to load the $\lambda R$ values in advance.

### 8.3.7 SAD reduction

SAD reduction for motion estimation has already been explained in detail in section 6.4.2.11. For this VLSI architecture the pre-SAD methodology is used, with $N_B$ being directly calculated within this architecture by the number of alpha-plane values which are equal to one.

### 8.3.8 Three Step Search (TSS)

For the three step search, basically the TSS-C9 approach, Fig. 6.34, is employed. PE0 ... PE2 calculate the SADs of the search positions of the first row, PE4 ... PE6 calculate the SADs of the search positions of the second row, and PE8 ... PE10 calculate the SADs of the search positions of the third row. A delay unit is required between PE2 and PE4, as well as between PE6 and PE8. For this VLSI implementation, PE3 and PE7 are acting as a delay element without calculating any SAD.

Compared to the alternative solution with programmable delay units, this approch offers the advantage that no VLSI area and control for an additional delay unit is required.

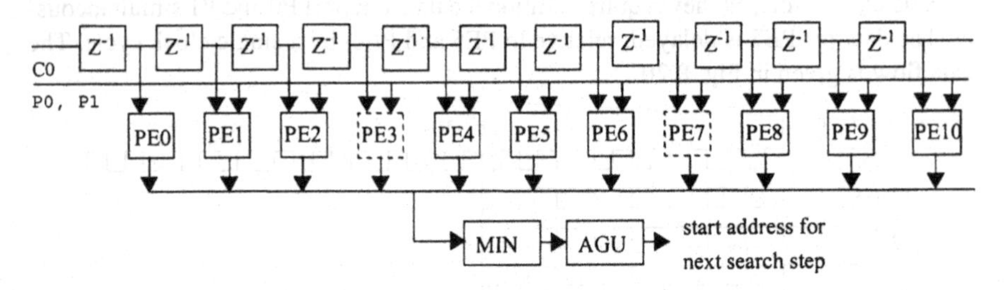


*Figure 8.21: Programmable delay units used for Three step search (TSS)*

## 8.4 MEMORY CONFIGURATIONS

### 8.4.1 Memory access and size

The mode for memory access for the search area RAM is 2 bytes per clock cycle for the full-search algorithm with/without 2:1, 4:1, 4:1 alternate pel subsampling. The modes for memory access for the current block RAM are a) 1 byte per clock cycle for the full-search algorithm within high maximum throughput constraints, and b) 2 bytes / clock for 2:1, 4:1, 4:1 alternate pel subsampling.

The minimum size of the search area memory for [-16, +15] pel search area is calculated to be 47 x 47 bytes = 2,209 Bytes, and for [-8, +7] pel search area the RAM size is 31 x 31 byte= 961 Bytes. Usually more RAM (higher DRAM/SRAM size or additional RAM banks) is used to allow simultaneous loading of the next current block data while performing motion estimation. The amount of the additional RAM depends primarily on system considerations, e.g. on task switching time of the main processor. The minimum size of current block memory is determined to be 256 bytes. The double size can be used to allow simultaneously loading of the next current block data while performing motion estimation. The amount of the additional RAM for current blocks depends primarily on system considerations, e.g. on the task switching time of the main processor.

## 8.4.2 Memory modules

For VLSI implementation the NEC CBC-10 0.25 μm [NEC CBC10] library was used. Based on the data of this library several RAM types are discussed.

### 8.4.2.1 HS2P: High speed synchronous dual-port RAM (1R +1 R/W)

Note: In terms of throughput this is the preferred solution, as no memory bank conflicts occur, which would have to be eliminated by providing additional cycles. However, the dual-port RAM is highly area and power consuming for some VLSI libraries. ($C_L$= load capacitance, A: operation rate: 100%=1, f: read operating frequency).

| RAM type | access time (max.) | $I_{read}$ / (f*A), typ. ($C_L$= 1pF) [mA/Mhz] | area [$mm^2$] |
|---|---|---|---|
| 1 * 2,209 byte for search area data | 3.28 ns | 0.275 | 1.43 |
| 1 * 256 byte for current block data | 3.28 ns | 0.170 | 0.35 |
| **sum** | | 0.448 | **1.78** |

### 8.4.2.2 HD2P: High density synchronous dual-port RAM (1R +1 W)

***Two RAM banks for cur and search area each (even and odd)***

Note: For this solution additional waitstates have to be inserted by the control logic, thus reducing the overall throughput.

| RAM type | access time (max.) | $I_{read}$ / (f*A), typ. ($C_L$= 1pF) [mA/Mhz] | area [$mm^2$] |
|---|---|---|---|
| 2 * 1,105 byte for search area data | 5.24 ns | 2* 0.0147= 0.0294 | 2 * 0.421 = 0.842 |
| 2 * 128 byte for current block data | 5.24 ns | 2* 0.0083 = 0.0166 | 2 * 0.105 = 0.205 |
| **sum** | | 0.046 | **1.047** |

***HD2P: One RAM bank for cur and search area each***

Note: For this solution additional waitstates have to be inserted by the control logic, thus reducing throughput. This solution is only viable for low throughput applications, or for solutions in which the RAM is accessed with twice the speed of the data path.

| RAM type | access time | $I_{read}$ / (f*A), typ. ($C_L$= 1pF) [mA/Mhz] | area [$mm^2$] |
|---|---|---|---|
| 1 * 2,209 byte for search area data | 5.24 ns | 0.0223 | 0.778 |
| 1 * 256 byte for current block data | 5.24 ns | 0.0077 | 0.146 |
| **sum** | | 0.030 | **0.924** |

#### 8.4.2.3 Alpha-plane RAM

For the alpha-plane RAM of size 256 bit several possible solutions were taken into consideration:

| RAM type | size [$mm^2$] |
|---|---|
| 16-bit word width | 0.114 |
| 8-bit word width | 0.074 |
| 1-bit word width | 0.069 |

As with this VLSI architecture, only a single binary alpha value per clock cycle is fed to one of the PEs in case a new SAD calculation begins, the RAM with one-bit word width is used. This solution also offers the benefit of smaller VLSI area requirements.

#### 8.4.2.4 Other RAM types

Note that this architecture can also be combined with off-chip memory of arbitrary size (within some limits). However, the throughput decreases but may be sufficient for some applications. With address mapping, the Search Engine II architecture can also work directly on the VOP-buffer.

## 8.5 RESULTS

### 8.5.1 Throughput

#### 8.5.1.1 Full-search

A memory access of 2 bytes per clock for search area (prev.) and 1 Byte per clock cycle for current block is required. One byte per clock is required for the search area data for prologue and epilogue until a $\eta$ of 100% is achieved.

| Search range | PE array size | $N_{cycles}$ clock cycles | Memory access search area (using dual port RAM) | Core Profile @ Level1 (2 QCIF 30 fps) | *for comparison:* Core @ L1: 16 PEs without data-path design |
|---|---|---|---|---|---|
| [-8, 7] | 16 x 1 | (16 * 16 * 16) + 15 = 4,111 | 7,939 Bytes / MB | 44.9 MByte/s | 372.6 Mbyte/s |
| [-16, 15] | 16 x 1 | (32 * 32 * 16) + 15 = 16,399 | 31,744 Bytes / MB | 179.8 Mbyte/s | 1,486.3 Mbyte/s |

*Figure 8.22: Clock cycles and memory bandwidth for the full-search motion estimation algorithm using the Search Engine II*

The advantage of this data path architecture is a memory bandwith reduction of about a factor 9. However, a special addressing scheme for search area memory and current block memory is required, and a higher fanout for prev, prev' and cur is required to drive the 16 MUX.

#### 8.5.1.2 Pel subsampling

The following table gives the number of clock cycles and the memory access bandwidth of various pel subsampling algorithms.

| Mode | [-16, +15] full-search range | | | | [-8, +7] full-search range | | | |
|---|---|---|---|---|---|---|---|---|
| | $N_{cycles}$ | Memory bandwith [Bytes/MB] | | | $N_{cycles}$ | Memory bandwidth [Bytes/MB] | | |
| | clock cycles | prev | cur | prev + cur | clock cycles | prev | cur | prev + cur |
| full-search | 16,399 | 31,744 | 16,399 | 48,153 | 4,111 | 7,936 | 4,111 | 12,047 |
| 2:1 subsampling | 8,200 | 15,360 | 16,400 | 31,760 | 2,056 | 3,840 | 4,112 | 7,952 |
| 4:1 subsampling | 4,104 | 7,680 | 8,202 | 15,882 | 1,032 | 1,920 | 2,064 | 3,984 |
| 4:1 alternate pel subsampling | 4,623 | 8,640 | 9,246 | 17,886 | 1,152 | 2,160 | 2,304 | 4,464 |

*Figure 8.23: Clocks and memory bandwidth of pel subsampling ME with Search Engine II*

#### 8.5.1.3 Advanced prediction mode and half-pel

The following table depicts the number of clock cycles and the memory bandwidth for the advanced prediction mode and the half-pel motion estimation in various combinations.

| Mode | $N_{cycles}$ | Memory bandwidth [Bytes/MB] | | |
|---|---|---|---|---|
| | clock cycles | prev | cur | prev + cur |
| advanced prediction mode [-2, +2] full-search range including half-pel | 1,222 | 2,240 | 1,536 | 3,776 |
| advanced prediction mode [-2, +2] full-search range | 830 | 1,600 | 1,280 | 2,880 |
| half-pel motion estimation (postprocessing of 16x16 integer pel MV) | 324 | 576 | 256 | 832 |
| half-pel motion estimation (postprocessing of one 8x8 integer pel MV) | 98 | 160 | 64 | 224 |

*Figure 8.24: Clocks and memory bandwidth of advanced prediction and half-pel ME with SE-II*

#### 8.5.1.4 Three step search (TSS)

The clock cycles and the memory bandwidth of the TSS search algorithm is depicted in Fig. 8.25.

| Mode | $N_{cycles}$ | Memory bandwith [Bytes/MB] | | |
|---|---|---|---|---|
| | clock cycles | prev | cur | prev+cur |
| Three Step Search: 1. step | 356 | 576 | 256 | 832 |
| Three Step Search: 2. step | 308 | 480 | 256 | 736 |
| Three Step Search: 3. step | 284 | 432 | 256 | 688 |
| **sum** | **948** | **1,488** | **768** | **2,256** |

*Figure 8.25: Clock cycles and memory bandwidth of the Three Step Search (TSS) algorithm*

## 8.5.2 VLSI area and speed

VHDL synthesis with the NEC CBC10 library (0.25 μm) results in 100 Mhz (min.) and these preliminary gate counts, Fig. 8.26. The total gate count is estimated to result into an VLSI area of about 0.5 $mm^2$ (without wiring and RAM).

| **Module** | **Gate Equivalents** |
|---|---|
| delay line | 1,621 |
| PE element: PE0 ... PE7 | 8 x 625 = 5,000 |
| PE element: PE8 ... PE15 | 8 x 671 = 5,368 |
| Address Generation Unit (AGU) | 4,114 |
| Coordinates Tracer (MV calculation) | 882 |
| PE control and MIN | 5,351 |
| **Total** | **22,340** |

*Figure 8.26: VLSI design results: gate counts*

Note for Fig. 8.26 that the PEs which perform the actual computational work are about two times this size, compared to the AGU and the PE control unit.

# 8.6 COMPARISON WITH RELATED WORK

For the flexible VLSI architectures proposed for TSS and the full-search algorithm [Dutta 96], [Dutta 98], and [Zheng 97], no detailed VLSI implementation data has been published so far. Both use 16 memory modules and a 16x16 interconnection network. [Cheng 97], who compared VLSI implementations of several motion estimation algorithms, 1.) did not present a VLSI architecture using more than a single motion estimation algorithm, and 2.) usesd estimated data for area and throughput.

| | Siemens AG<br><br>Germany<br>1995 | AT&T Bell Labs,<br>Holmdel,<br>USA<br>1996 | Univ. Science &<br>Tech. Anhui,<br>China<br>1997 | Search Engine II<br>TU Munich,<br>Germany<br>1998 |
|---|---|---|---|---|
| Number of PEs | 256 | 64 | 48 | 16 |
| Number of memory modules | 0<br>(external memory) | 8 | 16 | 2<br>(+ 1 for MPEG-4) |
| Clock Frequency | 72 Mhz | 66 Mhz | n.a. | 100 Mhz<br>(min.) |
| Design style | full custom | full custom | n.a. | VHDL synthesis |
| Area | 228 mm$^2$ | 4.2 x 6.6 mm$^2$<br>(incl. RAM) | 24kG + RAM | 0.5 mm$^2$ (22.3 kG)<br>+ 1.78 mm$^2$ RAM<br>+ wiring |
| Process | 0.6 μm CMOS, 2 ML | 0.5 μm CMOS<br>3ML | n.a. | 0.25 μm<br>NEC CBC10 |
| Processing Power | 18 GOPS<br>@ 72 MHz | 14 GOPS<br>@ 75 MHz | n.a. | 4.8 GOPS<br>@ 100 MHz |
| Power Consumption | n.a. | 0.3 W @ 66Mhz | n.a. | n.a. |
| Supply Voltage | n.a. | 3.3 V | n.a. | 2.5 V |
| On-Chip Memory | no | 768 Byte | n.a. | 2,209 + 256 + 16<br>= 2,481 Byte |
| Full-search | yes | yes | [-16, +15] | [-16, +15]<br>[-8, +7] |
| Search area subsampling | no | yes | pipelined TSS<br>(cf. 6.4.2.2) | TSS and other |
| Pel subsampling | yes (degraded efficiency) | no | no | 2:1, 4:1,<br>4:1 alternating |
| Mean | | yes | no | yes |
| Fractional-pel | yes | n.a. | no | 1/2 pel |
| Advanced prediction mode | no | no | no | yes |
| Arbitrary shapes (MPEG-4) | no | no | no | yes |
| References | [Vos 95:1], [Ber 97] | [Lin 96], [Lin 96a] | [Zhang 97] | this chapter of this book, [Kuhn 99] |

*Figure 8.27: Comparison with other flexible motion estimation chips (n.a.: not available)*

# 8.7 PROCESSOR - FLEXIBLE ASIC COMPARISON USING THE ATB METRIC

For the comparison of the processor implementation (software) and the flexible VLSI implementation (hardware) of the different motion estimation algorithms, the Area-Time-Bandwidth metric (ATB) as defined in 6.2.3.2 was used. For the Ultra-SPARC processor, a silicon area of 149 $mm^2$ at 0.35 μm CMOS technology was employed for this comparison [SPARC], after scaling down to 0.25 μm. For the Search Engine II a total area number of 2 $mm^2$ was used.

The data for the processor based implementation was gained using the iprof tool (c.f. chapter 3), where a single clock cycle was approximated by one RISC instruction. For the software implementation of the fast motion estimation algorithms on the Ultra-SPARC CPU no VIS instructions (cf. 6.4.5) were used. For Search Engine II, the memory bandwidth of the data transferred from the local memory to the motion-estimation data path was taken into account. Note that for this comparison only the number of magnitude is of interest.

Fig. 8.28a depicts the number of clock cycles required for [-16, +15] pel full-search motion estimation, the 4:1 pel alternate pel subsampling algorithm, and the TSS motion estimation. All algorithms are used together with advanced prediction mode [-2, +2] pel and half-pel pel resolution. Fig. 8.28b depicts the memory access bandwidth of these algorithms, both for processor based implementation and VLSI implementation using the Search Engine II architecture. Fig. 8.29 depicts the comparison of both solutions using the ATB metric. It is shown that the flexible VLSI implementation is more favourable in terms of the ATB metric by numbers of magnitude, compared to a processor based implementation.

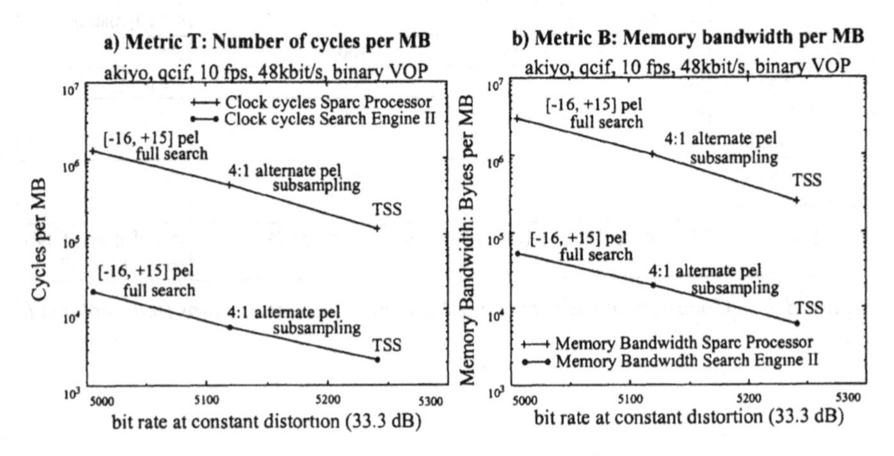


*Figure 8.28: Processor-ASIC comparison using the time (T) and the bandwidth (B) metric*

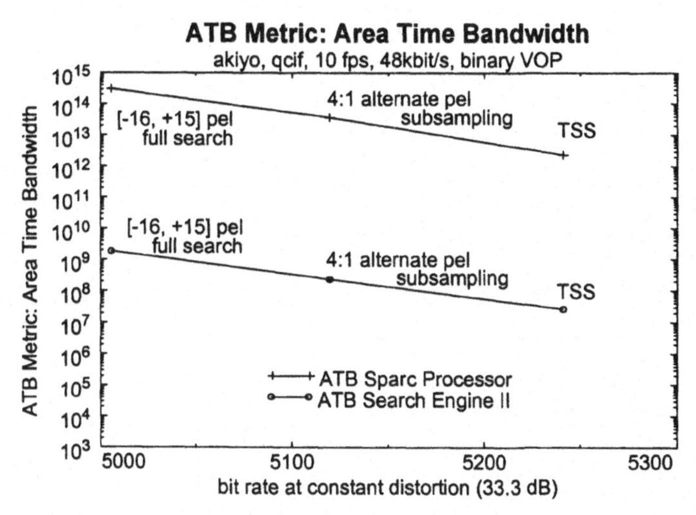


*Figure 8.29: Processor ASIC comparison using the Area-Time-Bandwidth (ATB) metric*

# 8.8 SUMMARY

This chapter presented a VLSI implementation of an 1D PE array using two dual-port RAMs for the search area data and the current block data. Various motion estimation algorithms are supported with high efficiency. The design aim were: 1.) to use only a small number of memory modules (with a maximum of two ports), and 2.) to achieve high efficiency with low memory bandwidth, as these factors are important for a low-power implementation.

The algorithm-VLSI mapping of various motion estimation algorithms was presented in this chapter: 1.) full-search ME with [-16, 15] and [-8, +7] pel search area, 2.) MPEG-4 motion estimation for arbitrarily-shaped objects, 3.) advanced prediction mode, 4.) 2:1 pel subsampling, 5.) 4:1 pel subsampling, 6.) 4:1 alternate pel subsampling as proposed by [Liu 93], 7.) preferring of the zero MV, 8.) R/D-optimized ME and 9.) half-pel ME. A general observation is that for the implementation of several motion estimation algorithms, the area required for the address generation unit and the result comparison unit, is about of half the area size as the PE array.

The VLSI architecture was implemented, using synthesizable VHDL and resulted into a size of 22.8 Kgates (without RAM), 100 MHz (min.) using a 0.25 μm commercial CMOS library. Finally, this architecture was compared with other architectures. It was shown that the flexible VLSI implementation depicted here is more favourable in terms of the ATB (Area-Time-Bandwidth) metric by numbers of magnitude, compared to a processor based implementation.

# 9

# SUMMARY

The emerging multimedia standard MPEG-4 combines interactivity, natural and synthetic digital video, audio and computer graphics. Typical applications are: video conferencing, mobile videophone, multimedia cooperative work, teleteaching, and, however, games. With MPEG-4, the next step from block-based video coding (ISO/IEC MPEG-1, MPEG-2, ITU-T H.261, H.263) to object-based video coding is taken. The step from block-based to object-based video coding requires a new methodology for VLSI design, as well as new VLSI architectures with considerable higher flexibility, where the motion estimation (ME) of the encoder is the computationally most demanding tool within the MPEG-4 standard.

Therefore, this book compares several MPEG-4 motion estimation algorithms implemented on a programmable processor with dedicated, but flexible VLSI architectures. It is shown by means of an Area-Time-Bandwidth (ATB) metric, that the flexible VLSI architectures developed within this book are favorable by magnitudes of order to processor based solutions, both for high throughput and for low-power constraints.

***Motion estimation***

Motion estimation has proven to be effective to exploit temporal redundancy of video sequences and is therefore a central part of the MPEG-1/MPEG-2/MPEG-4 and the H.261/H.263 video compression standards. These video compression schemes are based on a block-based hybrid coding concept, which was extended within the MPEG-4 standardization effort to support arbitrarily-shaped video objects.

Motion estimation algorithms have attracted much attention in research and industry, because of these reasons:

1. It is computationally the most demanding algorithm for a video encoder (about 60-80% of the total computation time) which limits the performance of the encoder in terms of encoding speed, and is therefore mainly responsible for high power consumption.

2. The motion estimation algorithm has a high impact on the visual performance of an encoder for a given bit rate.

3. The method to extract motion vectors from the video material is not standardized, thus being open to competition.

***MPEG-4 requirements***

While for previously standardized video coding schemes usually a single motion estimation algorithm has been employed, MPEG-4 now introduces the support of multiple arbitrarily-shaped video objects (VOs), in which for every single VO different quality constraints and, therefore, different motion estimation algorithms can be selected. This is especially useful for real-time applications, in which, for example, the coding options and the motion estimation algorithm can be optimized for low complexity for the background VO (e.g. a landscape). A high visual quality can be obtained for the foreground VO (e.g. a person talking) by spending more computational power on motion estimation.

Therefore, a low-power MPEG-4 motion estimation architecture is required to support

1. multiple VOs (video objects),
2. several motion estimation algorithms with different complexity/distortion trade-off, and
3. the alpha-plane defined within MPEG-4.

***Objectives***

As a high memory access bandwidth is mainly responsible for throughput limitations and high power consumption in various motion estimation algorithms, the aim was to minimize the memory bandwidth between the data path and the on-chip buffer memories.

The MPEG-4 encoder, with its support of several modes and video objects, suggests at first glance a programmable processor based implementation. The question arose, whether a programmable processor based or dedicated VLSI-based solution is better suited for low power motion estimation. Processors offer a high flexibility and reconfigurability through programmability, which has, however, to be paid for with higher power consumption, higher area demands and lower throughput, compared to dedicated ASICs (application-specific integrated circuits). Dedicated ASICs usually have the disadvantage of being limited in terms of flexibility. Therefore, within this work efforts have been made to develop VLSI architectures for motion estimation with low area, low computation time demands, and the flexibility to support several algorithms, hence being favorable to programmable processors.

Therefore, the general thesis of this book could be formulated as follows: The thesis of this book is to show that dedicated, but flexible VLSI architectures, supporting several motion estimation algorithms, are favorable within the MPEG-4 context, compared to programmable processors, in terms of Area (A), Computation Time (T), and memory bandwidth (B), and are therefore advantageous in terms of power consumption and throughput.

***Methodology and results***

The design flow and HW/SW comparison methodology used within this work is depicted in Fig. 9.1.

The advantage of dedicated, but flexible VLSI architectures over software based implementations is demonstrated within this work for several motion estimation algorithms. In the following, the methodology and the results are summarized.

Block-matching motion estimation algorithms usually depict relatively simple arithmetic operations, but require a high memory bandwidth. Therefore, for low-power and high throughput optimization of motion estimation at algorithm and system level, the minimization of the memory bandwidth from the data path to the local buffers is proposed in the introduction.

In the second chapter of this book, an overview on and classification of fast motion estimation algorithms for MPEG-4 is performed. In chapter 3, software complexity metrics and software based complexity analysis tools are discussed. As existing tools for software based computational complexity analysis are limited in terms of portability and analysis of memory bandwidth, a new portable tool, `iprof`, has been developed within this book.

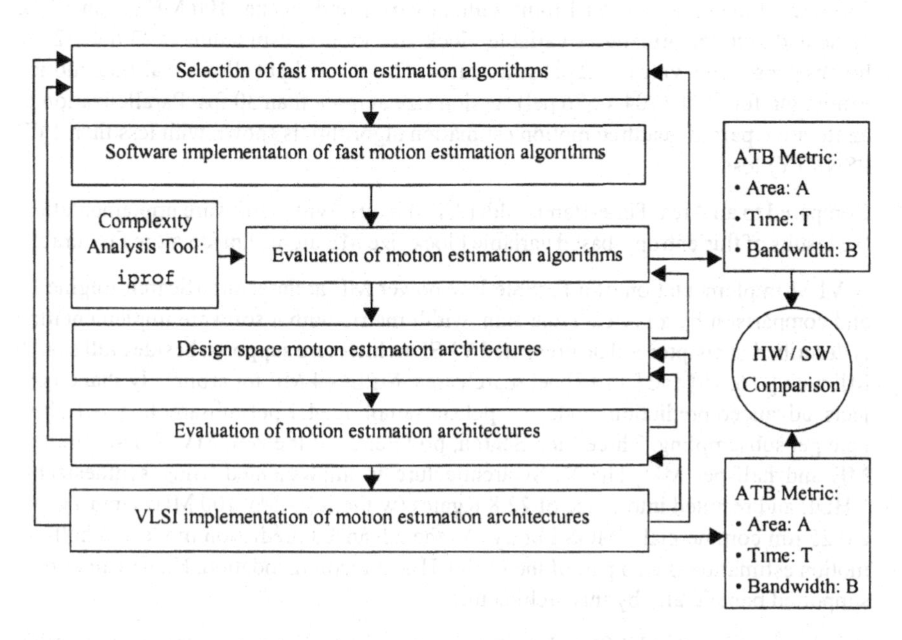


*Figure 9.1: Design flow and hardware/software comparison methodology*

A detailed time-dependent complexity analysis of a software implementation of the MPEG-4 video part, as well as a discussion on the impact of these results for VLSI architectures is described in chapter 4. A complexity and visual quality analysis of several fast motion estimation algorithms within the MPEG-4 framework is described in chapter 5.

An overview of and discussion on VLSI complexity metrics and the basic components of motion estimation (ME) architectures for MPEG-4 is given in chapter 6. Referring to the design space of these basic components, dedicated, flexible as well as programmable VLSI architectures are discussed for a number of motion estimation algorithms. Based on this discussion, the basic architectures for a high throughput MPEG-4 motion estimation architecture supporting optionally luminance-corrected full search variable block size motion estimation (Search Engine I) and a low-power motion estimation architecture for several fast motion estimation algorithms (Search Engine II) were derived.

A VLSI implementation of a flexible high throughput ME architecture (Search Engine I) is described in chapter 7. The VLSI architecture supports: 1.) fixed and 2.) variable block size full-search motion estimation, 3.) segment-matching, 4.) support of arbitrary-shaped video objects (MPEG-4) and 5.) luminance-corrected variable block size motion estimation. It is shown that, with this architecture, the additional chip size area costs for the support of block-matching with arbitrarilyy-shaped objects (MPEG-4) are very low, and luminance-corrected variable block size motion-estimation can be implemented with about 15 $mm^2$ additional chip size with 0.35 μm CMOS-technology. The total chip size area is 54.1 $mm^2$ without wiring and reaches 100 MHz (min.). The presented architecture allows variable block size motion estimation at 23,668 32x32 blocks per second with +/- 32 pel search area. This throughput allows real-time motion estimation for 4CIF (704 x 576 pel) session size at more than 30 fps. Parallelization of an iterative partial quadtree motion estimation algorithm is shown with less than 1 dB PSNR (Y) loss.

Compared to an Area-Time-Bandwidth (ATB) metric with a software implementation, the results of this entropy-based variable block size ME algorithm are more favourable.

A VLSI implementation of a flexible low-power ME architecture (Search Engine II) and comparison by an Area-Time-Bandwidth metric with a software implementation is described in chapter 8. The presented VLSI architecture supports, besides full search ME with [-16, 15] and [-8, +7] pel search area, MPEG-4 ME for arbitrarily shaped objects, advanced prediction mode, 2:1 pel subsampling, 4:1 pel subsampling, 4:1 alternate pel subsampling, Three Step Search, preference of the zero MV, R/D-optimized ME and half-pel ME. The VLSI architecture is implemented using synthesizable VHDL and resulted into a size of 22.8 Kgates (without RAM), 100 MHz (min.) using a 0.25 μm commercial CMOS library. As the advanced prediction mode and half-pel motion estimation is also part of the ITU-T H.263 recommendation, H.263 can also be supported beneficially by this architecture.

Finally, the presented VLSI architecture has proved to be superior, compared to other algorithm-specific ASICs and processor based solutions, using the above mentioned ATB metric.

# ANNEX

***Calculation of reference functions and PSNR distances***

In order to compare two different motion estimation algorithms in terms of subjective visual quality (PSNR) for a single test sequence, either the bit rate of the two sequences should be equal for a PSNR comparison or vice-versa. However, without using a rate-control algorithm the bit rates are difficult to match by only varying the inter quantizer parameter $Q_p$. Therefore the rate/distortion-curve for both algorithms was calculated by variation of $Q_p$ in the vicinity of the algorithm's linear working point. The bit rate and PSNR was calculated by these two R/D-curve approximations. For comparison, the resulting bit rate of a single simulation was calculated as a bit rate deviation to the original MPEG-4 verification model algorithm (i.e. "+6.34%"), e.q. (10.1).

$$BitrateDeviation = \left(\frac{bitrate}{bitrate\_of\_reference\_VM} - 1\right) \cdot 100\% \qquad (10.1)$$

Furthermore, the reference R/D-curves were calculated for every sequence using the original MPEG-4 verification model implementation of the motion estimation, Fig. 10.1.

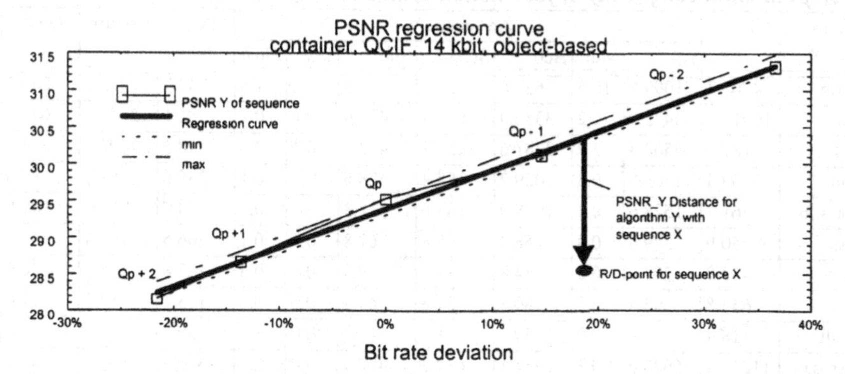


*Figure 10.1: Regression curve for the sequence „container", QCIF, 14kbit, object-based*

To obtain a useful approximation of the R/D-curve for a particular test sequence, additional simulations were performed with different $Q_p$ variations (-2, -1, 0, +1, +2) obtaining five R/D-points. Using these five points, a reference curve was calculated using linear regression as depicted in Fig. 10.1 for each of the 24 test sequences. With these reference curves, a *PSNRYdistance* was calculated for every simulation which was used to determine the subjective quality of an algorithm *alg* and a single sequence *seq*, e.q. (10.2).

$$PSNRYdistance(alg, seq) = PSNRY(alg, seq) - PSNRYref(BitrateDeviation(alg, seq)) \quad (10.2)$$

$$significance(alg,seq) = \frac{|PSNRYdistance(alg, seq)|}{reference\_dispersion(seq)} \quad (10.3)$$

Fig. 10.1 depicts the five calculated R/D-points and the resulting reference curve of the original MPEG-4 VM motion estimator. The distance of these points with respect to the reference curve has to be considered when interpreting the *PSNR_y distances*. This is expressed by the *significance* of the *PSNR_y_distance* of a simulation, eq. (10.3). For a significance near or below one, it was considered to be not sensible to use the *PSNRYdistance* as a ranking criterion.

***Fast motion estimation algorithms***

The following tables depict the instruction usage and the memory bandwidth for the sequence akiyo, qcif, at 10kbit/s (as used in chapter 5). The next tables give an overview on the MPEG-4 sequences and the parameters used for this evaluation. For this analysis rectangular VO ("frame") as well as binary shaped VO ("object") were used. The last table depicts the PSNR results of the sequences using the various fast motion estimation algorithms.

***Computational complexity of fast motion estimation algorithms***

| Algo-rithm | All | Arithmetic | | | | | | | Load Store | Control | Other |
|---|---|---|---|---|---|---|---|---|---|---|---|
| | | All | Mul | Add | Sub | Shift | Div | Float | | | |
| orig | 4765.7 | 1099.8 | 16.5 | 630.6 | 333.1 | 45.3 | 1.2 | 1.6 | 1744.7 | 1712.7 | 200.7 |
| fs | 10016.0 | 4422.9 | 0.2 | 3356.9 | 635.8 | 414.0 | 0.0 | 0.1 | 2292.1 | 2646.3 | 643.1 |
| rd | 10187.0 | 4508.4 | 3.9 | 3381.6 | 653.7 | 442.3 | 0.1 | 0.3 | 2318.5 | 2697.2 | 643.7 |
| liu | 3403.8 | 1563.0 | 0.3 | 1029.2 | 188.4 | 328.5 | 0.1 | 0.1 | 785.6 | 858.1 | 188.6 |
| pelsub | 3361.7 | 1528.6 | 0.3 | 1018.0 | 184.6 | 312.8 | 0.1 | 0.1 | 774.5 | 858.0 | 188.6 |
| tss | 830.9 | 379.6 | 0.4 | 258.4 | 45.8 | 67.8 | 0.1 | 0.1 | 206.5 | 196.4 | 46.3 |
| cote | 433.8 | 203.0 | 0.5 | 134.8 | 21.6 | 39.9 | 0.1 | 0.3 | 112.8 | 95.6 | 20.6 |
| hier | 633.8 | 296.8 | 0.3 | 189.3 | 29.8 | 69.4 | 0.0 | 0.1 | 163.1 | 140.2 | 31.0 |
| null | 118.4 | 62.9 | 0.0 | 39.1 | 1.5 | 17.7 | 0.0 | 0.1 | 38.7 | 14.9 | 1.6 |
| fsrbsad | 11281.1 | 5670.6 | 0.3 | 3365.1 | 637.7 | 417.9 | 0.0 | 0.1 | 2300.3 | 2653.6 | 644.9 |
| fsmme | 13755.0 | 5827.7 | 0.3 | 2749.1 | 637.9 | 1775.8 | 0.0 | 0.1 | 3576.3 | 3965.1 | 325.6 |
| fspdc | 12584.7 | 5448.1 | 0.2 | 3018.7 | 637.7 | 1775.5 | 0.0 | 0.1 | 2916.6 | 3886.9 | 321.5 |
| fsdpdc | 18975.4 | 9244.7 | 0.4 | 4626.0 | 1263.3 | 2052.8 | 0.0 | 0.1 | 3361.7 | 6269.1 | 35.2 |
| fsbbm | 7535.2 | 3749.6 | 152 | 2997.4 | 36.6 | 535.5 | 0.0 | 0.1 | 1382.9 | 2290.4 | 91.8 |
| proj | 1686.4 | 852.2 | 0.3 | 496.1 | 139.9 | 200.7 | 0.1 | 0.1 | 467.9 | 290.2 | 71.4 |

***Memory bandwidth of fast motion estimation algorithms***

| Algo-rithm | Total | Load | | | | Store | | | |
|---|---|---|---|---|---|---|---|---|---|
| | | Total | 16 bit | 32 bit | 64 bit | Total | 16 bit | 32 bit | 64 bit |
| orig | 3983.1 | 2802.7 | 1285.2 | 57.6 | 0.2 | 1180.5 | 2.8 | 293.6 | 0.1 |
| fs | 4318.9 | 4277.2 | 2055.0 | 41.8 | 0.0 | 41.8 | 8.4 | 6.2 | 0.0 |
| rd | 4408.1 | 4335.6 | 2060.3 | 53.7 | 0.0 | 72.4 | 8.6 | 13.7 | 0.0 |
| liu | 1518.4 | 1459.7 | 608.0 | 60.9 | 0.0 | 58.7 | 8.9 | 10.2 | 0.0 |
| pelsub | 1510.7 | 1452.0 | 604.1 | 61.0 | 0.0 | 58.7 | 8.8 | 10.2 | 0.0 |
| tss | 400.1 | 365.8 | 160.0 | 11.5 | 0.0 | 34.3 | 9.1 | 4.0 | 0.0 |
| cote | 221.5 | 191.0 | 81.5 | 7.0 | 0.0 | 30.5 | 8.4 | 3.4 | 0.0 |
| hier | 318.1 | 278.2 | 115.2 | 11.9 | 0.0 | 40.0 | 11.4 | 4.2 | 0.1 |
| null | 71.4 | 50.2 | 22.5 | 1.3 | 0.0 | 21.2 | 7.7 | 1.4 | 0.0 |
| fsrbsad | 4336.7 | 4292.0 | 2060.5 | 42.7 | 0.0 | 44.7 | 8.8 | 6.7 | 0.0 |
| fsmme | 4337.7 | 4293.3 | 2061.2 | 42.7 | 0.0 | 44.4 | 8.7 | 6.7 | 0.0 |
| fspdc | 4334.8 | 4290.6 | 2060.1 | 42.6 | 0.0 | 44.2 | 8.7 | 6.6 | 0.0 |
| fsdpdc | 4542.3 | 4472.6 | 2073.5 | 81.4 | 0.0 | 69.7 | 15.1 | 9.8 | 0.0 |
| fsbbm | 2909.9 | 2715.8 | 988.1 | 184.9 | 0.0 | 194.1 | 17.1 | 39.9 | 0.0 |
| proj | 928.8 | 855.1 | 326.3 | 50.6 | 0.0 | 73.7 | 17.6 | 9.6 | 0.0 |

***Test sequences***

| Nr | Test Sequence | Type | Size | Bitrate | VOPs / Frames | Frame-rate | Intra-quant. | Inter-quant. |
|---|---|---|---|---|---|---|---|---|
| 1 | mother_daughter | frame | QCIF | 10k | 299 | 7.5 | 14 | 14 |
| 2 | hall_monitor | frame | QCIF | 10k | 299 | 7.5 | 17 | 17 |
| 3 | container | frame | QCIF | 10k | 299 | 7.5 | 16 | 16 |
| 4 | mother_daughter | frame | QCIF | 24k | 299 | 10 | 8 | 8 |
| 5 | silent | frame | QCIF | 24k | 299 | 10 | 13 | 13 |
| 6 | container | frame | QCIF | 24k | 299 | 10 | 9 | 9 |
| 7 | foreman | frame | QCIF | 48k | 299 | 10 | 13 | 13 |
| 8 | coastguard | frame | QCIF | 48k | 299 | 10 | 13 | 13 |
| 9 | news | frame | CIF | 48k | 299 | 7.5 | 19 | 19 |
| 10 | news | frame | CIF | 112k | 299 | 15 | 11 | 11 |
| 11 | foreman | frame | CIF | 112k | 299 | 15 | 31 | 31 |
| 12 | coastguard | frame | CIF | 112k | 299 | 15 | 31 | 31 |
| 13 | bream | object | QCIF | 24k | 299 | 7.5 | 16 | 16 |
| 14 | bream | object | QCIF | 48k | 299 | 7.5 | 8 | 8 |
| 15 | children | object | QCIF | 24k | 299 | 10 | 20 | 20 |
| 16 | children | object | CIF | 48k | 299 | 7.5 | 29 | 29 |
| 17 | speaker (news) | object | CIF | 20k | 299 | 7.5 | 19 | 19 |
| 18 | monitor (news) | object | CIF | 34k | 149 | 7.5 | 19 | 19 |
| 19 | speaker (news) | object | CIF | 52k | 299 | 15 | 11 | 11 |
| 20 | monitor (news) | object | CIF | 80k | 149 | 15 | 11 | 11 |
| 21 | akiyo (weather) | object | QCIF | 24k | 299 | 10 | 11 | 11 |
| 22 | akiyo (weather) | object | QCIF | 48k | 299 | 10 | 6 | 6 |
| 23 | container (container) | object | QCIF | 10k | 299 | 7.5 | 12 | 12 |
| 24 | container (container) | object | QCIF | 14k | 299 | 7.5 | 9 | 9 |

***PSNR-distance [dB] of fast motion estimation algorithms to original at same bit rate***

| Nr. | RD | liu | tss | cote | hier | null | fsrbsad | fsmme | fspdc | fsdpdc | fsbbm | proj |
|---|---|---|---|---|---|---|---|---|---|---|---|---|
| 1 | 0.36 | 0.08 | -0.04 | 0.15 | -0.15 | -2.83 | 0.08 | 0.04 | -0.79 | -0.15 | -0.82 | 0.04 |
| 2 | 0.09 | 0.01 | 0.03 | 0.00 | -0.03 | -1.23 | 0.04 | -0.12 | -0.61 | -0.05 | -0.41 | 0.03 |
| 3 | 0.01 | 0.04 | 0.01 | 0.02 | -0.03 | -3.32 | 0.07 | 0.04 | -0.24 | 0.05 | -0.02 | 0.04 |
| 4 | 0.2 | 0.12 | 0.05 | 0.09 | -0.03 | -3.12 | 0.12 | 0.08 | -0.83 | -0.21 | -0.80 | 0.03 |
| 5 | 0.59 | 0.14 | 0.04 | 0.49 | -0.08 | -1.17 | 0.20 | 0.12 | -0.23 | 0.15 | -0.20 | 0.11 |
| 6 | 0.04 | -0.02 | -0.03 | 0.03 | -0.03 | -1.96 | -0.03 | -0.01 | -0.17 | -0.02 | -0.03 | -0.02 |
| 7 | 0.23 | -0.02 | -0.99 | -0.68 | -0.89 | -6.43 | -0.11 | -0.10 | -3.48 | -0.39 | -3.48 | -0.19 |
| 8 | 0.1 | 0.14 | 0.05 | -0.37 | 0.03 | -4.79 | 0.09 | -0.03 | -3.42 | -0.19 | -3.38 | -0.06 |
| 9 | 0.36 | -0.06 | -0.39 | 0.15 | -0.38 | -2.99 | 0.02 | -0.14 | -1.26 | -0.10 | -1.28 | -0.01 |
| 10 | 0.07 | -0.08 | -0.16 | -0.06 | -0.30 | -3.42 | -0.10 | -0.10 | -1.32 | -0.25 | -1.24 | -0.10 |
| 11 | 1.49 | 0.09 | -2.58 | 0.14 | -1.71 | -4.15 | 0.20 | 0.05 | -2.47 | -0.05 | -2.70 | 0.27 |
| 12 | 0.89 | 0.45 | 0.09 | -0.09 | 0.03 | -3.53 | 0.37 | 0.31 | -2.49 | 0.18 | -2.69 | 0.33 |
| 13 | 0.06 | 0.04 | -1.54 | -1.19 | -1.60 | -4.78 | 0.25 | 0.00 | -3.71 | -0.55 | -3.72 | -0.30 |
| 14 | -0.02 | -0.06 | -1.47 | -1.12 | -1.50 | -4.16 | 0.05 | -0.11 | -2.89 | -0.65 | -2.88 | -0.36 |
| 15 | -0.05 | 0.08 | -0.14 | -0.25 | -0.52 | -3.04 | 0.20 | -0.17 | -1.24 | 0.01 | -1.21 | 0.20 |
| 16 | 0.34 | 0.22 | -0.34 | -0.16 | -0.65 | -2.59 | 0.33 | 0.00 | -1.38 | 0.09 | -1.45 | 0.18 |
| 17 | -0.2 | 0.14 | -0.04 | 0.11 | -0.17 | -4.46 | 0.09 | 0.09 | -1.27 | 0.13 | -1.32 | 0.16 |
| 18 | 0.47 | -0.01 | -0.34 | 0.28 | -0.26 | -1.50 | 0.04 | -0.09 | -1.04 | -0.19 | -1.12 | 0.08 |
| 19 | -0.02 | 0.13 | 0.12 | 0.13 | 0.01 | -3.54 | 0.13 | 0.14 | -0.52 | 0.08 | -0.39 | 0.15 |
| 20 | 0.05 | 0.11 | 0.00 | -0.09 | -0.27 | -3.03 | -0.11 | 0.06 | -1.74 | -0.34 | -1.67 | -0.01 |
| 21 | 0 | -0.04 | -0.22 | -0.37 | -0.47 | -5.03 | 0.06 | -0.14 | -1.73 | -0.10 | -1.57 | 0.05 |
| 22 | 0.09 | -0.02 | -0.16 | -0.21 | -0.46 | -4.32 | 0.03 | -0.11 | -1.58 | -0.08 | -1.37 | 0.01 |
| 23 | -0.1 | 0.35 | 0.37 | 0.29 | 0.38 | -5.34 | 0.35 | 0.39 | 0.30 | 0.37 | 0.36 | 0.35 |
| 24 | 0.04 | 0.47 | 0.46 | 0.40 | 0.42 | -4.95 | 0.46 | 0.48 | 0.36 | 0.47 | 0.45 | 0.49 |

# BIBLIOGRAPHY

## CHAPTER 1: INTRODUCTION


[Itoh 95]: K. Itoh, K. Sasaki, Y. Nakagome: "Trends in low-power RAM circuit technologies", Proceedings of the IEEE, vol. 83, Apr. 1995, pp524-543

[Kaplan 94]: Jerry Kaplan: "Startup· a Silicon Valley adventure", Penguin books, Harmondsworth, Middlesex, England, 1994, 322p

[Kneip 98]: Kneip J., Bauer S., Vollmer J., Schmale B., Kuhn P.: „The MPEG-4 video coding standard - A VLSI point of view", IEEE Workshop on Signal Processing Systems (SiPS), Cambridge, MA, USA, Oct 1998

[Kuhn 98a] Kuhn, P., Stechele, W.: "Complexity Analysis of the Emerging MPEG-4 Standard as a Basis for VLSI Implementation", vol. SPIE 3309 Visual Communications and Image Processing, San Jose, Jan. 1998, pp. 498-509

[Liu 93:1]: B. Liu, A. Zaccarin: "New Fast Algorithms for the Estimation of Block Motion Vectors", IEEE Trans. on Circuits and Systems for Video Technology", Vol. 3, No. 2, April 1993, pp 148-157

[MPEG]: MPEG web site, internet: `http://www.cselt.it/mpeg`

[Mus 85]: Hans Georg Musmann, Peter Pirsch, Hans-Joachim Grallert: "Advances in Picture Coding", Proceedings of the IEEE, vol. 73 no. 4, Apr. 1985, pp 523-548

[N 2195]: "MPEG-4 Applications", ISO/IEC JTC1/SC29/WG11/N2195, March 1998, Tokyo, latest document publically available at `http://www.cselt.it/mpeg`

[Nacht 98]: Lode Nachtergaele, Francky Catthoor, Bhanu Kapoor, Stefan Jannsens, Dennis Moolenaar: "Low-Power Data Transfer and Storage exploration for H.263 video decoder system", IEEE Journal on Selected Areas in Communications, vol. 16, no. 1, Jan 1998, p 120-129

[Netra 88]: A. Netravali, B. Haskell: "Digital Pictures Representation and Compression, New York, Plenum 1988

[Per 96]: Fernando Pereira: "MPEG-4: A new Challenge for the representation of audio-visual information", Keynote at the International Picture Coding Symposium 96, Melbourne, Australia,1996, p1-10

[Pir 95:1]: Peter Pirsch, Nicolas Demassieux, Winfried Gehrke: "VLSI Architectures for Video Compression - A Survey", Proceedings of the IEEE, vol. 83, no. 2, feb. 1995, p. 220 - 246

[Puri 98]: A. Puri, R. L. Schmidt, B.G. Haskell:"Performance Evaluation of the MPEG-4 Visual Coding Standard", SPIE vol. 3309, Visual Communications and Image Processing, San Jose 1998, pp. 417-433

[Rab 96]: Jan M. Rabaey, Massoud Pedram: "Low Power Design Methodologies", Kluwer Academic Publishers, Boston / Dordrecht / London, 1996, 367p

[Sik 97a]: T. Sikora: "The MPEG-4 Video Standard Verification Model", IEEE Trans. Circuits and Systems for Video Technology, Vol. 7, No. 1, Feb 1997, pp 19 - 31

[Sik 97b]: T. Sikora: "MPEG Digital Video Coding Standards", IEEE Signal Processing Magazine, Vol. 14, No. 5, Sept. 1997, p 82 -100

[ZKuhn 97a]: Zeller M., Aicher W., Kuhn. P., Stechele W.: "Hard- und Softwareaspekte der Videocodierung - Teil 2: Videoprozessoren und Architekturen", F&M, Zeitschrift für Elektronik, Optik und Mikrosystemtechnik, Nr. 10, 105 Jahrgang, Carl Hanser Verlag, Muenchen, p698-702, Oct. 1997 (in german)

[ZKuhn 97b]: Zeller M., Kuhn P., Hutter A., Stechele W · "Hard- und Softwareaspekte der Videocodierung - Teil 1: Prinzipien und Standards", F&M, Zeitschrift für Elektronik, Optik und Mikrosystemtechnik, Jahrgang, 105, Carl Hanser Verlag, Muenchen, p338-342, May 1997 (in german)

# CHAPTER 2: MOTION ESTIMATION ALGORITHMS

### *Motion Estimation - General*

[Benz 97]: Ulrich Benzler: "Results of Core Experiment P8 (Motion and Aliasing Compensation Prediction)", ISO/IEC JTC1/SC29/WG11/M 2229, July 1997, Stockholm

[Berg 71]: Tobi Berger: "Rate Distortion Theory: A Mathematical Basis for Data Compression", Prentice-Hall Elec. Eng. Series, Englewood Cliffs, NJ, 1971

[Bha 95]: Vasudev Bhaskaran and Konstantinos Konstantinides: "Image and Video Compression Standards: Algorithms and Architectures", Kluwer Academic Publishers, 1995, 352pp.

[Duf 92]: F. Dufaux, M. Kunt: "Multigrid block matching motion estimation with an adaptive local mesh refinement", SPIE Proc. Visual Communications and Image Processing 1992, vol. 1818, Boston, MA, Nov. 1992, pp97-109

[Duf 95]: F. Dufaux and F. Moscheni: "Motion estimation techniques for digital TV: A review and a new contribution", Proc. IEEE, vol. 83, no. 6, pp. 858-876, 1995

[Gilge 90]: M. Gilge:" Regionenorientierte Transformationscodierung in der Bildkommunikation", Dissertation, Aachen, 1990 (in german)

[Gir 94]: B. Girod: "Rate constrained motion estimation", SPIE Proc. Visual Communications and Image Processing, vol 2308, pp 1026-1034, 1994

[H.261]: ITU-T Recommendation H.261: "Video Codec for Audiovisual Services at px64 kbit/s", Geneve 1990

[H.263]: ITU-T Recommendation H.263: "Video Coding For Low Bitrate Communication", Geneve 1996

[Hampson 96]: F. J. Hampson, R. E. H. Franich, J.-C. Pesquet, J. Biemont: "Pel-Recursive Motion Estimation in the Presence of Illumination Variations", ICIP'96, IEEE Intern. Conf. on Image Processing, 1996

[Kuhn 98b]: Kuhn P., et al.: "Complexity and PSNR-Comparison of several Fast Motion Estimation Algorithms for MPEG-4", vol. SPIE 3460 Applications of Digital Image Processing XXI, San Diego, July 1998, pp 486-499

[Kuhn 98a]: Kuhn, P., Stechele, W.: "Complexity Analysis of the Emerging MPEG-4 Standard as a Basis for VLSI Implementation", vol. SPIE 3309 Visual Communications and Image Processing, San Jose, Jan. 1998, pp. 498-509

[MPEG 4]: ISO/IEC JTC1/SC29/WG11/MPEG97/N1902: "Committee Draft of ISO/IEC 14496-2 (MPEG-4 Visual)", November 1997

[Mus 85]: Hans Georg Musmann, Peter Pirsch, Hans-Joachim Grallert: "Advances in Picture Coding", Proceedings of the IEEE, vol. 73 no. 4, Apr. 1985, pp 523-548

[Nega 89]: S. Negahdaripour, A. Shokrollahi, and M, Gennert: "Relaxing Brightness Constancy Assumption in Computing Optical Flow", ICIP' 89, IEEE International Conference on Image Processing, 1989

[Netra 88]: A. Netravali, B. Haskell: "Digital Pictures Representation and Compression, New York, Plenum 1988

[Nico 95]: H. Nicolas and C. Labit: "Motion and Illumination Variation Estimation Using a Hierarchy of Models: Application to Image Sequence Coding", Journal of Visual Communication and Image Representation, Vol. 6, No. 4 December, 1995, pp. 303-316

[Ohm 95]: Jens-R. Ohm: "Digitale Bildcodierung: Repraesentation, Kompression und Übertragung von Bildsignalen", Springer Verlag, 1995, 487 pp

[Pir 95:1]: Peter Pirsch, Nicolas Demassieux, Winfried Gehrke: "VLSI Architectures for Video Compression - A Survey", Proceedings of the IEEE, vol. 83, no. 2, feb. 1995, p. 220 - 246

[Press 92]: W.H. Press, S.A. Teukolsky, W.T. Vetterling, B. P. Flannery: "Numerical Recipes in C", Cambridge University Press, 1992, p 353

[Rao 90]: K.R. Rao, P. Yip: "Discrete Cosine Transform - Algorithms, Advantages, Applications", Academic Press, Boston / San Diego / New York / London / Sydney / Tokyo / Toronto, 1990, p 242ff

[TMN 9]: ITU-Telecommunication Standardization Sector-Study Group 16 Video Coding Experts Group: Video Codec Test Model Near Term Version 9

[Tek 95]: Murat A. Tekalp: "Digital Video Processing", Prentice Hall, 1995, 550pp

### *Fast Motion Estimation Algorithms*

[Armi 97]: Robert M. Armitano, Ronald W. Schafer, Frederick L. Kitson, Vasudev Bhaskaran: "Robust block-matching motion estimation technique for noisy sources", ICASSP 97

[Baek 96]: Yunju Baek, Hwang-Soek Oh and Heung-Kyu Lee: "Block-matching criterion for efficient VLSI implementation for motion estimation", IEE Electronics Letters, 20th June 1996, vol. 32, no. 13, Jun 1996, pp 1184-1185

[Bier 86]: M. Bierling: "Displacement Estimation by hierarchical Blockmatching", SPIE Vol. 1001 Visual Communications and Image Processing, May 1988, pp 942-951

[CCLin 97]: Chihfeng C. Lin, Daniel J. Pease: "An efficient block matching motion estimation algorithm on a valid assumption of the convex distortion", ICIP 97, 1997

[CHLee97]: Chang-Hsing Lee, Ling-Hwei Chen: "A Fast Motion Estimation Algorithm Based on the Block Sum Pyramid", IEEE Transactions on Image Processing, vol. 6, no. 11, Nov. 1997, pp 1587-1591

[CHLin 95]: Chung-Hung Lin, Ja-Ling Wu: "Fast motion estimation algorithm with adjustable search area", SPIE Visual Communications and Image Processing, vol. 2501, 1995, pp 1328-1336

[Chali 97]: Junavit Chalidabhongse, C.-C. Jay Kuo: "Fast Motion Vector Estimation using multiresolution-spatio-temporal correlations", IEEE Transactions on Circuits and Systems for Video Technology, vol. 7, no. 3, Jun. 1997, pp 477-488

[Chan 95]: Y.-L. Chan and W.-C. Siu: "Adaptive multiple-candidate hierarchical search for block matching algorithm", IEE Electronics Letters, vol. 31, no. 19, 14 th Sept. 1995, pp 1637-1639

[Chan 96]: Yui-Lam Chan, Wan-Chi Siu: "New adaptive pixel decimation for block motion vector estimation", IEEE Transactions on circuits and systems for video technology, vol. 6, no. 1, Feb. 1996, pp 113-118

[Chen 91]: Liang-Gee Chen, Wai-Ting Chen, Yeu-Shen Jehng, Tzi-Dar Chiueh: "An Efficient Parallel Motion Estimation Algorithm for Digital Image Processing", IEEE Transactions on Circuits and Systems for video technology, vol. 1, no. 4, Dec. 1991, pp 378-385

[Chen 95:1]: Mei-Juan Chen, Liang-Gee Chen, Tzi-Dar Chiueh, Yung-Pin Lee: "A new block-matching criterion for motion estimation and its implementation", IEEE Transactions on Circuits and Systems for Video Technology, vol. 5, no. 3, Jun. 1995, pp 231-236

[Cheng 96]: K.W. Cheng, S.C. Chan: "Fast block matching algorithms for motion estimation", ICASSP 96, 1996, p 2318ff

[Chow 93]: Keith Hung-Kei Chow, Ming L. Liu: "Genetic Motion Search Algorithm for Video Compression", IEEE Transactions on circuits and Systems for Video Technology, vol. 3, no. 6, Dec. 1993, pp 440-445

[Chu 96]: Wilson C. Chung, Faouzi Kossentini, Mark J. T Smith: "An efficient motion estimation technique based on a rate-distortion criterion", ICASSP 96, 1996

[Chun 94]: K.W. Chun, and J.B. Ra: "An improved block matching algorithm based on successive refinement on motion vector candidates", Signal Processing: Image Communication, vol. 6, 1994, pp 115-122

[Cob 97]: Muhammed Z. Coban, Russell M. Mersereau: "Computationally Efficient Exhaustive Search Algorithm for Rate-constrained Motion Estimation", IEEE International Conference on Image Processing, ICIP 97, Santa Barbara, Oct. 1997

[Cote 97]: Guy Cote, Michael Gallant, Faouzi Kossentini: "Efficient Motion Vector Estimation and Coding for H.263-based very low bit rate video compression", ITU-T SG 16, Q15-A-45, June 1997, 18 p

[Eck 95]. Stefan Eckart, Chad Fogg: "ISO/IEC MPEG-2 software video codec", SPIE Vol. 2419. Digital Video Compression: algorithms and Technologies, 1995, San Jose, CA, USA

[Fan 97]: Jiangping Fan, Fuxi Gan: "Motion Estimation based on Global and Local Uncompensability Analysis", IEEE Transactions on Image Processing, vol. 6, no. 11, Nov. 1997, pp 1584-1587

[Feng 95]: Jian Feng, Kwok-Tung LO, Hassan Mehrpour, A.E. Karbowiak: "Adaptive Block Matching Motion Estimation algorithm Using Bit-Plane Matching", IEEE International Conference on Image Processing, ICIP 95, 95

[Feng 95a]: Feng, J , Lo, K.T., Mehrpour, H. Karbowiak, A.E: "Adaptive block-matching motion estimation algorithm for video coding", IEE Electron. Letters, vol. 31, no. 18, 1995, pp 1542-1543

[Fuhrt 97]: Botho Furht, Joshua Greenberg, Raymond Westwater, "Motion Estimation Algorithms for Video Compression", Kluwer Academic Publishers, Boston / Dordrecht / London, 162 pp, 1997

[Gha 90]: M. Ghanbari: "The cross-search algorithm for motion estimation", IEEE Transactions on Communications, vol 38, no. 7, Jul. 1990, pp 950-953

[Girod 93]: Bernd Girod: "Motion-Compensating Prediction with Fractional Pel Accuracy", IEEE Transactions on Communications, vol. 41, no. 4, Apr. 1993, pp 604-612

[Haan 93]: Gerhard deHaan, Paul W.A.C. Biezen: "Sub-pixel motion estimation with 3-D recursive search block matching", Signal Processing: Image Communication", vol. 6, 1994, pp 229-239

[Haan 93a]: G. de Haan, P.W.A.C. Biezen, H Hijgen, O.A Ojo: "True motion estimation with 3-D recursive search block matching", IEEE Trans.Circuits and Systems for Video Technology, vol. 3, 1993, pp 350-367

[He 97]: Zhong-Li He, Kai-Keung Chan, Chi-Ying Tsui, Ming L. Liou: "Low Power Motion Estimation Design Using Adaptive Pixel Truncation", IEEE Proceedings of 1997 International Symposium on Low Power Electronics and Design, 1997, pp 167-171

[HeL 97]: Zhongli He, Ming L. Liou. "Design of Fast Motion Estimation Algorithm based on Hardware Consideration", IEEE Transactions on circuits and systems for video technology, vol 7, no. 5, oct. 1997, pp 819-823

[Hwang 97]· Wen-Jyi Hwang, Yeong-Cherng Lu, Yi-Chong Zeng: "Fast block-matching algorithm for video coding", IEE Electronics Letters, vol. 33, no. 10, 8th May 1997, pp 833-835

[JLu 97]: Jianhua Lu, Ming L. Liu: "A simple and efficient search algorithm for block-matching motion estimation", IEEE Transactions on Circuits and Systems for Video Technology, vol. 7, no. 2, Apr. 1997, pp 429-433

[Jain 81]: J. R. Jain, A.K. Jain: "Displacement Measurement and its application in interframe image coding", IEEE Transactions Commun. vol. COM-29, Dec. 1981, pp 1799-1808

[Jung 96]: Hae Mook Jung, Duck Dong Hwang, Choong Soo Park, Han Soo Kim: "An Annular Search Algorithm for Efficient Motion Estimation", International Picture Coding Symposium, PCS 96, 1996, pp 171-174

[Kap 85]: S. Kappagantula, K.R. Rao: "Motion compensated interframe image prediction", IEEE Transactions on Communications, 33(9), Sept 1985, pp 1011-1015

[Kim 92]: Joon-Seek Kim, Rae-Hong Park: "A fast feature-based block matching algorithm using integral projections", IEEE Journal on Selected areas in communications, vol. 10, no. 5, Jun. 1992, pp 968-971

[Kim 94:1]: M.K. Kim and J.K. Kim: "Efficient motion estimation algorithm for bidirectional prediction scheme", IEE Electronics Letters, vol. 30, no. 8, Apr. 1994, pp 632-633

[Koga 81]: T. Koga, K. Iinuma, A. Hirano, Y. Ijima, T. Ishiguro: "Motion compensated interframe coding for video conferencing", in Proc. NTC 81, pp. C 9.6.1 - 9.6.5

[Kos 97]: Faouzi Kossentini, Yuen-Wen Lee, Mark J. T. Smith, Rabab K. Ward: "Predictive RD optimized Motion Estimation for Very Low Bit-Rate Video Coding", IEEE Journal on Selected Areas in Communications, Vol. 15, No. 9, Dec 1997

[Kwan 97]: Chok-Kwan Cheung, Lai Man Po: "A hierarchical block motion estimation algorithm using partial distortion measures", IEEE International Conference on Image Processing, ICIP 97, Santa Barbara, Oct. 1997

[Lak 97]: Prasad Lakamsani: "An architecture for enhanced three step search generalized for hierarchical motion estimation algorithms", IEEE Trans. on Consumer Electronics, vol. 43, no. 2, May 1997, pp 221-227

[Leng 98]: Krisda Lengwehasatit, Antonio Ortega, Andrea Basso, Amy Reibmann: "A novel computationally scalable algorithm for motion estimation", SPIE 3309 VCIP Visual Communications and Image processing, San Jose, CA, Jan 1998, pp 68-79

[LKLiu 96]: Lurng-Kuo Liu, Ephraim Feig: "A Block-based Gradient Descent Search Algorithm for block-based motion estimation in video coding", IEEE Transactions on Circuits and Systems for Video Technology, vol. 6, no. 4, Aug. 1996, pp 419-422

[LKLiu 97]: Lurng-Kuo Liu: "Rate-constrained motion estimation algorithm for video coding", IEEE International Conference on Image Processing, ICIP 97, 1997

[LLuo 97]: Lijun Luo, Cairong Zou, Xiqi Gao, Zhenya He: "A new prediction search algorithm for block motion estimation in video coding", IEEE Trans on Consumer Electronics, vol. 43, no. 1, Feb. 1997, pp 56-61

[LWLS 93]: Liang-Wei Lee, Jhing-Fa Wang, Jau-Yien Lee, Jung-Dar Shie: "Dynamic Search-Window Adjustment and Interlaced search for block-matching algorithm", IEEE Transactions on circuits and systems for video technology, vol. 3, no. 1, Feb. 1993, pp 85-87

[Le 98]: Thinh M. Le, M. Snelgrove, S. Panchanatan: "Fast motion estimation using feature extraction and XOR operations", SPIE 3311 MHA Multimedia Hardware Architectures, San Jose, 1998, pp 108-118

[Lee 97]: Yuen-Wen Lee, Faouzi Kossentini, Mark J. T. Smith, Rabab Ward: "Prediction and Search Techniques for RD-Optimized motion estimation in a very low bit rate video coding framework", ICASSP 97, 1997, pp 2861 ff

[Lee 97a]: Yuen-Wen Lee, Faouzi Kossentini, Rabab Ward, Mark Smith: "Towards MPEG-4: An improved H.263-based video coder", Signal Processing: Image Communication, vol. 10, 1997, pp 143-158

[LeeC 96]: S. Lee and S.-I. Chae: "Motion Estimation algorithm using low resolution quantisation", IEE Electronic Letters, vol. 32, no. 7, 28 th. Mar. 1996, pp 647-648

[Lin 97]: Chun-Hung Lin, Ja-Ling Wu, Yi-Shin Tun: "DSRA: A blockmatching algorithm for near-real-time video encoding", IEEE Trans on Consumer Electronics, vol. 43, no. 2, May. 1997, pp 112-122

[Linz 96]: Elliot Linzer, Prasoon Tiwari, Mohammad Zubair: "High performance algorithms for MPEG motion estimation", ICASSP 96, 1996, pp 1935 ff

[Liu 93:1]: B. Liu, A. Zaccarin: "New Fast Algorithms for the Estimation of Block Motion Vectors", IEEE Trans. on Circuits and Systems for Video Technology", Vol. 3, No. 2, April 1993, pp 148-157

[Luo 97]: Li-Jun Luo, Cai-Rong Zou, Xi-Qi Gao, Zhen-Ya He: "Matching criterion function for block motion estimation: block feature matching function", IEE Electronics Letters, vol. 33, no. 11, 22nd May 1997, pp 929-931

[M3299]: Kai-Kuang Ma, Prabhudev I. Housr, Lei Huang: "Status Report of Core Experiment on Fast Block-Matching Motion Estimation", ISO/IEC JTC1/SC29/WG11, MPEG98/M3299, Tokyo, March 1998

[MCChen 96]: Michael C. Chen, Alan N. Willson, Jr.: "Rate-Distortion optimal motion estimation algorithm for video coding", ICASSP 96, 1996

[Mit 97]: Joan L. Mitchell, William B. Pennebaker, Chad E. Fogg, Didier J. LeGall: "MPEG video Compression standard, Chapter 13: Motion Estimation", Chapman & Hall, 1997, pp 283-311

[Miz 96]: Marcelo M. Mizuki, Ujjaval Y. Desai, Ichiro Masaki, Anantha Chandrakasan: " A binary block matching architecture with reduced power consumptation and silicon area requirement", ICASSP 96, 1996, p 3249ff

[Nam 95]: Kwon Moon Nam, Joon-Seek Kim Rae-Hong Park, Young Serk Shim: "A fast hierarchical motion vector estimation algorithm using mean pyramid", IEEE Transactions on Circuits and Systems for Video Technology, vol. 5, no. 4, Aug. 1995, pp 344-351

[Nat 96]: Balas Natarajan, Bhaskaran Vasudev, Konstantinos Konstantinides: "Low-complexity algorithm and architecture for block-based motion estimation via one-bit transforms", ICASSP 96, 1996, p 2345 ff

[Nat 97]: Balas Natarajan, Vasudev Bhaskaran, Konstantinos Konstantinides: "Low-Complexity Block-based motion estimation via One-Bit Transforms", IEEE Transactions on Circuits and Systems for Video Technology, vol. 7, no. 4, Aug. 1997, pp 702-706

[Oli 97]: Geraldo Cesar de Oliveira, Abraham Alcaim: "On fast motion compensation algorithms for video coding", International Picture Coding Symposium, PCS 97, Berlin Sept. 1997, pp 467-472

[Panus 97]: K. Panusopone and K.R. Rao: "Efficient Motion Estimation for Block based video compression", ICASSP 97, 1997, pp 2677 ff

[Pick 97]: Mark R. Pickering, John F. Arnold, Michael R. Frater: "An adaptive search length algorithm for block matching motion estimation", IEEE Transactions on Circuits and Systems for video-technology, vol. 7, no. 6, Dec. 1997, pp 906--912

[Po 96]: Lai-Man Po, Wing-chung Ma: "A novel Four-Step Search Algorithm for fast blockmatching", IEEE Transactions on Circuits and Systems for Video Technology, vol. 6, no. 3, Jun. 1996, pp. 313-317

[Puri 87]: A. Puri, H.M. Hang, D.L. Schilling: "An efficient blockmatching algorithm for motion compensated coding", Proc. IEEE ICASSP 1987, pp2.4.1.-25.4.4.

[RLi 94]: R. Li, B. Zeng, M.L. Liu: "A new three-step search algorithm for block motion estimation," IEEE Transactions on Circuits and Systems for Video Technology, vol. 4, no. 4, pp 438--442, Aug. 1994

[SKim 95]: Sungook Kim, Junavit Chalidabhongse, C.-C. Jay Kuo: "Fast motion vector estimation by using spatio-temporal correlation of motion field", SPIE Visual Communications and Image Processing VCIP 2501, 1995, pp 810-821

[SKim 96]: Sungook Kim, Junavit Chalidabhongse, C.-C. Jay Kuo: "A new stochastic block matching algorithm (SBMA) for video coding based on modified 3-step search", ICASSP 96, 1996

[Sauer 96]: Ken Sauer and Brian Schwartz: "Efficient Motion Estimation using Integral Projections", Transactions on Circuits and Systems for Video Technology, vol. 6, no. 5, Oct. 1996, pp 513-518

[Shi 97]: Y. Q. Shi, X. Xia: "A thresholding Multiresolution Block Matching Algorithm", IEEE Transactions on Circuits and Systems for Video Technology, vol. 7, no. 2, Apr. 1997, pp 437-440

[Sri 85]: Ram Srinivasan and K.R.Rao: "Predictive Coding Based on efficient motion estimation", IEEE Transactions on Communications, vol. COM-33, no. 8, Aug. 1985, pp 888-896

[Sri 98]: Parthasarathy Sriram, Subramania Sudharsanan: "Entropy Constrained Motion Estimation: Algorithm and Implementation", SPIE 3311 MHA Multimedia Hardware Architectures, 1998, San Jose, CA, pp 99-107

[Song 98]: Xudong Song, Ya-Qin Zhang, Tihao Chiang: "Hierarchical motion estimation using binary pyramids with 3-scale tilings", SPIE 3309 VCIP Visual Communications and Image Processing, 1998, San Jose, CA, pp 80-87

[Song 98a]: Byung Cheol Song, Jong Beom Ra: "A hierarchical block matching algorithm using partial distortion criteria", SPIE 3309 VCIP Visual Communications and Image processing, 1998, San Jose, CA, pp 88-95

[Su 97]: Jonathan K. Su, Russell M. Mersereau: "Non-iterative Rate-constrained motion estimation for OB-MC", IEEE International Conference on Image Processing, ICIP 97, Santa Barbara, 1997

[Tao 98]: Bo Tao, Michael T. Orchard: "Feature-Acceleratod Block Matching", SPIE 3309 VCIP Visual Communications and Image processing, San Jose, CA, 1998, pp 469-476

[Tzo 94]: D. Tzovaras, M. G. Strintzis, H. Sahinolu: "Evaluation of multiresolution block matching techniques for motion and disparity estimation", Signal Processing: Image Communication, vol. 6, 1994, pp 56-67

[WLi 95]: W. Li, E. Salari "Succesive elimination algorithm for motion estimation", IEEE Trans. Image Processing, vol. 4 pp 105-107, Jan 1995, pp 105-107

[XLee 96]: Xiaobing Lee, Ya-Qin Zhang: "A fast hierarchical Motion-Compensation Scheme for Video Coding using Block-feature Matching", IEEE Transactions on Circuits and Systems for Video Technology, vol. 6, no. 6, Dec. 1996, pp 627-635

[Xie 92]: Kan Xie, Luc Van Eycken, and Andre Oosterlinck: "A new block based motion estimation algorithm", Signal Proc: Image Communication, vol. 4, Nov. 1992, pp 507-517

[Xu 97]: Jie-Bin Xu, Lai-Man Po, Chok-Kwan Cheung: "A new prediction model search algorithm for fast block motion estimation", IEEE Internat. Conference on Image Processing, ICIP 97, Santa Barbara, 1997

[Yu 97]: Fengqi Yu and Alan N. Willson, Jr.: "A flexible hardware-oriented fast algorithm for motion estimation", ICASSP 97, 1997, pp 2681 ff

[Zeng 97]: Bing Zeng, Renxiang Li, Ming L. Liou: "Optimization of Fast Block motion estimation algorithms", IEEE Transactions on circuits and systems for video technology, vol. 7, no. 6, Dec. 1997, pp. 833-844

[ZHe 97]: Zhongli He, Ming L. Liou: "A high performance fast search algorithm for block matching motion estimation", IEEE Transactions on circuits and systems for video technology, vol. 7, no. 5, Oct. 1997, pp 826-828

[Zhong 96]: Sheng Zhong, Francis Chin, Y.S. Cheung, Doug Kwan: "Hierarchical Motion Estimation based on visual patterns for video coding", ICASSP 96, 1996

[Zhu 98]: Shan Zhu, Kai-Kuang Ma: "A new diamond search algorithm for fast block matching", submitted to IEEE Transactions on Image processing

### *Variable block size motion estimation*

[Cal 97]: S. Calzone, K. Chen, C.-C. Chuang, A. Divakaran, S. Dube, L. Hurd, J. Kari, G. Liang, F.-H. Lin, J. Muller, H.K. Rising III: "Video Compression by Mean-Corrected Motion Compensation of Partial Quadtrees", IEEE Trans. on Circuits and Systems for Video Technology, Vol. 7, No. 1, p 86ff, February 1997

[Chan 90]: M.H. Chan and Y.B. Yu and A.G. Constantinides: "Variable size block matching motion compensation with applications to video coding", IEE Proceedings, Vol 137, Pt. I, No. 4, August 1990

[Gis 96]: J. V. Gisladottir, K. Ramchandran and M. Orchard: " Motion-based representation of Video Sequences using Variable Block Sizes", SPIE Visual Communications and Image Processing,, Vol. 2727, 1996, Orlando, p 368-374

[Har 96]: Erich Haratsch: "Core Experiment P5 für MPEG-4: Entropy Constrained Variable Block Size Motion Estimation, Motion Compensation", Studienarbeit, Institute for Integrated Circuits, Technical University of Munich, Germany, 1996

[HBi 96]: Hao Bi, Wai-Yip Chan: "Rate-constrained hierarchical motion estimation using BFOS tree pruning", ICASSP 96, 1996, p 2317

[Kim 95a]: Jong Won Kim, Sang Uk Lee: "Rate-distortion optimization between hierarchical variable block size motion estimation and motion sequence coding", SPIE visual Communications and Image Processing (VCIP), vol. 2501, 1995, pp822-833

[Kim 96]: Jong Won Kim, Sang Uk Lee: "On the Hierarchical Variable Block Size Motion Estimation Technique for Motion Sequence Coding", SPIE Visual Communications and Image Processing, Vol. 2094, 1993, Cambridge, p 372-383

[Lee 95]: Jungwoo Lee, "Optimal Quadtree for Variable Block Size Motion Estimation", ICIP-95, Washington DC, Oct 1995

[Li 96]: W. Li and F. Dufaux: "Image Sequence Coding by Multigrid Motion Estimation and Segmentation based coding of prediction errors", SPIE Visual Communications and Image Processing, Vol. 2094, 1993, Cambridge, p 542-552

[M 0553]: S. Calzone, R. Chuang, A. Divakaran, S. Dube, L. Hurd, J. Kari, G. Liang, F.-H. Lin, J. Muller, H.K. Rising, S. Ye, M. Zeug: "Iterated Systems MPEG-4 Video Submission Technical Description", ISO/IEC JTC1/SC29/WG11 MPEG96/M0553, Munich, Germany, 1996

[M 1031]: R. Chuang, L. Hurd, S. Lyles, J. Muller, and M. Zeug: "Results of Core Experiment P5, Comparision of entropy constrained variable block size motion estimation motion compensation", ISO/IEC JTC1/SC29/WG11 MPEG96/M1031, Tampere, Finland, 1996

[M 1291]: T. Wiegand, M. Flierl: "Results of Core Experiment P5 (Entropy-Constrained Variable Block Size Coding)", ISO/IEC JTC1/SC29/WG11 MPEG96/M1291, Chicago, Illinois, 1996

[M 1294]: A. Hutter, E. Haratsch, S. Herrmann, P. Kuhn : "Results of Core Experiment P5 (Entropy Constrained Variable Block Size Coding)", ISO/IEC JTC1/SC29/WG11 MPEG96/M1294, Chicago, Illinois, 1996

[Puri 87]: A. Puri, H.-M. Hang and D.L. Schilling: "Interframe Coding with Variable Block-size Motion Compensation", Globecom 1987

[Schu 96]: Guido M. Schuster, Aggelos K. Katsaggelos: "Rate-Distortion Based Video Compression Optimal Video Frame Compression and Object Boundary Encoding", Kluwer Academic Publishers, Boston, 1996, 312 pp.

[Schu 97]: Guido M. Schuster, Aggelos K. Katsaggelos: "A video compression scheme with optimal bit allocation among segmentation, motion and residual error", IEEE Transactions on Image processing, vol. 6, no. 11, 1997, pp 1487-1502

[Su 91]. Gary J. Sullivan, Richard L. Baker: "Rate-Distortion optimized motion compensation for video compression using fixed an variable size blocks", IEEE Global Communication Conference (Globecom), Phoenix, Arizona, Dec 2-5, no. 3.3 1, 1991, pp 85-90

[Su 94]: Gary J. Sullivan and Richard L. Baker: "Efficient Quadtree Coding of Images and Video", IEEE Trans. Image Processing, vol. 3, no. 3, pp.327--331, May 1994

[Stro 91]: Peter Strobach: "Quadtree-Structured Recursive Plane Decomposition Coding of Images", IEEE Transactions on Signal Processing, vol. 39, no. 6, Jun 1991, pp 1380-1397

[Truo 96]: K. K. Truong and C. H. Richardson: "A hierarchical Video Coder with cache motion estimation", ICASSP 1996, p 1209-1212

[YHua 96]: Yan Huang, Xinhua Zhuang, Changseng Yang: "Two block-based motion compensation methods for videocoding", IEEE Transactions on circuits and systems for video technology, vol. 6, No. 1, Feb. 1996

# CHAPTER 3: METHODOLOGY FOR COMPLEXITY ANALYSIS

[Bak 91]: Nabajyoti Barkakati: Object-Oriented Programming in C++, SAMS, 1991

[Be 95]: Michael Bekerman and Avi Mendelson: „A Performance Analysis of Pentium Processor Systems“, IEEE Micro, October 1995, pp 72-83

[Bha 96]: Dilep P. Bhandarkar: Alpha Implementations and Architecture: Complete Reference and Guide, Digital Press, 1996

[Cha 91]: Steve Chamberlain: libbfd - The Binary File Descriptor Library, April 1991, documentation for the GNU libbfd library, in package ftp://prep.ai.mit.edu/pub/gnu/binutils-2.6.tar.gz

[Con 95]: Thomas M. Conte, Charles E. Girmac (Editors): Fast Simulation of Computer Architectures, Kluwer Academic Publishers, 1995

[Dall 94]: Sonia L.Q. Dall'Agnol, Abraham Alcaim, Jose Roberto B. de Marca: "Performance of LSF Vector Quantizers for VSELP Coders in Noisy Channels", European Transactions on Telecommunications and related technologies, Vo. 5, No. 5, Sept-Oct 1994, p559

[Fen 93]: Jay Fenlason and Richard M. Stallman: GNU gprof, Jan 1993, documentation for the GNU profiler, in package: ftp://prep.ai.mit.edu/pub/gnu/binutils-2.6.tar.gz

[Grah 82]: S. Graham, P. Kessler, M. McKusick, gprof: A Call Graph Execution Profiler, in Proceedings of the SIGPLAN82 Symposium on Compiler Construction, SIGPLAN Notices Vol.17, No.2, June 1982, pp 120-126

[GCT]: Brian Marick: Generic Coverage Tool (GCT): ftp://cs.uiuc.edu/pub/testing

[GNU CC]: The GNU C compiler: `ftp://prep.ai.mit.edu/pub/gnu/gcc-2.7.2.tar.gz`

[GNU MAN]: Richard M. Stallman: Using and Porting GNU CC, manual for the GNU C compiler, ftp://prep.ai.mit edu/pub/gnu

[Grah 83]: S. Graham, P. Kessler, M. McKusick, An Execution Profiler for Modular Programs, Software—Practice and Experience, 1993, Vol 13, p671-685

[Hen 90]: John L. Hennesey, David A. Patterson: Computer architecture a quantitative approach, Kaufmann, (german translation used, publisher: Vieweg, 1994)

[iprof]: `iprof`-Software available under GNU licence terms. For software location and information send email to: `Peter.Kuhn@computer.org`

[Itoh 95]: K. Itoh, K. Sasaki, Y. Nakagome: "Trends in low-power RAM circuit technologies", Proceedings of the IEEE, vol. 83, Apr. 1995, pp524-543

[Kuhn 98a] Kuhn, P., Stechele, W.: "Complexity Analysis of the Emerging MPEG-4 Standard as a Basis for VLSI Implementation", vol. SPIE 3309 Visual Communications and Image Processing, San Jose, Jan. 1998, pp. 498-509

[Kuhn 98b]: Kuhn P., et al.: "Complexity and PSNR-Comparison of several Fast Motion Estimation Algorithms for MPEG-4", vol. SPIE 3460 Applications of Digital Image Processing XXI, San Diego, July 1998

[Larus 90]: James R Larus: Abstract Execution: A Technique for Efficiently Tracing Programs, Software Practice and Expierience, Volume 20, Number 12, Dec. 1990, pp 1241-1258

[Larus 93]: James R. Larus: Efficient Program Tracing, IEEE Computer, Volume 26, Number 5, May 1993, pp 52-60

[Larus 94]: James R. Larus, Thomas Ball. Rewriting Executable Files to Measure Program Behavior, Software Practice and Expierience, Volume 24, Number 2, Feb. 1994, pp 197-218

[Larus 95]: James R. Larus and Eric Schnarr, EEL: Machine-Independent Executable Editing, Proceedings of the SIGPLAN '95 Conference on Programming Language Design and Implementation (PLDI), June 1995, pp 291-300

[M 2863]: Kuhn P.: "A Complexity Analysis Tool: iprof (version 0.3)", ISO/IEC JTC1/SC29/WG11/M2863, Fribourg (CH), Switzerland, October 1997

[M 1056]: Kuhn, P.. "A portable Instruction Level Profiler for Complexity Analysis - Software``, ISO/IEC JTC1/ SC29/WG11 MPEG96/M1056, Tampere, Finland, 1996

[M 0921]: Kuhn, P.: "A portable Instruction Level Profiler for Complexity Analysis - Documentation``, ISO/IEC JTC1/ SC29/WG11 MPEG96/M0921, Tampere, Finland, 1996

[M 0838]: Kuhn, P.: "Instrumentation Tools and Methods for MPEG-4 VM: Review and a new Proposal``, ISO/IEC JTC1/SC29/WG11 MPEG96/M0838, Firence, Italy, 1996

[Marca 94]: Jose Roberto B. de Marca: "An LSF Quantizer for the North-American Half-Rate Speech Coder", IEEE Transactions on Vehicular Technology, Vol. 43, No. 3, Aug. 1994

[Meh 96]: Huzefa Mehta, Robert Michael Owens, Mary Jane Irwin: "Instruction Level Power Profiling", ICASSP 96, p3327

[Nacht 98]: Lode Nachtergaele, Francky Catthoor, Bhanu Kapoor, Stefan Jannsens, Dennis Moolenaar: "Low-Power Data Transfer and Storage exploration for H.263 video decoder system", IEEE Journal on Selected Areas in Communications, vol. 16, no. 1, Jan 1998, pp 120-129

[Oli 97]: Geraldo Cesar de Oliveira, Abraham Alcaim: "On fast motion compensation algorithms for video coding", International Picture Coding Symposium, PCS 97, Berlin Sept. 1997, pp 467-472

[Sri 94]: Amitabh Srivastava, Alan Eustace: ATOM: A system for building customized program analysis tools, Proceedings of the SIGPLAN 1994 Conference on Programming Language Design an Implementation (PLDI), Orlando (Florida, USA), Jun. 1994, pp. 196-205

[Strou 91]: Bjarne Stroustroup: The C++ Programming Language, 2nd Edition, Addison Wesley, 1991

[Tiw 96]: Vivek Tiwari, Sharad Malik, Andrew Wolfe, Mike Tien-Chien Lee: "Instruction Level Power Analysis and Optimization of Software", Journal of VLSI Signal Processing Systems, vol. 13, 1996, pp 223-238.

[Tremb 95]: Marc Tremblay, Guillermo Maturana, Atsushi Inoue, Les Kohn: A fast and flexible performance simulator for micro-architecture trade-off analysis on UltraSPARC-I, 32th Design Automation Conference, DAC 95, 1995.

[Wea 94]: David L. Weaver, Tom Germond: „The Sparc Architecture Manual“, 1994, Prentice Hall

# CHAPTER 4: COMPLEXITY ANALYSIS OF MPEG-4 VISUAL

[Bha 95]: V. Bhaskaran, K. Konstantinides: "Image and Video Compression Standards: Algorithms and Architectures", Kluwer Academic Publishers, Boston / Dordrecht / London, 1995, p208 ff

[Gut 92]: K. Guttag, R J. Gove, J.R. van Aken: "A single-chip multiprocessor for multimedia: the MVP", IEEE Computer Graphics and Applications, 12 (6), Nov. 1992, p53 -64

[H.261]: ITU-T Recommendation H.261: "Video Codec for Audiovisual Services at px64 kbit/s", Geneve 1990

[H.263]: ITU-T Recommendation H.263: "Video Coding For Low Bitrate Communication", Geneve 196

[iprof]: `iprof`-Software available under GNU licence terms. For location and information email to: `Peter.Kuhn@computer.org`

[Kap 98]: Bhanu Kapoor: "An analysis of memory bandwidth requirement for the h.263 video codec", SPIE 3309 Visual Communications and Image Processing, San Jose, Jan. 1998, pp525-534

[Kuhn 98a] Kuhn, P., Stechele, W.: "Complexity Analysis of the Emerging MPEG-4 Standard as a Basis for VLSI Implementation", vol. SPIE 3309 Visual Communications and Image Processing, San Jose, Jan. 1998, pp. 498-509

[Kuhn 98b]. Kuhn P., et al.: "Complexity and PSNR-Comparison of several Fast Motion Estimation Algorithms for MPEG-4", vol. SPIE 3460 Applications of Digital Image Processing XXI, San Diego, July 1998

[M 3204]: Kuhn P.: "Complexity Analysis of single video tools of the MPEG-4 verification Model", ISO/IEC JTC1/SC29/WG11/M3204, San Jose, USA, Jan. 1998

[M 2862]: Kuhn P., Diebel G.: "Complexity Analysis of the MPEG-4 Video VM 8.0", ISO/IEC JTC1/SC29/WG11/ M2862, Fribourg, Switzerland, October 1997

[M 1257]· Kuhn, P.: "Complexity Analysis of the MPEG-4 Video Verification Model Decoder``, ISO/IEC JTC1/ SC29/WG11 MPEG96/M1257, Chicago, Illinois, 1996

[M 0920]: Kuhn, P.: "Complexity Analysis of the MPEG-4 Video Verfication Model Encoder using Profiling Tools``, ISO/IEC JTC1/SC29/WG11 MPEG96/M0920, Tampere, Finland, 1996

[MomVM]. ISO/IEC JTC1/SC29/WG11/MPEG97/M2915: "Momusys Implementation of the VM (VM8-971021)", Fribourg, October 1997

[MPEG 1]: ISO/IEC 11172-2 Information Technology - Coding of Moving Pictures and Associated Audio for digital storage media at up to 1.5 Mbit/s. Part2: Video

[MPEG 2]: ISO/IEC 13818-2 Information Technology - Generic Coding of Moving Pictures and Associated Audio Information. Part2 Video

[MPEG 4]: ISO/IEC JTC1/SC29/WG11/MPEG97/N1902: "Committee Draft of ISO/IEC 14496-2 (MPEG-4 Visual)", November 1997

[Sik 97a]: T. Sikora: "The MPEG-4 Video Standard Verification Model", IEEE Trans. Circuits and Systems for Video Technology, Vol. 7, No. 1, pp 19 - 31, Feb 1997

[Sik 97b]: T. Sikora: "MPEG Digital Video Coding Standards", IEEE Signal Processing Magazine, Vol. 14, No. 5, p 82 -100, Sept. 1997

[TMN 9]: ITU-Telecommunication Standardization Sector-Study Group 16 Video Coding Experts Group: Video Codec Test Model Near Term Version 9

[VM 8.0]: ISO/IEC JTC1/SC29/WG11/MPEG97/N1796: "MPEG-4 Video Verification Model Version 8.0", Stockholm, July 1997

[Zhou 95]: C.G. Zhou, L. Khon, D. Rice, I. Kabir, A. Jabbi, X.-P. Hu: "MPEG Video Decoding with the UltraSPARC Visual Instruction Set", Digest of Papers COMPCON Spring 95, p470 - 475, IEEE, March 1995

# CHAPTER 5: ANALYSIS OF FAST MOTION ESTIMATION ALGORITHMS

[Baek 96]: Yunju Baek, Hwang-Soek Oh and Heung-Kyu Lee: „Block-matching criterion for efficient VLSI implementation for motion estimation“, IEE Electronics Letters, 20th June 1996, vol. 32, no. 13, Jun 1996, pp 1184-1185

[Chen 95:1]: Mei-Juan Chen, Liang-Gee Chen, Tzi-Dar Chiueh, Yung-Pin Lee: „A new block-matching criterion for motion estimation and its implementation“, IEEE Transactions on Circuits and Systems for Video Tech nology, vol. 5, no. 3, Jun. 1995, pp 231-236

[Cote 97]: Guy Cote, Michael Gallant, Faouzi Kossentini: „Efficient Motion Vector Estimation and Coding for H.263-based very low bit rate video compression“, ITU-T SG 16, Q15-A-45, June 1997, 18 p

[Ghar 90]: H. Gharavi, M. Mills: „Block-Matching Motion Estimation Algorithms - New Results“, IEEE Transactions on Circuits and Systems“, Vol 37, No. 5, May 1990, p649-651

[LeeC 96]: S. Lee and S.-I. Chae: „Motion Estimation algorithm using low resolution quantisation“, IEE Electronic Letters, vol. 32, no. 7, 28 th. Mar. 1996, pp 647-648

[Kim 92]: Joon-Seek Kim, Rae-Hong Park: „A fast feature-based block matching algorithm using integral projections“, IEEE Journal on Selected areas in communications, vol. 10, no. 5, Jun. 1992, pp 968-971

[Koga 81]: T. Koga, K. Iinuma, A. Hirano, Y. Ijima, T. Ishiguro: „Motion compensated interframe coding for video conferencing“, in Proc. NTC 81, pp. C 9.6.1 - 9.6.5

[Kuhn 98a] Kuhn, P., Stechele, W.: "Complexity Analysis of the Emerging MPEG-4 Standard as a Basis for VLSI Implementation", vol. SPIE 3309 Visual Communications and Image Processing, San Jose, Jan. 1998, pp. 498-509

[Kuhn 98b]: Kuhn P., et al.: "Complexity and PSNR-Comparison of several Fast Motion Estimation Algorithms for MPEG-4", vol. SPIE 3460 Applications of Digital Image Processing XXI, San Diego, July 1998, pp 486-499

[Liu 93:1]: B. Liu, A. Zaccarin: „New Fast Algorithms for the Estimation of Block Motion Vectors“, IEEE Trans. on Circuits and Systems for Video Technology“, Vol. 3, No. 2, April 1993, pp 148-157

[Nam 95]: Kwon Moon Nam, Joon-Seek Kim Rae-Hong Park, Young Serk Shim: „A fast hierarchical motion vector estimation algorithm using mean pyramid“, IEEE Transactions on Circuits and Systems for Video Technology, vol. 5, no. 4, Aug. 1995, pp 344-351

[Nat 97]: Balas Natarajan, Vasudev Bhaskaran, Konstantinos Konstantinides: „Low-Complexity Block-based motion estimation via One-Bit Transforms“, IEEE Transactions on Circuits and Systems for Video Technology, vol. 7, no. 4, Aug. 1997, pp 702-706

[MomVM]: ISO/IEC JTC1/SC29/WG11/MPEG97/M2915: „Momusys Implementation of the VM (VM8-971021)“, Fribourg, October 1997

[M 3551]: Kuhn P.: „A Complexity Analysis Tool· iprof (version 0.41)“, ISO/IEC JTC1/SC29/WG11/M3551, Dublin, July 1998

[M 3204]: Kuhn P.: „Complexity Analysis of single video tools of the MPEG-4 verification Model“, ISO/IEC JTC1/SC29/WG11/M3204, San Jose, USA, Jan. 1998

[M 2863]: Kuhn P.: „A Complexity Analysis Tool: iprof (version 0.3)“, ISO/IEC JTC1/SC29/WG11/M2863, Fribourg (CH), Switzerland, October 1997

[M 1056]: Kuhn, P.· „A portable Instruction Level Profiler for Complexity Analysis - Software“, ISO/IEC JTC1/SC29/WG11 MPEG96/M1056, Tampere, Finland, 1996

[M 0921]: Kuhn, P.: „A portable Instruction Level Profiler for Complexity Analysis Documentation“, ISO/IEC JTC1/SC29/WG11 MPEG96/M0921, Tampere, Finland, 1996

[M 0838]: Kuhn, P.: „Instrumentation Tools and Methods for MPEG-4 VM: Review and a new Proposal", ISO/IEC JTC1/SC29/WG11 MPEG96/M0838, Firence, Italy, 1996

# CHAPTER 6: VLSI ARCHITECTURES FOR MOTION ESTIMATION

[Bhas 95]: Vasudev Bhaskaran, Konstantinos Konstantinides: "Image and Video Compression Standards", Kluwer Academic Publishers, 1995, p237-247

[Bha 95:1]: V. Bhaskaran, K. Konstanitides, R.B. Lee, J. P. Beck: Algorithmic and Architectural Enhancements for Real-Time MPEG-1 Decoding on a general Purpose RISC workstation", IEEE Trans. on circuits and systems for video technology, vol. 5, no. 5, Oct. 1995

[Boxer 95]: Aaron Boxer: "Where Buses cannot go", IEEE Spectrum, vol. 32, no. 2, 1995, p 41-45

[Carl 97]: David A. Carlson, Ruben W. Castelino, Robert O. Mueller: "Multimedia Extensions for a 550-MHz RISC Microprocessor", IEEE Journal for Solid-State Circuits, Vol. 32, No. 11, November 1997, p1618-1624

[Ding 98]: Wei Ding, Alex Z.J. Mou and Daniel Rice: "A standard-based software-only video conference codec on ultra-sparc", SPIE Vol. 3309, Visual Communications and Image Processing, San Jose 1998, p535-542

[He 96]: Zhong L. He, Ming L. Liou: "Design Trade-Offs for Real-Time Block-Matching Motion Estimation", Lecture Notes on Computer Science, Image Analysis and Processing, 1996, pp 129-138

[Kohn 95]: L. Kohn, G. Maturana, M. Tremblay, A. Prabhu, G. Zyner: " The visual instruction Set (VIS) on UltraSPARC", IEEE Compcon, 1995, p462-469

[Kung 88]: S.Y. Kung: "VLSI Array Processors", Prentice Hall, Englewood Cliffs, 1988

[Kala 96]: Paul Kala: "Hardware/Software Interactions on the MPACT media processor", Hot Chips VIII, Stanford, California, august 19-20, 1996, p 179-191

[Knebel 93]: P. Knebel, B. Arnold, M. Bass, W. Kever, J.D. Lamb, R.b. Lee, P. L. Perez, S. Undy, W. Walker: " HP's PA7100LC: A low-cost superscalar PA-RISC processor", IEEE Compcon, 1993, p441-447

[Kneip 98]: Kneip J., Bauer S., Vollmer J., Schmale B., Kuhn P.: „The MPEG-4 video coding standard - A VLSI point of view", IEEE Workshop on Signal Processing Systems (SiPS), Cambridge, MA, USA, Oct 1998

[Mad 95]: Vijay K. Madisetti: "VLSI Digital Signal Processors - An Introduction to Rapid Prototyping and Design Synthesis", Butterworth Heinemann/IEEE Press, 1994, p446-494

[Mou 96]: A. Mou, D. Rice, W. Ding: "VIS-based native video processing on Ultra SPARC", Proc. IEEE International Conference on Image Processing, vol II, Sep 1996, p 153-156

[MPEG]: MPEG web site, internet: `http://www.cselt.it/mpeg`

[NEC 98]: NEC Electronics: "CBC-C10 Family, 0.25 mm Cell-Based IC - preliminary user's manual", n.p., January 1998

[RLee 95]: R. Lee: "Accelerating Multimedia with Enhanced Microprocessors", IEEE Micro, April 1995, p 22-32

[Schmeck 95]: Hartmut Schmeck: "Analyse von VLSI-Algorithmen", Spektrum-Verlag, Heidelberg, Berlin, Oxford, 1995, chapter 3 (in german)

[TMN 9]: ITU-Telecommunication Standardization Sector-Study Group 16 Video Coding Experts Group: Video Codec Test Model Near Term Version 9

[Yasu 95]: Kazunori Yasuda, Tsuyoshi Oda, Kouichi Uchide: "Apparatus for detecting motion vector and encoding picture sign", European Patent Application, EP 0695 096 A1, 1996, 17p

### *Motion Estimation VLSI*

[Alves 94:1]: Lirida Alves de Barros, Nicolas Demassieux: "Real-time architecture for large displacements estimation", SPIE Proceedings Vol. 2308: 'Visual Communications and Image Processing '94, 1994, pp 1765-1776

[Baek 96]: Yunju Baek, Hwang-Soek Oh and Heung-Kyu Lee: "Block-matching criterion for efficient VLSI implementation for motion estimation", IEE Electronics Letters, 20th June 1996, vol. 32, no. 13, Jun 1996, pp 1184-1185

[Ber 96]: J. P. Berns, T.G. Noll: "A flexible Motion Estimation Chip for Variable Size Block Matching", ASAP' 96, International Conference on Application-Specific Systems, Architectures and Processors, Chicago, 1996

[Ber 96b]: J. P. Berns, T.G. Noll: "A cascadable 200 GOPS Motion Estimation Chip for HDTV Application", IEEE Custom Integrated Circuits Conference, San Diego, May 5-8, 1996, p355-358

[Ber 97]: J. P. Berns, T.G. Noll. "200 GOPS Prozesor für die objektbasierte Bewegungsschätzung im digital-

HDTV", GMM Fachbericht Mikroelektronik, 1997, p143-148

[Bra 94]: J. Bracamonte, I. Defilippis, M. Ansorge, F. Pellandini: "Bit-serial parallel processing VLSI architecture for a blockmatching motion estimation algorithm", International Picture Coding Symposium, PCS 94, 1994, p22-25

[BugW 97]: Alexander Bugeja, Woodward Yang: "A reconfigurable VLSI coprocessing system for the blockmatching algorithm", IEEE Transactions on very large scale integration (VLSI) systems, vol. 5, no. 3, sept. 1997, p329-337

[Chan 93]: Eric Chan, Sethuraman Panchananthan: "Motion Estimation Architecture for Video compression", IEEE Trans. on Consumer Electronics, vol. 39, no. 3, aug. 1993, pp. 292-297

[Chang 95]: Shifan Chang, Juin-Haur Hwang, Chein-Wei Jen: "Scalable Array Architecture Design for Full Search Block Matching", IEEE Transactions on Circuits and Systems for Video Technology, vol. 5, no. 4, aug. 1995, pp. 332--343

[Charlot 95:1]: D. Charlot, J.-M. Bard, B. Canfield, C. Cuney, A. Graf, A. Pirson, D. Teichner, F. Yassa: "A RISC Controlled Motion Estimation Processor for MPEG-2 and HDTV Encoding", ICASSP 95, vol. 5, p 3287

[Chen 93]: Sau-Gee Chen: "An area/time-efficient motion estimation micro core", IEEE Transactions on Consumer Electronics, vol. 39, no. 3, aug 1993, p 298-303

[Chen 95:1]: Mei-Juan Chen, Liang-Gee Chen, Tzi-Dar Chiueh, Yung-Pin Lee: "A new block-matching criterion for motion estimation and its implementation", IEEE Transactions on Circuits and Systems for Video Technology, vol. 5, no. 3, Jun. 1995, pp 231-236

[Cheng 96]: Sheu-Chih Cheng and Hsueh-Ming Hang: "A Comparison of Block-Matching Algorithms for VLSI-Implementations", SPIE Proceedings Vol. 2727, Visual Communications and Image Processing, 1996, pp 994-1005

[Cheng 97]: Sheu-Chih cheng, Hsueh-Ming Hang: "A Comparison of Block-Matching Algorithms Mapped to Systolic-Array Implementation", IEEE Transactions on circuits and systems for video technology, vol. 7, no. 5, oct 1997, pp. 741-757

[Cheng 97a]: Hsueh-Ming Hang, Yung-Ming Chou, Sheu-Chih Cheng: "Motion Estimation for Video Coding Standards", Journal of VLSI Signal Processing, 1997 Kluwer Academic Publishers, Netherlands, pp. 113-136

[Cmar 97]: R. Cmar, S. Vernalde: "Highly scalable parallel parametrizable architecture of the motion estimator", "IEEE ED & TC 97 (European Design & Test Conference), Paris, France, Mar. 1997, pp. 208--212

[Col 91]: Oswald Colavin, Alain Artieri, Jean-Francois Naviner, Renaud Pacalet: "A dedicated circuit for real time motion estimation", EUROASICs 1991

[Costa 95]: Alessandra Costa, Alessandro De Gloria, Paolo Faraboschi, Filippo Passagio: "A VLSI Architecture for hierarchical motion estimation", IEEE Trans. on Consumer Electronics, vol. 41, no. 2, may 1995, p 248

[Dogi 97]: S. Dogimont, M. Gumm, F. Mombers, D. Mlynek, A. Torielli: "Conception and Design of a RISC CPU for the use as embedded controller within a parallel multimedia architecture", IEEE International Conference on Application-specific Systems, Architectures and Processors (ASAP 97), Zurich, Switzerland, July 1997, pp 412-421

[Dut 96]: Santanu Dutta and Wayne Wolf: "A Flexible Parallel Architecture Adapted to Block-Matching Motion-Estimation Algorithms", IEEE Transactions on Circuits and Systems for Video Technology, vol. 6, no. 1, feb 1996, pp74-86

[Dut 98]: Santanu Dutta, Wayne Wolf, Andrew Wolfe: "A methodology to evaluate memory architecture design tradeoffs for video signal processors", IEEE Transactions on Circuits and Systems for Video Technology, vol. 8, no. 1, feb 1998, pp. 36-53

[Fri 92]: Emmanuel D. Frimout, Johannes N. Driessen, Ed F. Deprettere: "Parallel Architecture for a Pel-Recursive Motion Estimation Algorithm", IEEE Transactions on Circuits and systems for video technology, vol. 2, no. 2, Jun. 1992, p159-168

[Greef 95]: E. De Greef, F. Catthoor, H. De Man: "A memory efficient, programmable multi-processor architecture for real-time motion estimation type algorithms", Algorithms and Parallel VLSI Architectures III, M. Moonen and F.Catthoor (Editors), Elsevier Publishers, vol. III, 1995, p191-202

[Gupta 95]: Gagan Gupta and Chaitali Chakrabarti: "Architectures for Hierarchical and other Block Matching Algorithms", "IEEE Transactions on Circuits and Systems for Video Technology, vol. 5, no. 6, dec 1995, p 477-489

[Hans 96]: E. Hanssens and J.-D. Legat: "A parallel Processor for Motion Estimation", SPIE Vol. 2727, 1996, pp 1006-1016

[He 96]: Zhong L. He, Ming L. Liou: "Design Trade-Offs for Real-Time Block-Matching Motion Estimation", Lecture Notes in Computer Science: Recent developments in Computer Vision, 1996, pp. 129-138

[He 97]: Zhong-Li He, Kai-Keung Chan, Chi-Ying Tsui, Ming L. Liou: "Low Power Motion Estimation Design Using Adaptive Pixel Truncation", IEEE Proceedings of 1997 International Symposium on Low Power Electronics and Design, 1997, pp 167-171

[He 97a]: Zhong L. He, Ming L. Liou: "Cost effective VLSI architectures for full-search block-matching motion estimation algorithm", Journal of VLSI Signal Processing, no. 17, Kluwer Academic Publishers, Netherlands, 1997, pp. 225-240

[Hsieh 92]: Chaur-Heh Hsieh, Ting-Pang Lin: "VLSI Architecture for block-matching motion estimation algorithm", IEEE Circuits and Systems for Video technology, vol. 2, no. 2, jun 1992, pp. 169--175

[Huang 94:2]: Shih-Yu Huang and Kuen-Rong Hsieh and Jia-Shung Wang: "Very large scale integration (VLSI) architecture for motion estimation and vector quantization", SPIE Proceedings Vol. 2308, Visual Communications and Image Processing, 1994, pp. 1742-1752

[Ishi 95]: Kazuya Ishihara, Shinichi Masuda, Shinichi Hattori, Hirofumi Nishikawa, Yoshihide Ajioka, Tsuyoshi Yamada, Hiroyuki Amishiro, Masahiko Yoshimoto: "A half-pel Precision MPEG2 Motion-Estimation Processor with Concurrent Three-Vector Search", ISSCC 95, 1995, pp. 288-189

[Ishi 95b]: Kazuya Ishihara, S. Masuda, S. Hattori, H. Nishikawa, Y. Ajioka, T. Yamada, H. Amishiro, S. Uramoto, M. Yoshimoto, T. Sumi: "A Half-pel precision MPEG-2 Motion Estimation Processor with concurrent Three-vector search", IEEE Journal of Solid-state Circuits, vol. 30, no. 12, dec 1995, pp. 1502-1509

[Jehng 93]: Yeu-Shen Jehng, Liang-Gee Chen, Tzi-Dar Chiueh: "An efficient and simple VLSI Tree architecture for motion estimation algorithms", IEEE Transactions on Signal Processing, vol. 41, no. 2, feb 1993, pp. 889-900

[Jong 94]: Her-Ming Jong and Liang-Gee Chen and Tzi-Dar Chiueh: "Parallel Architectures for 3-Step Hierarchical Search Block-Matching Algorithm", vol. 4, no. 4, aug. 1994, pp. 407-415

[Jung 95]: Hae-Kwan Jung, C-P Hong, J-S Choi, Y-H Ha: "A VLSI Architecture for the alternative subsampling-based blockmatching algorithm", IEEE Transactions on Consumer Electronics, vol. 41, no. 2, may 1995, pp 239-347

[Kom 89]: T. Komarek, P. Pirsch: "Array Architectures for blockmatching algorithms", IEEE Trans. Circuits and SystemsVol. 36, No. 10, Oct. 1989, pp1301-1308

[Kom 93]: T. Komarek: "VLSI-Architekturen für Displacementschätzverfahren auf der Basis von Blockmatching-Algorithmen", Dissertation, Hannover, 1993, VDI Verlag (in german)

[Kuhn 97a]: Kuhn P., Stechele W.: "VLSI architecture for variable block size motion estimation with luminance correction", vol. SPIE 3162 Advanced Signal Processing: Algorithms, Architectures and Implementations, San Diego, July 1997, pp. 497-508

[Kuhn 97b]: Kuhn, P., Weisgerber, A., Poppenwimmer, R. Stechele, W.: "A flexible VLSI architecture for Variable Block Size Segment Matching with Luminance Correction", ASAP 97, IEEE International Conference on Application-specific Systems, Architectures and Processors, Zurich, July 14 - 16, 1997

[Kuhn 97c]: Kuhn P., Eiermann M., Stechele W.: "A flexible Segment Matching Processor for Motion and Illumination Estimation", PCS 97, International Picture Coding Symposium, Berlin, Sept. 1997

[Kuhn 97d]: Kuhn, P. , Weisgerber A. , Poppenwimmer R., Stechele W.: "Eine flexible Architektur für moderne Verfahren der Bewegungsschätzung", 7. Dortmunder Fernsehseminar, Dortmund, Sept. 1997 (in german)

[Kuhn 97e]: Kuhn, P., Eiermann M., Weisgerber W. , Poppenwimmer R., Stechele, W.:,,VLSI Implementation of Mean-Corrected Block-Matching Motion Estimation of Partial Quadtrees", VLBV 97, Workshop for Very Low Bitrate Video Coding, Linköping, Sweden, July 1997

[Kwak 95]: Jinsuk Kwak, Jinwoong Kim, Kichul Kim: "A field and Frame-based Motion Estimator with a very flexible search range", SPIE Proceedings Vol. 2727, Visual Communications and Image Processing, mar. 1996, p394-402

[Lai 97]: Yeong-Kang Lai, Liang-Gee Chen, Yung-Pin Lee: "A flexible Data-interlacing Architecture for full-search block-matching algorithm", IEEE International Conference on Application-specific Systems, Architectures and Processors (ASAP 97), Zurich, Switzerland, July 14 - 16, jul. 1997, pp. 96-104

[Lai 97a]: Yeong-Kang Lai, Liang-Gee Chen, Tsung-Han Tsai, Po-Cheng Wu: "A flexible high-throughput VLSI architecture with 2-D data-reuse for full-search motion estimation", IEEE ICIP 97, 1997, p1101

[Lak 97]: Prasad Lakamsani: "An architecture for enhanced three step search generalized for hierarchical motion estimation algorithms", IEEE Trans. on Consumer Electronics, vol. 43, no. 2, may 1997, pp. 221-227

[LeeC 96]: S. Lee and S.-I. Chae: "Motion Estimation algorithm using low resolution quantisation", IEE Electronic Letters, vol. 32, no. 7, 28 th. Mar. 1996, pp 647-648

[Lee 94a]: Chen-Yi Lee, Shih-Chou Juan, Wen-Wei Yang: "A parallel Bit-Level Maximum/Minimum Selector for Digital and Video Signal Processing", IEEE Transactions on Circuits and Systems-II: Analog and Digital Signal Processing, vol. 41, no. 10, oct. 1994, pp. 693-695

[LeeDL 94]: Jong Hwa Lee, Myeong Kyoo Doh, Choong Woong Lee: "A VLSI Chip for Motion Estimation of HDTV dignals", IEEE Trans on Consumer Electronics, vol. 40, no. 2, may 1994, pp. 154-160

[LeeLu 97]: Chen-Yi Lee, Mei-Cheng Lu: "An efficient VLSI architecture for full-search block matching algorithms", Journal of VLSI Signal Processing, vol. 15, pp. 272-282

[Le 98]: Thinh M. Le, M. Snelgrove, S. Panchanatan: "Fast motion estimation using feature extraction and

XOR operations", SPIE 3311 MHA Multimedia Hardware Architectures, San Jose, 1998, pp 108-118

[Li 96]: Xiaoming Li, Cesar Gonzales: "A locally quadratic model of the motion estimation error criterion function and its application to subpixel interpolations", IEEE Transactions on Circuits and Systems for Video Technology, vol. 6, no. 1, feb. 1996

[Lin 96]: Hong-Dar Lin and Alex Anesko and Brian Petryna: "A 14-Gops Programmable Motion Estimator For H.26X Video Coding", IEEE Journal of Solid State Circuits, vol. 31, no. 11, Nov. 1996, pp. 1096-1075

[Lin 96a]: H.-D. Lin, A. Anesko, B. Petryna: "A 14 GOPS Programmable Motion Estimator for H.26X Video Coding", 1996 IEEE International Solid-State Circuits Conference (ISSCC), 1996, p. 246-247

[Lu 95:1]: Mei-Cheng Lu, Chen-Yi Lee: "Semi-Systolic Array Based Motion Estimation Processor Design", ICASSP 95, vol. 5, 1995, pp. 3299

[Meng 91]: Teresa H.-Y. Meng, Andy C. Hung: "Parallel Array Architectures for Motion Estimation", ASAP 91, Proc. of the International Conference on Application Specific Array Processors, Sept. 2-4, 1991, Barcelona, Spain, pp. 214--235

[Miz 96]: Marcelo M. Mizuki and Ujjaval Y. Desai and Ichiro Masaki and Anantha Chandrakasan: "A binary block matching architecture with reduced power consumption and silicon area requirement", ICASSP 96, 1996, p3249

[Mosh 97]: Vasily G. Moshnyaga, Keikichi Tamaru: "A memory efficient array architecture for real-time motion estimation", IPPS 97, Proceedings of the 11 th International Parallel processing Symposium, April 1 -5, 1997, Geneva, Switzerland, CDROM

[Mosh 97a]: Vasily G. Moshnyaga, Keikichi Tamaru: "A memory array architecture for full-search block matching algorithm", ICASSP 97, p. 4109-4112

[Nam 94]: Seung Hyun Nam, Jong Seob Baek, Moon Key Lee: "Flexible VLSI Architecture of Full Search motion estimation for video applications", IEEE Trans. on Consumer Electronics, vol. 40, no. 2, may 1994, p176-184

[Nam 96]: Seung Hyun, Moon Key Lee: "Flexible VLSI Architecture for Motion Estimation for Video Image Compression", IEEE Transactions on Circuits and Systems II: Analog and Digital Signal Processing, vol. 43, no. 6, jun 1996, p. 467-470

[Nat 96]: Balas, Natarajan and Bhaskaran Vasudev and Konstantinos Konstantinides: "Low-complexity Algorithm and Architecture for Block-based Motion Estimation via one-bit transforms", ICASSP 96, 1996, p. 3245

[Nat 97]: Balas Natarajan, Vasudev Bhaskaran, Konstantinos Konstantinides: "Low-Complexity Block-based motion estimation via One-Bit Transforms", IEEE Transactions on Circuits and Systems for Video Technology, vol. 7, no. 4, Aug. 1997, p 702

[OOI 97]: Yasushi Ooi, O. Ohnishi, Y. Yokohama, Y. Katayama, M. Mizuno, M. Yamashina, H. Takano, N. Hayashi, I. Tamitani: "An MPEG-2 Encoder Architecture based on a single-chip dedicated LSI with a control MPU", ICASSP 97, 1997, p. 599

[Ogura 95]: Eiji Ogura, Yuuichirou Ikenaga, Y. Iida. Y. Hosoya, M. Takashima, K. Yamashita: "A cost efective motion estimation processor LSI using a simple and efficient algorithm", IEEE Transactions on Consumer Electronics, vol. 41, no. 3, Aug. 1995, p690-697

[Pan 96]: Sung Bum Pan, Seung Soo Chae, Rae-Hong Park: "VLSI architectures for block matching algorithms using systolic arrays", IEEE Transactions on Circuits and Systems for video technology, vol. 6, no. 1, feb. 1996, p67-73

[Pir 95:1]: Peter Pirsch, Nicolas Demassieux, Winfried Gehrke: "VLSI Architectures for Video Compression - A Survey", Proceedings of the IEEE, vol. 83, no. 2, feb. 1995, p. 220 - 246

[Pirson 95:1]: Alain Pirson, Fathy Yassa, Philippe Paul, Barth Canfield, Friedrich Rominger, Andreas Graf, Detlef Teichner: "A Programmable Motion Estimation Processor for Full Search Block Matching", ICASSP 95, vol. 5, 1995, p. 3283

[Reven 93]: C.V. Reventlow, M. Talmi, S. Wolf, M. Ernst, K. Mueller, C. Stoffers: "System Considerations and the System Level Design of a Chip Set for Real-Time TV and HDTV motion estimation", Journal of VLSI processing, Kluwer Academic Publishers, vol. 5, 1993, p. 237-248

[Ross 91]: J. Rosseel, F. Catthoor, H. De Man: "The systematic design of motion estimation array architecture", ASAP 91, Proc. of the International Conference on Application Specific Array Processors (ASAP 91), Sept. 2-4, 1991, Barcelona, SpainSept. 1991, p. 40-54

[Saha 95]: Arindam Saha, Raja Neogi: "Parallel programmable algorithm and architecture for real-time motion estimation of various video applications", IEEE Trans. on Consumer Electronics, vol. 41, no. 4, Nov. 1995, p 1069--1079

[Skim 97]: Sangjoong Kim, Yonggil Kim, Kangbin Yim, Hwaja Chung, Kyunghee Choi, Yongdeak Kim, Gihyun Jung: "A fast motion estimator for real-time systems", IEEE Trans. on Consumer Electronics, vol. 43, no. 1, Feb. 1997, pp. 24-33

[ST 94]: SGS-Thomson Microelectronics: "STi 3220 Motion Estimation Processor", datasheet, January 1994, 24p

[Su 96]: Kazuhito Suguri et al: "A Real-Time motion Estimation and Compensation LSI with Wide Search

Range for MPEG2 Video Coding", IEEE Journal of Solid State Circuits, vol. 31, no. 11, Nov. 1996, p. 1733--1741

[Sun 93]: Ming-Ting Sun: "Algorithms and VLSI architectures for motion estimation", LSI Implementations for Image Communications, P. Pirsch (Editor), Elsevier Science Publishers B.V., 1993, pp 251-282

[Ura 94]: Shin-ichi Uramotom A. Takabatake, M. Suzuki, H. Sakurai, M. Yoshimoto: "A Half-Pel Precision Motion Estimation Processor for NTSC-Resolution Video", IEICE Trans. Electronics, vol. E77-C, no. 12, Dec. 1994, pp. 1930-1936

[Vos 89]: Luc De Vos, M. Stegherr: "Paramaterizable VLSI architectures for the full-search block-matching algorithm," IEEE Trans. Circuits Syst., Vol. 36, Oct. 1989, p 1309-1316

[Vos 90]: Luc De Vos: "VLSI-architectures for the hierarchical block-matching algorithm for HDTV applications", SPIE Vol. 1360 Visual Communications and Image Processing, 1990, pp. 398-409

[Vos 94]: Luc de Vos: "Dedizierte VLSI-Architekturen für Block Matching Algorithmen", VDI-Verlag, Düsseldorf 1994, Reihe 10 Informatik/Kommunikationstechnik, Nr. 326 (dissertation, in german), 171 p

[Vos 95:1]: Luc De Vos, M. Schöbinger: "VLSI Architecture for a Flexible Block Matching Processor", IEEE Transactions on Circuits and Systems for Video Technology, Vol. 5, No. 5, October 1995, pp. 417-428

[Wang 94]: Bor-Min Wang and Jui-Cheng and Shyang Chang: "Zero Waiting-Cycle Hierarchical Block Matching Algorithm and its Array Architectures", IEEE Trans. on Circuits and Systems for Video Technology, vol. 4, no. 1, Feb. 1994, p. 18-27

[Wang 95:1]: Chin-Liang Wang, Ker-Min Chen, Jin-Min Hsiung: "A High-Throughput, Flexible VLSI Architecture for Motion Estimation", ICASSP 95, 1995, pp. 3295

[WLi 95]: W. Li, E. Salari: "Succesive elimination algorithm for motion estimation", IEEE Trans. Image Processing, vol. 4 pp 105-107, Jan 1995, pp 105-107

[Wu 93]: Chen-Mie Wu, Ding-Kuen Yeh: "A VLSI Motion Estimator for Video image compression", IEEE Trans. on Consumer Electronics, vol. 39, no. 4, Nov. 1993, pp. 837-846

[Wu 95:1]: An-Yeu Wu, K.J. Ray Liu: "Algorithm-Based Low-Power Transform Coding Architectures", ICASSP 95, vol. 5, 1995, p. 3267-3272

[XLee 96]: Xiaobing Lee, Ya-Qin Zhang: "A fast hierarchical Motion-Compensation Scheme for Video Coding using Block-feature Matching", IEEE Transactions on Circuits and Systems for Video Technology, vol. 6, no. 6, Dec. 1996, pp 627-635

[Yang 89]: Kun-Min Yang, Ming-Ting Sun, Lancelot Wu: "A family of VLSI designs for the motion compensation block-matching algorithm", IEEE Transactions on circuits and systems, vol. 36, no. 10, Oct. 1989, pp. 1317-1325

[Yeh 97]: Yuan-Hau Yeh, Chen-Yi Lee: "Buffer Size Optimization for Full-Search Block Matching Algorithms", IEEE International Conference on Application-specific Systems, Architectures and Processors (ASAP 97), Zurich,Switzerland, July 14 - 16, 1997, pp. 76-85

[Yeo 95]: Hangu Yeo and Yu Hen Hu: "A Novel Modular Systolic Array Architecture for Full-Search Block Matching Motion Estimation", IEEE Transactions on Circuits and Systems for Video technology, vol. 5, no. 5, oct. 1995, pp. 407-416

[Yeo 95:1]: Hangu Yeo, Yu Hen Hu: "A Novel Modular Systolic Array Architecture for Full-Search Block Matching Motion Estimation", ICASSP 95, vol. 5, 1995, p 3303

[YeoHu 96]: Hangu Yeo, Yu Hen Hu: "A high-throughput modular architecture for three-step search block matching motion estimation", ICASSP 96, 1996, p. 2305

### *Adress Generation Units*

[Coh 97]: Ben Cohen: "VHDL: Answers to Frequently Asked Questions", Kluwer, Academic Publishers, Boston, Dordrecht, London, 1997

[Hart 97]: Reiner W. Hartenstein, Juergen Becker, Michael Herz, Ulrich Nageldinger: "A novel sequencer hardware for application specific computing", IEEE International Conference on Application-specific Systems, Architectures and Processors (ASAP 97), Zurich, Switzerland, July 14 - 16, 1997, pp 392-401

[Grant 94]: D.M. Grant, J. van Meerbergen, P.E.R. Lippens: "Optimization of Address Generator Hardware", Proccedings of the 5 th ACM/ IEEE Europ. Design and Test Conference, pp 325-329, 1994

[Kita 91]: Kazukuni Kitagaki, Takeshi Oto, Tatsuhiko Demura, Yoshitsugu Araki, and Tomoji Takad: "A new address generation unit architecture for video signal processing", SPIE Vol. 1606 Visual Communications and Image Processing 91: Image Processing, pp 891-900

[Liem 97]: Clifford Liem: "Retargetable Compilers for Embedded Core Processors", Kluwer Academic Publishers, Boston, Dordrecht, London, 1997

[Mira 97]: Miguel Miranda, Martin Kaspar, Francky Catthor, Hugo De Man: "Architectural Exploitation and Optimization for Counter Based Hardware Address Generation", European Design and Test Conference 97, Paris, France, March 17-20, 1997, pp293-298

[Nel 93]: Mark Nelson: "Datenkomprimierung - Effiziente Algorithmen in C", Heinz Heise Verlag, Hannover, 1993, pp475

[Sud 97]: Ashok Sudarsanam, Stan Liao, Srinivas Devadas: "Analysis and Evaluation of Address Arithmetic Capabilities in Custom DSP Architectures", IEEE/ACM Design Automation Conference (DAC), Anaheim, CA, 1997, pp 287-292

[Syn 98]: Synopsys: "VHDL Compiler Reference Manual", Document Order Number: 1US01-10430, WWW: http://www.synopsys.com

[Wess 97]: Bernhard Wess, Martin Gotschlich: "Optimierungstechniken für Adressrecheneinheiten in DSPs", DSP Deutschland 97 - Grundlagen, Architekturen, Tools, Applikationen, 30.9.-1.10.97, Munich, Germany (in german)

# CHAPTER 7: VLSI IMPLEMENTATION: SEARCH ENGINE I (2D-ARRAY)

[Kuhn 97a]: Kuhn P., Stechele W.: "VLSI architecture for variable block size motion estimation with luminance correction", vol. SPIE 3162 Advanced Signal Processing: Algorithms, Architectures and Implementations, San Diego, July 1997, pp. 497-508

[Kuhn 97b]: Kuhn, P., Weisgerber, A., Poppenwimmer, R. Stechele, W.: "A flexible VLSI architecture for Variable Block Size Segment Matching with Luminance Correction", ASAP 97, IEEE International Conference on Application-specific Systems, Architectures and Processors, Zurich, July 14 - 16, 1997

[Kuhn 97c]: Kuhn P., Eiermann M., Stechele W.: "A flexible Segment Matching Processor for Motion and Illumination Estimation", PCS 97, International Picture Coding Symposium, Berlin, Sept. 1997

[Kuhn 97d]: Kuhn, P., Weisgerber A., Poppenwimmer R., Stechele W.: "Eine flexible Architektur für moderne Verfahren der Bewegungsschätzung", 7. Dortmunder Fernsehseminar, Dortmund, Sept. 1997 (in german)

[Kuhn 97e]: Kuhn, P., Eiermann M., Weisgerber W., Poppenwimmer R., Stechele, W.:„VLSI Implementation of Mean-Corrected Block-Matching Motion Estimation of Partial Quadtrees", VLBV 97, Workshop for Very Low Bitrate Video Coding, Linköping, Sweden, July 1997

# CHAPTER 8: VLSI IMPLEMENTATION: SEARCH ENGINE II (1D-ARRAY)

[Cheng 96]: K.W. Cheng, S.C. Chan: "Fast block matching algorithms for motion estimation", ICASSP 96, 1996, p 2318ff

[Dut 96]: Santanu Dutta and Wayne Wolf: "A Flexible Parallel Architecture Adapted to Block-Matching Motion-Estimation Algorithms", IEEE Transactions on Circuits and Systems for Video Technology, vol. 6, no. 1, feb 1996, pp74-86

[Dut 98]: Santanu Dutta, Wayne Wolf, Andrew Wolfe: "A methodology to evaluate memory architecture design tradeoffs for video signal processors", IEEE Transactions on Circuits and Systems for Video Technology, vol. 8, no. 1, feb 1998, pp. 36-53

[Kuhn 99]: Kuhn P., Niedermeier U., Chao L.-F., Stechele W.: "A flexible low power VLSI architecture for MPEG-4 motion estimation", vol. SPIE 3653 Visual Communications and Image Processing, San Jose, Jan. 1999

[NEC 98]: NEC Electronics: "CBC-C10 Family, 0.25 µm Cell-Based IC - preliminary user's manual", n.p., January 1998

[SPARC]: Internet: `http://www.sun.com/microelectronics/products/microproc.html`

# INDEX